T0182131

Springer Optimization and Its Applications

VOLUME 133

Aims and Scope
Optimization has been expanding in all directions at an astonishing rate during the last few decades. New algorithmic and theoretical techniques have been developed, the diffusion into other disciplines has proceeded at a rapid pace, and our knowledge of all aspects of the field has grown even more profound. At the same time, one of the most striking trends in optimization is the constantly increasing emphasis on the interdisciplinary nature of the field. Optimization has been a basic tool in all areas of applied mathematics, engineering, medicine, economics and other sciences.

The series *Springer Optimization and Its Applications* publishes undergraduate and graduate textbooks, monographs and state-of-the-art expository works that focus on algorithms for solving optimization problems and also study applications involving such problems. Some of the topics covered include nonlinear optimization (convex and nonconvex), network flow problems, stochastic optimization, optimal control, discrete optimization, multi-objective programming, description of software packages, approximation techniques and heuristic approaches.

More information about this series at http://www.springer.com/series/7393

Jan A. Snyman · Daniel N. Wilke

Practical Mathematical Optimization

Basic Optimization Theory and Gradient-Based Algorithms

Second Edition

 Springer

Jan A. Snyman
Department of Mechanical
 and Aeronautical Engineering
University of Pretoria
Pretoria
South Africa

Daniel N. Wilke
Department of Mechanical
 and Aeronautical Engineering
University of Pretoria
Pretoria
South Africa

Additional material to this book can be downloaded from http://extras.springer.com.

ISSN 1931-6828   ISSN 1931-6836   (electronic)
Springer Optimization and Its Applications
ISBN 978-3-030-08486-8   ISBN 978-3-319-77586-9   (eBook)
https://doi.org/10.1007/978-3-319-77586-9

Mathematics Subject Classification (2010): 65K05, 90C30, 90C26, 90C59

Printed on acid-free paper

This Springer imprint is published by the registered company Springer International Publishing AG
part of Springer Nature
The registered company address is: Gewerbestrasse 11, 6330 Cham, Switzerland

To our wives and friends
Alta and Ella

Preface to the second edition

The first edition (2005) of Practical Mathematical Optimization has proved to be a rigorous yet practical introduction to the fundamental principles of mathematical optimization. As stated in the preface to the first edition also included in this new edition, the aim of the text was to equip the reader with the basic theory and algorithms to allow for the solution of practical problems with confidence and in an informed way.

However, since the publication of the first edition more than a decade ago, the complexity of and computing requirements for the solution of mathematical optimization problems have significantly increased. The accessibility and definition of computing platforms have expanded by huge proportions. The fundamental physical limitations on speeding up central processing units have spurred on the advancement of multi-core computing environments that now regularly include graphical processing units. The diversity of software platforms is ever expanding with new domain and computing specific software platforms being released weekly. This edition addresses these recent advancements together with novel ideas already touched on in the first edition that have since matured considerably. They include the handling of noise in objective functions and gradient-only optimization strategies introduced and discussed in the first edition. In order to assist in the coverage of further developments in this area, and in particular of recent work in the application of mainly gradient-only methods to piecewise smooth discontinuous objective functions, it is a pleasure to welcome as co-author for this edition the younger colleague, Daniel N. Wilke.

This second edition of Practical Mathematical Optimization now takes account of the above recent developments and aims to bring this text up to date. Thus, this book now includes a new and separate chapter dedicated to advanced gradient-only formulated solution strategies for optimizing noisy objective functions, specifically piecewise smooth discontinuous objective functions, for which solution formulations and strategies are thoroughly covered. A comprehensive set of alternative solution strategies are presented that include gradient-only line search methods and gradient-only approximations. The application of these strategies is illustrated by application to well-motivated example problems. Also new to this edition is a dedicated chapter on the construction of surrogate models using only zero-order information, zero- and first-order information, and only first-order information. The latter approach being particularly effective in constructing smooth surrogates for discontinuous functions.

A further addition is a chapter dedicated to numerical computation which informs students and practicing scientists and engineers on ways to easily setup and solve problems without delay. In particular, the scientific computing language *Python* is introduced, which is available on almost all computing platforms ranging from dedicated servers and desktops to smartphones. Thus, this book is accompanied by a *Python* module `pmo`, which makes all algorithms presented in this book easily accessible as it follows the well-known `scipy.optimize.minimize` convention. The module is designed to allow and encourage the reader to include their own optimization strategies within a simple, consistent, and systematic framework. The benefit to graduate students and researchers is evident, as various algorithms can be tested and compared with ease and convenience.

To logically accommodate the new material, this edition has been restructured into two parts. The basic optimization theory that covers introductory optimization concepts and definitions, search techniques for unconstrained minimization, and standard methods for constrained optimization is covered in the first five chapters to form Part I. This part contains a chapter of detailed worked-out example problems, while other chapters in Part I are supplemented by example problems and exercises that can be done by hand using only pen, paper, and a calculator. In Part II, the focus shifts to computer applications of relatively new and mainly gradient-based numerical strategies and algorithms that are covered over four chapters. A dedicated computing chapter using *Python* is included as the final chapter of Part II, and the reader is encouraged to consult this chapter as required to complete the exercises in the preceding three chapters. The chapters in Part II are also supplemented by numerical exercises that are specifically designed so as to encourage the students to plan, execute, and reflect on numerical investigations. In summary, the twofold purpose of these questions is to allow the reader, in the first place, to gain a deeper understanding of the conceptual material presented and, secondly, to assist in developing systematic and scientific numerical investigative skills that are so crucial for the modern-day researcher, scientist, and engineer.

Jan Snyman and Nico Wilke
Pretoria
30 January 2018

Preface to the first edition

It is intended that this book is used in senior- to graduate-level semester courses in optimization, as offered in mathematics, engineering, computer science, and operations research departments. Hopefully, this book will also be useful to practicing professionals in the workplace.

The contents of this book represent the fundamental optimization material collected and used by the author, over a period of more than twenty years, in teaching Practical Mathematical Optimization to undergraduate as well as graduate engineering and science students at the University of Pretoria. The principal motivation for writing this work has not been the teaching of mathematics per se, but to equip students with the necessary fundamental optimization theory and algorithms, so as to enable them to solve practical problems in their own particular principal fields of interest, be it physics, chemistry, engineering design, or business economics. The particular approach adopted here follows from the author's own personal experiences in doing research in solid-state physics and in mechanical engineering design, where he was constantly confronted by problems that can most easily and directly be solved via the judicious use of mathematical optimization techniques. This book is, however, not a collection of case studies restricted to the above-mentioned specialized research areas, but is intended to convey the basic optimization principles and algorithms to a general audience in such a way that, hopefully, the application to their own practical areas of interest will be relatively simple and straightforward.

Many excellent and more comprehensive texts on practical mathematical optimization have of course been written in the past, and I am much indebted to many of these authors for the direct and indirect influence their work has had in the writing of this monograph. In the text, I have tried as far as possible to give due recognition to their contributions. Here, however, I wish to single out the excellent and possibly underrated book of D. A. Wismer and R. Chattergy (1978), which served to introduce the topic of nonlinear optimization to me many years ago, and which has more than casually influenced this work.

With so many excellent texts on the topic of mathematical optimization available, the question can justifiably be posed: Why another book and what is different here? Here, I believe, for the first time in a relatively brief and introductory work, due attention is paid to certain inhibiting difficulties that can occur when fundamental and classical gradient-based algorithms are applied to real-world problems. Often students, after having mastered the basic theory and algorithms, are disappointed to find that due to real-world complications (such as the presence of noise and discontinuities in the functions, the expense of function evaluations, and an excessive large number of variables), the basic algorithms they have been taught are of little value. They then discard, for example, gradient-based algorithms and resort to alternative non-fundamental methods. Here, in Chapter 4 (now Chapter 6) on new gradient-based methods, developed by the author and his co-workers, the above-mentioned inhibiting real-world difficulties are discussed, and it is shown how these optimization difficulties may be overcome without totally discarding the fundamental gradient-based approach.

The reader may also find the organization of the material in this book somewhat novel. The first three chapters present the basic theory, and classical unconstrained and constrained algorithms, in a straightforward manner with almost no formal statement of theorems and presentation of proofs. Theorems are of course of importance, not only for the more mathematically inclined students, but also for practical people interested in constructing and developing new algorithms. Therefore, some of the more important fundamental theorems and proofs are presented separately in Chapter 6 (now Chapter 5). Where relevant, these theorems are referred to in the first three chapters. Also, in order to prevent cluttering, the presentation of the basic material in Chapters 1 to 3 is interspersed with very few worked-out examples. Instead, a generous number of worked-out example problems are presented separately in Chapter 5 (now Chapter 4), in more or less the same order as the presentation of the corresponding theory given in Chapters 1 to 3. The separate presentation of the example problems may also be convenient for students who have to prepare for the inevitable tests and examinations. The instructor may also use these examples as models to easily formulate similar problems as additional exercises for the students, and for test purposes.

Although the emphasis of this work is intentionally almost exclusively on gradient-based methods for nonlinear problems, this book will not be complete if only casual reference is made to the simplex method for solving linear programming (LP) problems (where of course use is also made of gradient information in the manipulation of the gradient vector **c** of the objective function, and the gradient vectors of the constraint functions contained in the matrix **A**). It was therefore decided to include, as Appendix A, a short introduction to the simplex method for LP problems. This appendix introduces the simplex method along the lines given by Chvatel (1983) in his excellent treatment of the subject.

The author gratefully acknowledges the input and constructive comments of the following colleagues to different parts of this work: Nielen Stander, Albert Groenwold, Ken Craig, and Danie de Kock. A special word of thanks goes to Alex Hay. Not only did he significantly contribute to the contents of Chapter 4 (now Chapter 6), but he also helped with the production of most of the figures and in the final editing of the manuscript. Thanks also to Craig Long who assisted with final corrections and to Alna van der Merwe who typed the first LaTeX draft.

Jan Snyman
Pretoria
31 May 2004

Contents

PART II: GRADIENT-BASED ALGORITHMS

Table of notation

$\mathbb{R}^n$	n-dimensional Euclidean (real) space
T	(superscript only) transpose of a vector or matrix
$\mathbf{x}$	column vector of variables, a point in $\mathbb{R}^n$ $\mathbf{x} = [x_1, x_2, \ldots, x_n]^T$
$\in$	element in the set
$f(\mathbf{x}), f$	objective function
$\mathbf{x}^*$	local optimizer
$\mathbf{x}_g^*$	non-negative associated gradient projection point
$f(\mathbf{x}^*)$	optimum function value
$g_j(\mathbf{x}), g_j$	j^{th} inequality constraint function
$\mathbf{g}(\mathbf{x})$	vector of inequality constraint functions
$h_j(\mathbf{x}), h_j$	j^{th} equality constraint function
$\mathbf{h}(\mathbf{x})$	vector of equality constraint functions
C^1	set of continuous differentiable functions
C^2	set of continuous and twice continuous differentiable functions
$\min, \min_{\mathbf{x}}$	minimize w.r.t. $\mathbf{x}$
$\mathbf{x}^0, \mathbf{x}^1, \ldots$	vectors corresponding to points 0,1,...
$\{\mathbf{x} \mid \ldots\}$	set of elements $\mathbf{x}$ such that ...
$\dfrac{\partial f}{\partial x_i}$	first partial derivative w.r.t. x_i
$\dfrac{\partial \mathbf{h}}{\partial x_i}$	$= [\dfrac{\partial h_1}{\partial x_i}, \dfrac{\partial h_2}{\partial x_i}, \ldots, \dfrac{\partial h_r}{\partial x_i}]^T$
$\dfrac{\partial \mathbf{g}}{\partial x_i}$	$= [\dfrac{\partial g_1}{\partial x_i}, \dfrac{\partial g_2}{\partial x_i}, \ldots, \dfrac{\partial g_m}{\partial x_i}]^T$
∇	first derivative operator
∇_A	first associated derivative operator

$\boldsymbol{\nabla} f(\mathbf{x})$
gradient vector $= \left[\dfrac{\partial f}{\partial x_1}(\mathbf{x}), \dfrac{\partial f}{\partial x_2}(\mathbf{x}), \ldots, \dfrac{\partial f}{\partial x_n}(\mathbf{x}) \right]^T$

$\boldsymbol{\nabla}^2$
second derivative operator (elements $\dfrac{\partial^2}{\partial x_i \partial x_j}$)

$\mathbf{H}(\mathbf{x}) = \boldsymbol{\nabla}^2 f(\mathbf{x})$ Hessian matrix (second derivative matrix)

$\left. \dfrac{df(\mathbf{x})}{d\lambda} \right|_{\mathbf{u}}$
directional derivative at $\mathbf{x}$ in the direction $\mathbf{u}$

$\subset, \subseteq$ subset of

$|\cdot|$ absolute value

$\|\cdot\|$ Euclidean norm of vector

$\cong$ approximately equal

$F(\ )$ line search function

$F[,]$ first order divided difference

$F[,,]$ second order divided difference

$(\mathbf{a}, \mathbf{b})$ scalar product of vector $\mathbf{a}$ and vector $\mathbf{b}$

$\mathbf{I}$ identity matrix

θ_j j^{th} auxiliary variable

L Lagrangian function

λ_j j^{th} Lagrange multiplier

$\boldsymbol{\lambda}$ vector of Lagrange multipliers

$\exists$ exists

$\Rightarrow$ implies

$\{\cdots\}$ set

$V[\mathbf{x}]$ set of constraints violated at $\mathbf{x}$

ϕ empty set

$\mathcal{L}$ augmented Lagrange function

$\langle a \rangle$ maximum of a and zero

$\dfrac{\partial \mathbf{h}}{\partial \mathbf{x}}$
$n \times r$ Jacobian matrix $= [\boldsymbol{\nabla} h_1, \boldsymbol{\nabla} h_2, \ldots, \boldsymbol{\nabla} h_r]$

$\dfrac{\partial \mathbf{g}}{\partial \mathbf{x}}$
$n \times m$ Jacobian matrix $= [\boldsymbol{\nabla} g_1, \boldsymbol{\nabla} g_2, \ldots, \boldsymbol{\nabla} g_m]$

s_i slack variable

$\mathbf{s}$ vector of slack variables

D determinant of matrix $\mathbf{A}$ of interest in $\mathbf{A}\mathbf{x} = \mathbf{b}$

D_j determinant of matrix $\mathbf{A}$ with j^{th} column replaced by $\mathbf{b}$

$\lim\limits_{i \to \infty}$ limit as i tends to infinity

Part I
Basic optimization theory

Chapter 1

INTRODUCTION

1.1 What is mathematical optimization?

Formally, *Mathematical Optimization* is the process of

(i) the *formulation* and

(ii) the *solution* of a constrained optimization problem of the general mathematical form:

$$\underset{\text{w.r.t. } \mathbf{x}}{\text{minimize}}\ f(\mathbf{x}),\ \mathbf{x} = [x_1, x_2, \ldots, x_n]^T \in \mathbb{R}^n$$

subject to the constraints:

$$
\begin{aligned}
g_j(\mathbf{x}) &\le 0, \quad j = 1,\ 2,\ \ldots,\ m \\
h_j(\mathbf{x}) &= 0, \quad j = 1,\ 2,\ \ldots,\ r
\end{aligned}
\tag{1.1}
$$

where $f(\mathbf{x})$, $g_j(\mathbf{x})$ and $h_j(\mathbf{x})$ are scalar functions of the real *column vector* $\mathbf{x}$.

The continuous components x_i of $\mathbf{x} = [x_1, x_2, \ldots, x_n]^T$ are called the (*design*) *variables*, $f(\mathbf{x})$ is the *objective function*, $g_j(\mathbf{x})$ denotes the respective *inequality constraint functions* and $h_j(\mathbf{x})$ the *equality constraint functions*.

© Springer International Publishing AG, part of Springer Nature 2018
J.A. Snyman and D.N. Wilke, *Practical Mathematical Optimization*,
Springer Optimization and Its Applications 133,
https://doi.org/10.1007/978-3-319-77586-9_1

The optimum vector $\mathbf{x}$ that solves problem (1.1) is denoted by $\mathbf{x}^*$ with corresponding optimum function value $f(\mathbf{x}^*)$. If no constraints are specified, the problem is called an *unconstrained* minimization problem.

Mathematical Optimization is often also called *Nonlinear Programming*, *Mathematical Programming* or *Numerical Optimization*. In more general terms Mathematical Optimization may be described as the science of determining the *best* solutions to mathematically defined problems, which may be models of physical reality or of manufacturing and management systems. In the first case solutions are sought that often correspond to minimum energy configurations of general structures, from molecules to suspension bridges, and are therefore of interest to Science and Engineering. In the second case commercial and financial considerations of economic importance to Society and Industry come into play, and it is required to make decisions that will ensure, for example, maximum profit or minimum cost.

The history of the Mathematical Optimization, where functions of many variables are considered, is relatively short, spanning roughly only 70 years. At the end of the 1940s the very important simplex method for solving the special class of linear programming problems was developed. Since then numerous methods for solving the general optimization problem (1.1) have been developed, tested, and successfully applied to many important problems of scientific and economic interest. There is no doubt that the advent of the computer was essential for the development of these optimization methods. However, in spite of the proliferation of optimization methods, there is no universal method for solving all optimization problems. According to Nocedal and Wright (1999): "...there are numerous algorithms, each of which is tailored to a particular type of optimization problem. It is often the user's responsibility to choose an algorithm that is appropriate for the specific application. This choice is an important one; it may determine whether the problem is solved rapidly or slowly and, indeed, whether the solution is found at all." In a similar vein Vanderplaats (1998) states that "The author of each algorithm usually has numerical examples which demonstrate the efficiency and accuracy of the method, and the unsuspecting practitioner will often invest a great deal of time and effort in programming an algorithm, only to find that it will not in fact solve the particular problem being attempted. This often leads to disenchantment with these techniques that can be avoided if the user is knowledgeable in the basic concepts of numerical optimization." With these representative and authoritative opinions in mind, and also taking into account the present authors' personal experiences in developing algorithms and applying them to design problems in mechanics, this text has been written to provide a brief but unified introduction to optimization concepts and methods. In addition, an overview of a set of novel algorithms, developed by the authors and their students at the University of Pretoria over the past thirty years, is also given.

The emphasis of this book is almost exclusively on gradient-based methods. This is for two reasons. (i) The authors believe that the introduction to the topic of mathematical optimization is best done via the classical gradient-based approach and (ii), contrary to the current popular trend of using non-gradient methods, such as genetic algorithms (GA's), simulated annealing, particle swarm optimization and other evolutionary methods, the authors are of the opinion that these search methods are, in many cases, computationally too expensive to be viable. The argument that the presence of numerical noise and multiple minima disqualify the use of gradient-based methods, and that the only way out in such cases is the use of the above mentioned non-gradient search techniques, is not necessarily true. It is the experience of the authors that, through the judicious use of gradient-based methods, problems with numerical noise and multiple minima may be solved, and at a fraction of the computational cost of search techniques such as genetic algorithms. In this context Chapter 6, dealing with the new gradient-based methods developed by the first author and gradient-only methods developed by the authors in Chapter 8, are especially important. The presentation of the material is not overly rigorous, but hopefully correct, and should provide the necessary information to allow scientists and engineers to select appropriate optimization algorithms and to apply them successfully to their respective fields of interest.

Many excellent and more comprehensive texts on practical optimization can be found in the literature. In particular the authors wish to acknowledge the works of Wismer and Chattergy (1978), Chvatel (1983), Fletcher (1987), Bazaraa et al. (1993), Arora (1989), Haftka and Gürdal (1992), Rao (1996), Vanderplaats (1998), Nocedal and Wright (1999) and Papalambros and Wilde (2000).

1.2   Objective and constraint functions

The values of the functions $f(\mathbf{x})$, $g_j(\mathbf{x})$ and $h_j(\mathbf{x})$ at any point $\mathbf{x} = [x_1, x_2, \ldots, x_n]^T$, may in practice be obtained in different ways:

(i) from *analytically* known *formulae*, e.g. $f(\mathbf{x}) = x_1^2 + 2x_2^2 + \sin x_3$;

(ii) as the *outcome* of some complicated *computational process*, e.g. $g_1(\mathbf{x}) = a(\mathbf{x}) - a_{\max}$, where $a(\mathbf{x})$ is the stress, computed by means of a finite element analysis, at some point in a structure, the design of which is specified by $\mathbf{x}$; or

(iii) from *measurements* taken of a *physical process*, e.g. $h_1(\mathbf{x}) = T(\mathbf{x}) - T_0$, where $T(\mathbf{x})$ is the temperature measured at some specified point in a reactor, and $\mathbf{x}$ is the vector of operational settings.

The first two ways of function evaluation are by far the most common. The optimization principles that apply in these cases, where computed function values are used, may be carried over directly to also be applicable to the case where the function values are obtained through physical measurements.

Much progress has been made with respect to methods for solving different classes of the general problem (1.1). Sometimes the solution may be obtained *analytically*, i.e. a closed-form solution in terms of a *formula* is obtained.

In general, especially for $n > 2$, solutions are usually obtained *numerically* by means of suitable *algorithms* (computational recipes).

Expertise in the *formulation* of appropriate optimization problems of the form (1.1), through which an optimum decision can be made, is gained from *experience*. This exercise also forms part of what is generally known as the *mathematical modelling* process. In brief, attempting to solve real-world problems via mathematical modelling requires the cyclic performance of the four steps depicted in Figure 1.1. The main steps are: 1) the observation and study of the real-world situation associated with a practical problem, 2) the abstraction of the problem by the construction of a mathematical model, that is described in terms of

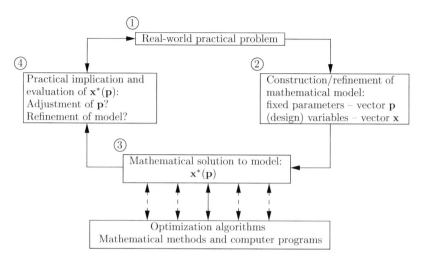

Figure 1.1: The mathematical modelling process

preliminary fixed model parameters $\mathbf{p}$, and variables $\mathbf{x}$, the latter to be determined such that model performs in an acceptable manner, 3) the solution of a resulting purely mathematical problem, that requires an analytical or numerical parameter dependent solution $\mathbf{x}^*(\mathbf{p})$, and 4) the evaluation of the solution $\mathbf{x}^*(\mathbf{p})$ and its practical implications. After step 4) it may be necessary to adjust the parameters and refine the model, which will result in a new mathematical problem to be solved and evaluated. It may be required to perform the modelling cycle a number of times, before an acceptable solution is obtained. More often than not, the mathematical problem to be solved in 3) is a *mathematical optimization problem*, requiring a numerical solution. The *formulation* of an appropriate and consistent optimization problem (or model) is probably the most important, but unfortunately, also the *most neglected* part of Practical Mathematical Optimization.

This book gives a very brief introduction to the *formulation* of optimization problems, and deals with different *optimization algorithms* in greater depth. Since no algorithm is generally applicable to all classes of problems, the emphasis is on providing sufficient information to allow for the selection of appropriate algorithms or methods for different specific problems.

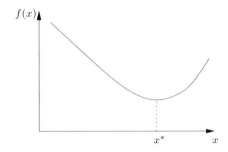

Figure 1.2: Function of single variable with optimum at x^*

1.3 Basic optimization concepts

1.3.1 Simplest class of problems: Unconstrained one-dimensional minimization

Consider the minimization of a smooth, i.e. continuous and twice continuously differentiable (C^2) function of a single real variable, i.e. the problem:

$$\underset{x}{\text{minimize}} \, f(x), \, x \in \mathbb{R}, \, f \in C^2. \tag{1.2}$$

With reference to Figure 1.2, for a strong local minimum, it is required to determine a x^* such that $f(x^*) < f(x)$ for all x.

Clearly x^* occurs where the slope is zero, i.e. where

$$f'(x) = \frac{df(x)}{dx} = 0,$$

which corresponds to the first order necessary condition. In addition *non-negative curvature* is necessary at x^*, i.e. it is required that the second order condition

$$f''(x) = \frac{d^2 f(x)}{dx^2} > 0$$

must hold at x^* for a strong local minimum.

A simple *special case* is where $f(x)$ has the simple *quadratic form*:

$$f(x) = ax^2 + bx + c. \tag{1.3}$$

Since the minimum occurs where $f'(x) = 0$, it follows that the closed-form solution is given by

$$x^* = -\frac{b}{2a}, \text{ provided } f''(x^*) = 2a > 0. \qquad (1.4)$$

If $f(x)$ has a *more general form*, then a closed-form solution is in general not possible. In this case, the solution may be obtained numerically via the *Newton-Raphson algorithm*:

Given an approximation x^0, iteratively compute:

$$x^{i+1} = x^i - \frac{f'(x^i)}{f''(x^i)}; \ i = 0, \ 1, \ 2, \ \ldots \qquad (1.5)$$

Hopefully $\lim_{i \to \infty} x^i = x^*$, i.e. the iterations converge, in which case a sufficiently accurate numerical solution is obtained after a finite number of iterations.

1.3.2   Contour representation of a function of two variables $(n = 2)$

Consider a function $f(\mathbf{x})$ of two variables, $\mathbf{x} = [x_1, x_2]^T$. The locus of all points satisfying $f(\mathbf{x}) = c = $ constant, forms a contour in the $x_1 - x_2$ plane. For each value of c there is a corresponding different contour.

Figure 1.3 depicts the contour representation for the example $f(\mathbf{x}) = x_1^2 + 2x_2^2$.

In three dimensions $(n = 3)$, the contours are *surfaces of constant function* value. In more than three dimensions $(n > 3)$ the contours are, of course, impossible to visualize. *Nevertheless, the contour representation in two-dimensional space will be used throughout the discussion of optimization techniques to help visualize the various optimization concepts.*

Other examples of 2-dimensional objective function contours are shown in Figures 1.4 to 1.6.

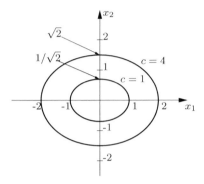

Figure 1.3: Contour representation of the function $f(\mathbf{x}) = x_1^2 + 2x_2^2$

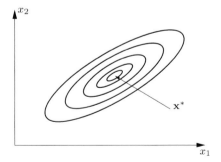

Figure 1.4: General quadratic function

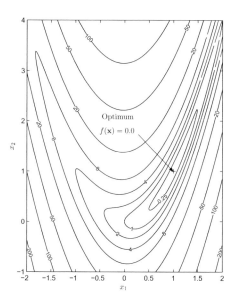

Figure 1.5: The two-dimensional Rosenbrock function $f(\mathbf{x}) = 10(x_2 - x_1^2)^2 + (1 - x_1)^2$

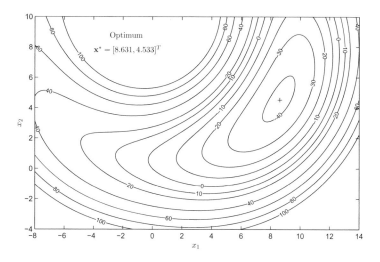

Figure 1.6: Potential energy function of a spring-force system (Vanderplaats 1998)

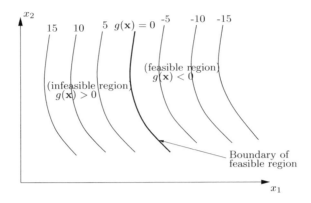

Figure 1.7: Contours within feasible and infeasible regions

1.3.3   Contour representation of constraint functions

1.3.3.1   Inequality constraint function $g(\mathbf{x})$

The contours of a typical inequality constraint function $g(\mathbf{x})$, in $g(\mathbf{x}) \leq$ 0, are shown in Figure 1.7. The contour $g(\mathbf{x}) = 0$ divides the plane into a *feasible region* and an *infeasible region*.

More generally, the boundary is a surface in three dimensions and a so-called "hyper-surface" if $n > 3$, which of course cannot be visualised.

1.3.3.2   Equality constraint function $h(\mathbf{x})$

Here, as shown in Figure 1.8, only the line $h(\mathbf{x}) = 0$ is a feasible contour.

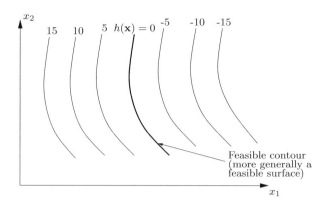

Figure 1.8: Feasible contour of equality constraint

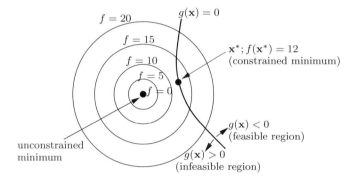

Figure 1.9: Contour representation of inequality constrained problem

1.3.4 Contour representations of constrained optimization problems

1.3.4.1 Representation of inequality constrained problem

Figure 1.9 graphically depicts the inequality constrained problem:

$$\min f(\mathbf{x})$$
$$\text{such that } g(\mathbf{x}) \leq 0.$$

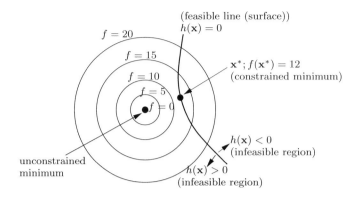

Figure 1.10: Contour representation of equality constrained problem

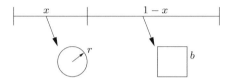

Figure 1.11: Wire divided into two pieces with $x_1 = x$ and $x_2 = 1 - x$

1.3.4.2   Representation of equality constrained problem

Figure 1.10 graphically depicts the equality constrained problem:

$$\min f(\mathbf{x})$$
$$\text{such that } h(\mathbf{x}) = 0.$$

1.3.5   Simple example illustrating the formulation and solution of an optimization problem

Problem: A length of wire 1 meter long is to be divided into two pieces, one in a circular shape and the other into a square as shown in Figure 1.11. What must the individual lengths be so that the total area is a minimum?

Formulation 1

Set length of first piece $= x$, then the area is given by $f(x) = \pi r^2 + b^2$. Since $r = \frac{x}{2\pi}$ and $b = \frac{1-x}{4}$ it follows that

$$f(x) = \pi \left(\frac{x^2}{4\pi^2} \right) + \frac{(1-x)^2}{16}.$$

The problem therefore reduces to an unconstrained minimization problem:

$$\text{minimize } f(x) = 0.1421x^2 - 0.125x + 0.0625.$$

Solution of Formulation 1

The function $f(x)$ is quadratic, therefore an analytical solution is given by the formula $x^* = -\frac{b}{2a}$ $(a > 0)$:

$$x^* = -\frac{-0.125}{2(0.1421)} = 0.4398 \text{ m},$$

and

$$1 - x^* = 0.5602 \text{ m with } f(x^*) = 0.0350 \text{ m}^2.$$

Formulation 2

Divide the wire into respective lengths x_1 and x_2 $(x_1 + x_2 = 1)$. The area is now given by

$$f(\mathbf{x}) = \pi r^2 + b^2 = \pi \left(\frac{x_1^2}{4\pi^2} \right) + \left(\frac{x_2}{4} \right)^2 = 0.0796x_1^2 + 0.0625x_2^2.$$

Here the problem reduces to an *equality constrained* problem:

$$\begin{aligned} \text{minimize } f(\mathbf{x}) &= 0.0796x_1^2 + 0.0625x_2^2 \\ \text{such that } h(\mathbf{x}) &= x_1 + x_2 - 1 = 0. \end{aligned}$$

Solution of Formulation 2

This constrained formulated problem is more difficult to solve. The closed-form analytical solution is not obvious and special constrained optimization techniques, such as the *method of Lagrange multipliers* to be discussed later, must be applied to solve the constrained problem analytically. The graphical solution is sketched in Figure 1.12.

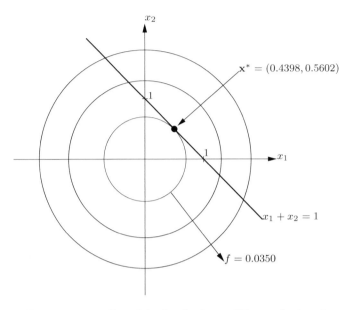

Figure 1.12: Graphical solution of Formulation 2

1.3.6   Maximization

The maximization problem: $\max_{\mathbf{x}} f(\mathbf{x})$ can be cast in the standard form (1.1) by observing that $\max_{\mathbf{x}} f(\mathbf{x}) = -\min_{\mathbf{x}}\{-f(\mathbf{x})\}$ as shown in Figure 1.13. Therefore in applying a minimization algorithm set $F(\mathbf{x}) = -f(\mathbf{x})$.

Also if the inequality constraints are given in the non-standard form: $g_j(\mathbf{x}) \geq 0$, then set $\tilde{g}_j(\mathbf{x}) = -g_j(\mathbf{x})$. In standard form the problem then becomes:

$$\text{minimize } F(\mathbf{x}) \text{ such that } \tilde{g}_j(\mathbf{x}) \leq 0.$$

Once the minimizer $\mathbf{x}^*$ is obtained, the maximum value of the original maximization problem is given by $-F(\mathbf{x}^*)$.

1.3.7   The special case of Linear Programming

A very important special class of the general optimization problem arises when both the objective function and all the constraints are linear func-

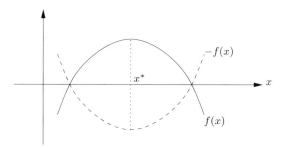

Figure 1.13: Maximization problem transformed to minimization problem

tions of $\mathbf{x}$. This is called a *Linear Programming* problem and is usually stated in the following form:

$$\min_{\mathbf{x}} f(\mathbf{x}) = \mathbf{c}^T \mathbf{x}$$

such that $\qquad\qquad\qquad\qquad\qquad$ (1.6)

$$\mathbf{A}\mathbf{x} \leq \mathbf{b}; \ \mathbf{x} \geq \mathbf{0}$$

where $\mathbf{c}$ is a real n-vector and $\mathbf{b}$ is a real m-vector, and $\mathbf{A}$ is a $m \times n$ real matrix. A linear programming problem in two variables is graphically depicted in Figure 1.14.

Special methods have been developed for solving linear programming problems. Of these the most famous are the simplex method proposed by Dantzig in 1947 (Dantzig 1963) and the interior-point method (Karmarkar 1984). A short introduction to the simplex method, according to Chvatel (1983), is given in Appendix A.

1.3.8 Scaling of design variables

In formulating mathematical optimization problems, great care must be taken to ensure that the scale of the variables are more or less of the same order. If not, the formulated problem may be relatively insensitive to the variations in one or more of the variables, and any optimization algorithm will struggle to converge to the true solution, because of extreme distortion of the objective function contours as result of the poor scaling. In particular it may lead to difficulties when selecting step lengths and

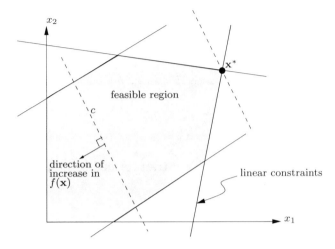

Figure 1.14: Graphical representation of a two-dimensional linear programming problem

calculating numerical gradients. Scaling difficulties often occur where the variables are of different dimension and expressed in different units. Hence it is good practice, if the variable ranges are very large, to scale the variables so that all the variables will be dimensionless and vary between 0 and 1 approximately. For scaling the variables, it is necessary to establish an approximate range for each of the variables. For this, take some estimates (based on judgement and experience) for the lower and upper limits. The values of the bounds are not critical. Another related matter is the scaling or normalization of constraint functions. This becomes necessary whenever the values of the constraint functions differ by large magnitudes.

1.4   Further mathematical concepts

1.4.1   Convexity

A line through the points $\mathbf{x}^1$ and $\mathbf{x}^2$ in $\mathbb{R}^n$ is the set

$$L = \{\mathbf{x}|\mathbf{x} = \mathbf{x}^1 + \lambda(\mathbf{x}^2 - \mathbf{x}^1),\ \text{for all}\ \lambda \in \mathbb{R}\}. \qquad (1.7)$$

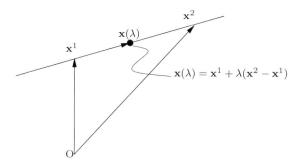

Figure 1.15: Representation of a point on the straight line through $\mathbf{x}^1$ and $\mathbf{x}^2$

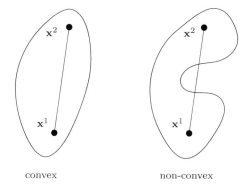

Figure 1.16: Examples of a convex and a non-convex set

Equivalently for any point $\mathbf{x}$ on the line there exists a λ such that $\mathbf{x}$ may be specified by $\mathbf{x} = \mathbf{x}(\lambda) = \lambda\mathbf{x}^2 + (1 - \lambda)\mathbf{x}^1$ as shown in Figure 1.15.

1.4.1.1   Convex sets

A set X is convex if for all $\mathbf{x}^1$, $\mathbf{x}^2 \in X$ it follows that

$$\mathbf{x} = \lambda\mathbf{x}^2 + (1 - \lambda)\mathbf{x}^1 \in X \text{ for all } 0 \le \lambda \le 1.$$

If this condition does not hold the set is non-convex (see Figure 1.16).

1.4.1.2   Convex functions

Given two points $\mathbf{x}^1$ and $\mathbf{x}^2$ in $\mathbb{R}^n$, then any point $\mathbf{x}$ on the straight line connecting them (see Figure 1.15) is given by

$$\mathbf{x} = \mathbf{x}(\lambda) = \mathbf{x}^1 + \lambda(\mathbf{x}^2 - \mathbf{x}^1), \ 0 < \lambda < 1. \tag{1.8}$$

A function $f(\mathbf{x})$ is a *convex function* over a *convex set* X if for all $\mathbf{x}^1$, $\mathbf{x}^2$ in X and for all $\lambda \in [0, 1]$:

$$f(\lambda \mathbf{x}^2 + (1 - \lambda)\mathbf{x}^1) \leq \lambda f(\mathbf{x}^2) + (1 - \lambda)f(\mathbf{x}^1). \tag{1.9}$$

The function is strictly convex if $<$ applies. Concave functions are similarly defined.

Consider again the line connecting $\mathbf{x}^1$ and $\mathbf{x}^2$. Along this line, the function $f(\mathbf{x})$ is a function of the single variable λ:

$$F(\lambda) = f(\mathbf{x}(\lambda)) = f(\mathbf{x}^1 + \lambda(\mathbf{x}^2 - \mathbf{x}^1)). \tag{1.10}$$

This is equivalent to $F(\lambda) = f(\lambda \mathbf{x}^2 + (1 - \lambda)\mathbf{x}^1)$, with $F(0) = f(\mathbf{x}^1)$ and $F(1) = f(\mathbf{x}^2)$. Therefore (1.9) may be written as

$$F(\lambda) \leq \lambda F(1) + (1 - \lambda)F(0) = F_{int}$$

where F_{int} is the linearly interpolated value of F at λ as shown in Figure 1.17.

Graphically $f(\mathbf{x})$ is *convex* over the convex set X if $F(\lambda)$ has the convex form shown in Figure 1.17 for any two points $\mathbf{x}^1$ and $\mathbf{x}^2$ in X.

1.4.2   Gradient vector of $f(\mathbf{x})$

For a function $f(\mathbf{x}) \in C^2$ there exists, at any point $\mathbf{x}$ a vector of first order partial derivatives, or gradient vector:

$$\nabla f(\mathbf{x}) = \begin{bmatrix} \dfrac{\partial f}{\partial x_1}(\mathbf{x}) \\[2ex] \dfrac{\partial f}{\partial x_2}(\mathbf{x}) \\[2ex] \vdots \\[2ex] \dfrac{\partial f}{\partial x_n}(\mathbf{x}) \end{bmatrix}. \tag{1.11}$$

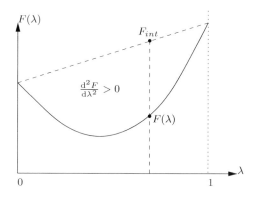

Figure 1.17: Convex form of $F(\lambda)$

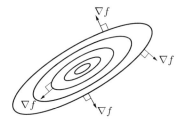

Figure 1.18: Directions of the gradient vector

It can easily be shown that if the function $f(\mathbf{x})$ is smooth, then at the point $\mathbf{x}$ the gradient vector $\boldsymbol{\nabla} f(\mathbf{x})$ is always perpendicular to the contours (or surfaces of constant function value) and is in the *direction of maximum increase* of $f(\mathbf{x})$, as depicted in Figure 1.18.

1.4.3   Hessian matrix of $f(\mathbf{x})$

If $f(\mathbf{x})$ is twice continuously differentiable then at the point $\mathbf{x}$ there exists a matrix of second order partial derivatives or *Hessian matrix*:

$$\mathbf{H}(\mathbf{x}) \;=\; \left\{ \frac{\partial^2 f}{\partial x_i \partial x_j}(\mathbf{x}) \right\} = \boldsymbol{\nabla}^2 f(\mathbf{x}) \tag{1.12}$$

$$= \begin{bmatrix} \dfrac{\partial^2 f}{\partial x_1^2}(\mathbf{x}) & \dfrac{\partial^2 f}{\partial x_1 \partial x_2}(\mathbf{x}) & \cdots \\[2ex] \dfrac{\partial^2 f}{\partial x_2 \partial x_1}(\mathbf{x}) & & \\[1ex] \vdots & & \\[1ex] \dfrac{\partial^2 f}{\partial x_n \partial x_1}(\mathbf{x}) & \cdots & \dfrac{\partial^2 f}{\partial x_n^2}(\mathbf{x}) \end{bmatrix} .$$

Clearly $\mathbf{H}(\mathbf{x})$ is a $n \times n$ symmetrical matrix.

1.4.3.1   Test for convexity of $f(\mathbf{x})$

If $f(\mathbf{x}) \in C^2$ is defined over a convex set X, then it can be shown (see Theorem 5.1.3 in Chapter 5) that if $\mathbf{H}(\mathbf{x})$ is positive-definite for all $\mathbf{x} \in X$, then $f(\mathbf{x})$ is strictly convex over X.

To test for convexity, i.e. to determine whether $\mathbf{H}(\mathbf{x})$ is positive-definite or not, apply Sylvester's Theorem or any other suitable numerical method (Fletcher 1987). For example, a convenient numerical test for positive-definiteness at $\mathbf{x}$ is to show that all the eigenvalues for $\mathbf{H}(\mathbf{x})$ are positive.

1.4.4   The quadratic function in $\mathbb{R}^n$

The quadratic function in n variables may be written as

$$f(\mathbf{x}) = \tfrac{1}{2}\mathbf{x}^T \mathbf{A}\mathbf{x} + \mathbf{b}^T \mathbf{x} + c \tag{1.13}$$

where $c \in \mathbb{R}$, $\mathbf{b}$ is a real n-vector and $\mathbf{A}$ is a $n \times n$ real matrix that can be chosen in a non-unique manner. It is usually chosen symmetrical in which case it follows that

$$\nabla f(\mathbf{x}) = \mathbf{A}\mathbf{x} + \mathbf{b}; \quad \mathbf{H}(\mathbf{x}) = \mathbf{A}. \tag{1.14}$$

The function $f(\mathbf{x})$ is called positive-definite if $\mathbf{A}$ is positive-definite since, by the test in Section 1.4.3.1, a function $f(\mathbf{x})$ is convex if $\mathbf{H}(\mathbf{x})$ is positive-definite.

1.4.5 The directional derivative of $f(\mathbf{x})$ in the direction u

It is usually assumed that $\|\mathbf{u}\| = 1$. Consider the differential:

$$df = \frac{\partial f}{\partial x_1}dx_1 + \cdots + \frac{\partial f}{\partial x_n}dx_n = \nabla^T f(\mathbf{x})d\mathbf{x}. \tag{1.15}$$

A point $\mathbf{x}$ on the line through $\mathbf{x}'$ in the direction $\mathbf{u}$ is given by $\mathbf{x} = \mathbf{x}(\lambda) = \mathbf{x}' + \lambda\mathbf{u}$, and for a small change $d\lambda$ in λ, $d\mathbf{x} = \mathbf{u}d\lambda$. Along this line $F(\lambda) = f(\mathbf{x}' + \lambda\mathbf{u})$ and the differential at any point $\mathbf{x}$ on the given line in the direction $\mathbf{u}$ is therefore given by $dF = df = \nabla^T f(\mathbf{x})\mathbf{u}d\lambda$. It follows that the *directional derivative* at $\mathbf{x}$ in the *direction* $\mathbf{u}$ is

$$\frac{dF(\lambda)}{d\lambda} = \frac{df(\mathbf{x})}{d\lambda}\bigg|_{\mathbf{u}} = \nabla^T f(\mathbf{x})\mathbf{u}. \tag{1.16}$$

1.5 Unconstrained minimization

In considering the unconstrained problem: $\min_{\mathbf{x}} f(\mathbf{x})$, $\mathbf{x} \in X \subseteq \mathbb{R}^n$, the following questions arise:

(i) what are the conditions for a minimum to exist,

(ii) is the minimum unique,

(iii) are there any relative minima?

Figure 1.19 (after Farkas and Jarmai 1997) depicts different types of minima that may arise for functions of a single variable, and for functions of two variables in the presence of inequality constraints. Intuitively,

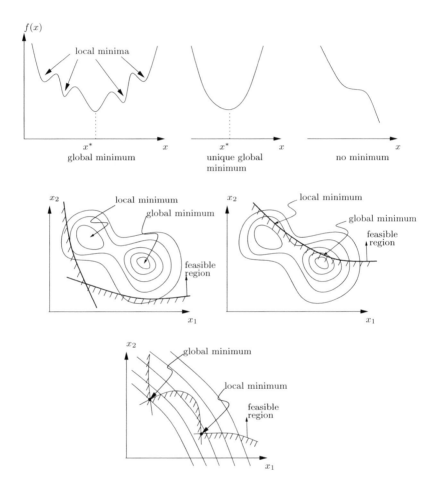

Figure 1.19: Types of minima

with reference to Figure 1.19, one feels that a general function may have a single unique global minimum, or it may have more than one local minimum. The function may indeed have no local minimum at all, and in two dimensions the possibility of saddle points also comes to mind. Thus, in order to answer the above questions regarding the nature of any given function more analytically, it is necessary to give more precise meanings to the above mentioned notions.

1.5.1 Global and local minima; saddle points

1.5.1.1 Global minimum

$\mathbf{x}^*$ is a global minimum over the set X if $f(\mathbf{x}) \geq f(\mathbf{x}^*)$ for all $\mathbf{x} \in X \subset \mathbb{R}^n$.

1.5.1.2 Strong local minimum

$\mathbf{x}^*$ is a strong local minimum if there exists an $\varepsilon > 0$ such that

$$f(\mathbf{x}) > f(\mathbf{x}^*) \text{ for all } \{\mathbf{x} \,|\, \|\mathbf{x} - \mathbf{x}^*\| < \varepsilon\}$$

where $\|\cdot\|$ denotes the Euclidean norm. This definition is sketched in Figure 1.20.

1.5.1.3 Test for unique local global minimum

It can be shown (see Theorems 5.1.4 and 5.1.5 in Chapter 5) that if $f(\mathbf{x})$ is strictly convex over X, then a strong local minimum is also the global minimum.

The global minimizer can be difficult to find since the knowledge of $f(\mathbf{x})$ is usually only local. Most minimization methods seek only a local minimum. An approximation to the global minimum is obtained in practice by the multi-start application of a local minimizer from randomly selected different starting points in X. The lowest value obtained after a sufficient number of trials is then taken as a good approximation to the global solution (see Snyman and Fatti 1987; Groenwold and Snyman

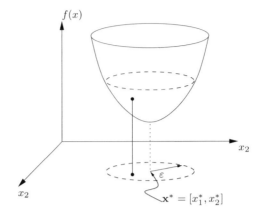

Figure 1.20: Graphical representation of the definition of a local minimum

2002). If, however, it is known that the function is strictly convex over X, then only one trial is sufficient since only one local minimum, the global minimum, exists.

1.5.1.4   Saddle points

$f(\mathbf{x})$ has a saddle point at $\overline{\mathbf{x}} = \begin{bmatrix} \mathbf{x}^0 \\ \mathbf{y}^0 \end{bmatrix}$ if there exists an $\varepsilon > 0$ such that for all $\mathbf{x}$, $\|\mathbf{x} - \mathbf{x}^0\| < \varepsilon$ and all $\mathbf{y}$, $\|\mathbf{y} - \mathbf{y}^0\| < \varepsilon$: $f(\mathbf{x}, \mathbf{y}^0) \leq f(\mathbf{x}^0, \mathbf{y}^0) \leq f(\mathbf{x}^0, \mathbf{y})$.

A contour representation of a saddle point in two dimensions is given in Figure 1.21.

1.5.2   Local characterization of the behaviour of a multi-variable function

It is assumed here that $f(\mathbf{x})$ is a smooth function, i.e., that it is a twice continuously differentiable function ($f(\mathbf{x}) \in C^2$). Consider again the line $\mathbf{x} = \mathbf{x}(\lambda) = \mathbf{x}' + \lambda \mathbf{u}$ through the point $\mathbf{x}'$ in the direction $\mathbf{u}$.

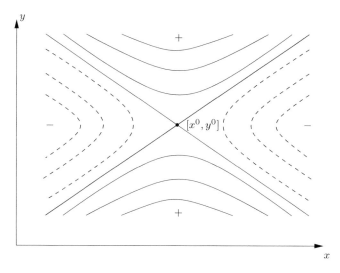

Figure 1.21: Contour representation of saddle point

Along this line a single variable function $F(\lambda)$ may be defined:

$$F(\lambda) = f(\mathbf{x}(\lambda)) = f(\mathbf{x}' + \lambda\mathbf{u}).$$

It follows from (1.16) that

$$\frac{dF(\lambda)}{d\lambda} = \frac{df(\mathbf{x}(\lambda))}{d\lambda}\bigg|_{\mathbf{u}} = \boldsymbol{\nabla}^T f(\mathbf{x}(\lambda))\mathbf{u} = g(\mathbf{x}(\lambda)) = G(\lambda)$$

which is also a single variable function of λ along the line $\mathbf{x} = \mathbf{x}(\lambda) = \mathbf{x}' + \lambda\mathbf{u}$.

Thus similarly it follows that

$$\begin{aligned}
\frac{d^2 F(\lambda)}{d\lambda^2} = \frac{dG(\lambda)}{d\lambda} = \frac{dg(\mathbf{x}(\lambda))}{d\lambda}\bigg|_{\mathbf{u}} &= \boldsymbol{\nabla}^T g(\mathbf{x}(\lambda))\mathbf{u} \\
&= \boldsymbol{\nabla}^T \left(\boldsymbol{\nabla}^T f(\mathbf{x}(\lambda))\mathbf{u}\right)\mathbf{u} \\
&= \mathbf{u}^T \mathbf{H}(\mathbf{x}(\lambda))\mathbf{u}.
\end{aligned}$$

Summarising: the first and second order derivatives of $F(\lambda)$ with respect to λ at any point $\mathbf{x} = \mathbf{x}(\lambda)$ on any line (any $\mathbf{u}$) through $\mathbf{x}'$ is given by

$$\frac{dF(\lambda)}{d\lambda} = \boldsymbol{\nabla}^T f(\mathbf{x}(\lambda))\mathbf{u}, \qquad (1.17)$$

$$\frac{d^2 F(\lambda)}{d\lambda^2} = \mathbf{u}^T \mathbf{H}(\mathbf{x}(\lambda))\mathbf{u} \qquad (1.18)$$

where $\mathbf{x}(\lambda) = \mathbf{x}' + \lambda\mathbf{u}$ and $F(\lambda) = f(\mathbf{x}(\lambda)) = f(\mathbf{x}' + \lambda\mathbf{u})$.

These results may be used to obtain Taylor's expansion for a multi-variable function. Consider again the single variable function $F(\lambda)$ defined on the line through $\mathbf{x}'$ in the direction $\mathbf{u}$ by $F(\lambda) = f(\mathbf{x}' + \lambda\mathbf{u})$. It is known that the Taylor expansion of $F(\lambda)$ about 0 is given by

$$F(\lambda) = F(0) + \lambda F'(0) + \tfrac{1}{2}\lambda^2 F''(0) + \ldots \qquad (1.19)$$

With $F(0) = f(\mathbf{x}')$, and substituting expressions (1.17) and (1.18) for respectively $sF'(\lambda)$ and $F''(\lambda)$ at $\lambda = 0$ into (1.19) gives

$$F(\lambda) = f(\mathbf{x}' + \lambda\mathbf{u}) = f(\mathbf{x}') + \boldsymbol{\nabla}^T f(\mathbf{x}')\lambda\mathbf{u} + \tfrac{1}{2}\lambda\mathbf{u}^T \mathbf{H}(\mathbf{x}')\lambda\mathbf{u} + \ldots$$

Setting $\boldsymbol{\delta} = \lambda\mathbf{u}$ in the above gives the expansion:

$$f(\mathbf{x}' + \boldsymbol{\delta}) = f(\mathbf{x}') + \boldsymbol{\nabla}^T f(\mathbf{x}')\boldsymbol{\delta} + \tfrac{1}{2}\boldsymbol{\delta}^T \mathbf{H}(\mathbf{x}')\boldsymbol{\delta} + \ldots \qquad (1.20)$$

Since the above applies for any line (any $\mathbf{u}$) through $\mathbf{x}'$, it represents the general Taylor expansion for a multi-variable function about $\mathbf{x}'$. If $f(\mathbf{x})$ is fully continuously differentiable in the neighbourhood of $\mathbf{x}'$ it can be shown that the truncated second order Taylor expansion for a multi-variable function is given by

$$f(\mathbf{x}' + \boldsymbol{\delta}) = f(\mathbf{x}') + \boldsymbol{\nabla}^T f(\mathbf{x}')\boldsymbol{\delta} + \tfrac{1}{2}\boldsymbol{\delta}^T \mathbf{H}(\mathbf{x}' + \theta\boldsymbol{\delta})\boldsymbol{\delta} \qquad (1.21)$$

for some $\theta \in [0, 1]$. This expression is important in the analysis of the behaviour of a multi-variable function at any given point $\mathbf{x}'$.

1.5.3 Necessary and sufficient conditions for a strong local minimum at $\mathbf{x}^*$

In particular, consider $\mathbf{x}' = \mathbf{x}^*$ a strong local minimizer. Then for any line (any $\mathbf{u}$) through $\mathbf{x}'$ the behaviour of $F(\lambda)$ in a neighbourhood of $\mathbf{x}^*$ is as shown in Figure 1.22, with minimum at at $\lambda = 0$.

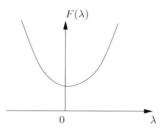

Figure 1.22: Behaviour of $F(\lambda)$ near $\lambda = 0$

Clearly, a *necessary first order condition* that must apply at $\mathbf{x}^*$ (corresponding to $\lambda = 0$) is that

$$\frac{dF(0)}{d\lambda} = \boldsymbol{\nabla}^T f(\mathbf{x}^*)\mathbf{u} = 0, \quad \text{for all } \mathbf{u} \neq \mathbf{0}. \tag{1.22}$$

It can easily be shown that this condition also implies that necessarily $\boldsymbol{\nabla} f(\mathbf{x}^*) = \mathbf{0}$.

A *necessary second order condition* that must apply at $\mathbf{x}^*$ is that

$$\frac{d^2 F(0)}{d\lambda^2} = \mathbf{u}^T \mathbf{H}(\mathbf{x}^*)\mathbf{u} > 0, \quad \text{for all } \mathbf{u} \neq \mathbf{0}. \tag{1.23}$$

Conditions (1.22) and (1.23) taken together are also *sufficient conditions* (i.e. those that imply) for $\mathbf{x}^*$ to be a strong local minimum if $f(\mathbf{x})$ is continuously differentiable in the vicinity of $\mathbf{x}^*$. This can easily be shown by substituting these conditions in the Taylor expansion (1.21).

Thus in summary, the *necessary and sufficient conditions* for $\mathbf{x}^*$ to be a strong local minimum are:

$$\begin{aligned} \boldsymbol{\nabla} f(\mathbf{x}^*) &= \mathbf{0} \\ \mathbf{H}(\mathbf{x}^*) &\text{ positive-definite.} \end{aligned} \tag{1.24}$$

In the argument above it has implicitly been assumed that $\mathbf{x}^*$ is an unconstrained minimum *interior* to X. If $\mathbf{x}^*$ lies on the *boundary* of X (see Figure 1.23) then

$$\frac{dF(0)}{d\lambda} \geq 0, \text{ i.e. } \boldsymbol{\nabla}^T f(\mathbf{x}^*)\mathbf{u} \geq 0 \tag{1.25}$$

for all *allowable* directions $\mathbf{u}$, i.e. for directions such that $\mathbf{x}^* + \lambda\mathbf{u} \in X$ for arbitrary small $\lambda > 0$.

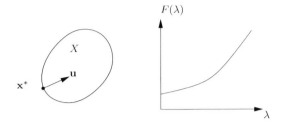

Figure 1.23: Behaviour of $F(\lambda)$ for all allowable directions of $\mathbf{u}$

Conditions (1.24) for an unconstrained strong local minimum play a very important role in the construction of practical algorithms for unconstrained optimization.

1.5.3.1   Application to the quadratic function

Consider the quadratic function:

$$f(\mathbf{x}) = \tfrac{1}{2}\mathbf{x}^T \mathbf{A}\mathbf{x} + \mathbf{b}^T \mathbf{x} + c.$$

In this case the first order necessary condition for a minimum implies that

$$\nabla f(\mathbf{x}) = \mathbf{A}\mathbf{x} + \mathbf{b} = \mathbf{0}.$$

Therefore a candidate solution point is

$$\mathbf{x}^* = -\mathbf{A}^{-1}\mathbf{b}. \tag{1.26}$$

If the second order necessary condition also applies, i.e. if $\mathbf{A}$ is positive-definite, then $\mathbf{x}^*$ is a unique minimizer.

1.5.4   General indirect method for computing $\mathbf{x}^*$

The general indirect method for determining $\mathbf{x}^*$ is to solve the system of equations $\nabla f(\mathbf{x}) = \mathbf{0}$ (corresponding to the first order necessary condition in (1.24)) by some numerical method, to yield all stationary points. An obvious method for doing this is Newton's method. Since in general the system will be non-linear, multiple stationary points are possible. These stationary points must then be further analysed in order to determine whether or not they are local minima.

1.5.4.1 Solution by Newton's method

Assume $\mathbf{x}^*$ is a local minimum and $\mathbf{x}^i$ an approximate solution, with associated unknown error $\boldsymbol{\delta}$ such that $\mathbf{x}^* = \mathbf{x}^i + \boldsymbol{\delta}$. Then by applying Taylor's theorem and the first order necessary condition for a minimum at $\mathbf{x}^*$ it follows that

$$0 = \nabla f(\mathbf{x}^*) = \nabla f(\mathbf{x}^i + \boldsymbol{\delta}) = \nabla f(\mathbf{x}^i) + \mathbf{H}(\mathbf{x}^i)\boldsymbol{\delta} + O\|\boldsymbol{\delta}\|^2.$$

If $\mathbf{x}^i$ is a good approximation then $\boldsymbol{\delta} \doteq \boldsymbol{\Delta}$, the solution of the linear system $\mathbf{H}(\mathbf{x}^i)\boldsymbol{\Delta} + \nabla f(\mathbf{x}^i) = \mathbf{0}$, obtained by ignoring the second order term in $\boldsymbol{\delta}$ above. A better approximation is therefore expected to be $\mathbf{x}^{i+1} = \mathbf{x}^i + \boldsymbol{\Delta}$ which leads to the Newton iterative scheme: Given an initial approximation $\mathbf{x}^0$, compute

$$\mathbf{x}^{i+1} = \mathbf{x}^i - \mathbf{H}^{-1}(\mathbf{x}^i)\nabla f(\mathbf{x}^i) \qquad (1.27)$$

for $i = 0,\ 1,\ 2,\ \ldots$ Hopefully $\lim_{i\to\infty} \mathbf{x}^i = \mathbf{x}^*$.

1.5.4.2 Example of Newton's method applied to a quadratic problem

Consider the unconstrained problem:

$$\text{minimize } f(\mathbf{x}) = \tfrac{1}{2}\mathbf{x}^T \mathbf{A}\mathbf{x} + \mathbf{b}^T\mathbf{x} + c.$$

In this case the first iteration in (1.27) yields

$$\mathbf{x}^1 = \mathbf{x}^0 - \mathbf{A}^{-1}(\mathbf{A}\mathbf{x}^0 + \mathbf{b}) = \mathbf{x}^0 - \mathbf{x}^0 - \mathbf{A}^{-1}\mathbf{b} = -\mathbf{A}^{-1}\mathbf{b}$$

i.e. $\mathbf{x}^1 = \mathbf{x}^* = -\mathbf{A}^{-1}\mathbf{b}$ in a single step (see (1.26)). This is to be expected since in this case no approximation is involved and thus $\boldsymbol{\Delta} = \boldsymbol{\delta}$.

1.5.4.3 Difficulties with Newton's method

Unfortunately, in spite of the attractive features of the Newton method, such as being quadratically convergent near the solution, the basic Newton method as described above does not always perform satisfactorily. The main difficulties are:

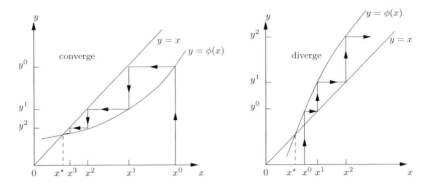

Figure 1.24: Graphical representation of Newton's iterative scheme for a single variable

(i) the method is not always convergent, even if $\mathbf{x}^0$ is close to $\mathbf{x}^*$, and

(ii) the method requires the computation of the Hessian matrix at each iteration.

The first of these difficulties may be illustrated by considering Newton's method applied to the one-dimensional problem: solve $f'(x) = 0$. In this case the iterative scheme is

$$x^{i+1} = x^i - \frac{f'(x^i)}{f''(x^i)} = \phi(x^i), \text{ for } i = 0,\ 1,\ 2,\ \ldots \qquad (1.28)$$

and the solution corresponds to the fixed point x^* where $x^* = \phi(x^*)$. Unfortunately in some cases, unless x^0 is chosen to be exactly equal to x^*, convergence will not necessarily occur. In fact, convergence is dependent on the nature of the fixed point function $\phi(x)$ in the vicinity of x^*, as shown for two different ϕ functions in Figure 1.24. With reference to the graphs Newton's method is: $y^i = \phi(x^i)$, $x^{i+1} = y^i$ for $i = 0,\ 1,\ 2,\ \ldots$. Clearly in the one case where $|\phi'(x)| < 1$ convergence occurs, but in the other case where $|\phi'(x)| > 1$ the scheme diverges.

In more dimensions the situation may be even more complicated. In addition, for a large number of variables, difficulty (ii) mentioned above becomes serious in that the computation of the Hessian matrix represents a major task. If the Hessian is not available in analytical form, use can be made of automatic differentiation techniques to compute it, or it can be estimated by means of finite differences. It should also be noted that in computing the Newton step in (1.27) a $n \times n$ linear system must be solved. This represents further computational effort. Therefore in practice the simple basic Newton method is not recommended. To avoid the convergence difficulty use is made of a modified Newton method, in which a more direct search procedure is employed in the direction of the Newton step, so as to ensure descent to the minimum $\mathbf{x}^*$. The difficulty associated with the computation of the Hessian is addressed in practice through the systematic update, from iteration to iteration, of an approximation of the Hessian matrix. These improvements to the basic Newton method are dealt with in greater detail in the next chapter.

1.6   Test functions

The efficiency of an algorithm is studied using standard functions with standard starting points $\mathbf{x}^0$. The total number of functions evaluations required to find the minimizer $\mathbf{x}^*$ is usually taken as a measure of the efficiency of the algorithm.

1.6.1   Unconstrained

Some classical unconstrained minimization test functions from (Rao 1996) are listed below.

1. Rosenbrock's parabolic valley:

$$f(\mathbf{x}) = 100(x_2 - x_1^2)^2 + (1 - x_1)^2; \ \mathbf{x}^0 = \begin{bmatrix} -1.2 \\ 1.0 \end{bmatrix} \ \mathbf{x}^* = \begin{bmatrix} 1 \\ 1 \end{bmatrix}.$$

2. Quadratic function:

$$f(\mathbf{x}) = (x_1 + 2x_2 - 7)^2 + (2x_1 + x_2 - 5)^2; \ \mathbf{x}^0 = \begin{bmatrix} 0 \\ 0 \end{bmatrix} \ \mathbf{x}^* = \begin{bmatrix} 1 \\ 3 \end{bmatrix}.$$

3. Powell's quartic function:

$$\begin{aligned} f(\mathbf{x}) &= (x_1 + 10x_2)^2 + 5(x_3 - x_4)^2 + (x_2 - 2x_3)^4 + 10(x_1 - x_4)^4; \\ \mathbf{x}^0 &= [3, -1, 0, 1]^T; \ \mathbf{x}^* = [0, 0, 0, 0]^T. \end{aligned}$$

4. Fletcher and Powell's helical valley:

$$f(\mathbf{x}) = 100\left((x_3 - 10\theta(x_1, x_2))^2 + \left(\sqrt{x_1^2 + x_2^2} - 1\right)^2\right) + x_3^2;$$

$$\text{where } 2\pi\theta(x_1, x_2) = \begin{cases} \arctan\dfrac{x_2}{x_1} & \text{if } x_1 > 0 \\ \pi + \arctan\dfrac{x_2}{x_1} & \text{if } x_1 < 0 \end{cases}$$

$$\mathbf{x}^0 = [-1, 0, 0]^T; \quad \mathbf{x}^* = [1, 0, 0]^T.$$

5. A non-linear function of three variables:

$$f(\mathbf{x}) = -\frac{1}{1 + (x_1 - x_2)^2} - \sin\left(\frac{1}{2}\pi x_2 x_3\right) - \exp\left(-\left(\frac{x_1 + x_3}{x_2} - 2\right)^2\right);$$

$$\mathbf{x}^0 = [0, 1, 2]^T; \quad \mathbf{x}^* = [1, 1, 1]^T.$$

6. Freudenstein and Roth function:

$$f(\mathbf{x}) = (-13 + x_1 + ((5 - x_2)x_2 - 2)x_2)^2$$
$$+(-29 + x_1 + ((x_2 + 1)x_2 - 14)x_2)^2;$$
$$\mathbf{x}^0 = [0.5, -2]^T; \quad \mathbf{x}^* = [5, 4]^T; \quad \mathbf{x}^*_{\text{local}} = [11.41\ldots, -0.8968\ldots]^T.$$

7. Powell's badly scaled function:

$$f(\mathbf{x}) = (10\,000 x_1 x_2 - 1)^2 + (\exp(-x_1) + \exp(-x_2) - 1.0001)^2;$$
$$\mathbf{x}^0 = [0, 1]^T; \quad \mathbf{x}^* = [1.098\ldots \times 10^{-5}, 9.106\ldots]^T.$$

8. Brown's badly scaled function:

$$f(\mathbf{x}) = (x_1 - 10^6)^2 + (x_2 - 2 \times 10^{-6})^2 + (x_1 x_2 - 2)^2;$$
$$\mathbf{x}^0 = [1, 1]^T; \quad \mathbf{x}^* = [10^6, 2 \times 10^{-6}]^T.$$

9. Beale's function:

$$f(\mathbf{x}) = (1.5 - x_1(1 - x_2))^2 + (2.25 - x_1(1 - x_2^2))^2$$
$$+(2.625 - x_1(1 - x_2^3))^2;$$
$$\mathbf{x}^0 = [1, 1]^T; \quad \mathbf{x}^* = [3, 0.5]^T.$$

10. Wood's function:

$$
\begin{aligned}
f(\mathbf{x}) &= 100(x_2 - x_1^2)^2 + (1 - x_1)^2 + 90(x_4 - x_3^2)^2 + (1 - x_3)^2 \\
&\quad + 10(x_2 + x_4 - 2)^2 + 0.1(x_2 - x_4)^2 \\
\mathbf{x}^0 &= [-3, -1, -3, -1]^T; \quad \mathbf{x}^* = [1, 1, 1, 1]^T.
\end{aligned}
$$

1.6.2 Constrained

Some classical constrained minimization test problems from Hock and Schittkowski (1981) are listed below.

1. Hock & Schittkowski Problem 1:

$$
f(\mathbf{x}) = 100(x_2 - x_1^2)^2 + (1 - x_1)^2
$$
such that
$$
x_2 \geq -1.5
$$

$$
\mathbf{x}^0 = \begin{bmatrix} -2 \\ 1 \end{bmatrix} \quad \mathbf{x}^* = \begin{bmatrix} 1 \\ 1 \end{bmatrix} \quad \lambda^* = 0
$$

2. Hock & Schittkowski Problem 2:

$$
f(\mathbf{x}) = 100(x_2 - x_1^2)^2 + (1 - x_1)^2
$$
such that
$$
x_2 \geq 1.5
$$

$$
\mathbf{x}^0 = \begin{bmatrix} -2 \\ 1 \end{bmatrix} \quad \mathbf{x}^* = \begin{bmatrix} 1.2243707487363527 \\ 1.5000000000000000 \end{bmatrix} \quad \lambda^* = 200
$$

3. Hock & Schittkowski Problem 6:

$$
f(\mathbf{x}) = (1 - x_1)^2
$$
such that
$$
10(x_2 - x_1^2) = 0
$$

$$
\mathbf{x}^0 = \begin{bmatrix} -1.2 \\ 1 \end{bmatrix} \quad \mathbf{x}^* = \begin{bmatrix} 1 \\ 1 \end{bmatrix} \quad \lambda^* = 0.4
$$

4. Hock & Schittkowski Problem 7:

$$f(\mathbf{x}) = \ln(1 + x_1^2) - x_2$$
such that
$$(1 + x_1^2)^2 + x_2^2 - 4 = 0$$

$$\mathbf{x}^0 = \begin{bmatrix} 2 \\ 2 \end{bmatrix} \quad \mathbf{x}^* = \begin{bmatrix} 0 \\ \sqrt{3} \end{bmatrix} \quad \lambda^* = 3.15$$

5. Hock & Schittkowski Problem 10:

$$f(\mathbf{x}) = x_1 - x_2$$
such that
$$-3x_1^2 + 2x_1x_2 - x_2^2 + 1 \geq 0$$

$$\mathbf{x}^0 = \begin{bmatrix} -10 \\ 10 \end{bmatrix} \quad \mathbf{x}^* = \begin{bmatrix} 0 \\ 1 \end{bmatrix} \quad \lambda^* = 1.0$$

6. Hock & Schittkowski Problem 18:

$$f(\mathbf{x}) = 0.01x_1^2 + x_2^2$$
such that
$$x_1x_2 - 25 \geq 0$$
$$x_1^2 + x_2^2 - 25 \geq 0$$
$$2 \leq x_1 \leq 50$$
$$2 \leq x_2 \leq 50$$

$$\mathbf{x}^0 = \begin{bmatrix} 2 \\ 2 \end{bmatrix} \quad \mathbf{x}^* = \begin{bmatrix} \sqrt{250} \\ \sqrt{2.5} \end{bmatrix} \quad \lambda^* = 0.079$$

7. Hock & Schittkowski Problem 27:

$$f(\mathbf{x}) = 0.01(x_1 - 1)^2 + (x_2 - x_1^2)^2$$
such that
$$x_1 + x_3^2 + 1 = 0$$

$$\mathbf{x}^0 = \begin{bmatrix} 2 \\ 2 \\ 2 \end{bmatrix} \quad \mathbf{x}^* = \begin{bmatrix} -1 \\ 1 \\ 0 \end{bmatrix} \quad \lambda^* = 2$$

8. Hock & Schittkowski Problem 42:

$$f(\mathbf{x}) = (x_1 - 1)^2 + (x_2 - 2)^2 + (x_3 - 3)^2 + (x_4 - 4)^2$$

such that

$$x_1 - 2 = 0$$
$$x_3^2 + x_4^2 - 2 = 0$$

$$\mathbf{x}^0 = \begin{bmatrix} 1 \\ 1 \\ 1 \\ 1 \end{bmatrix} \quad \mathbf{x}^* = \begin{bmatrix} 2 \\ 2 \\ 0.6\sqrt{2} \\ 0.8\sqrt{2} \end{bmatrix} \quad \lambda_{max}^* = 7.07, \ \lambda_{min}^* = 3.54$$

9. Hock & Schittkowski Problem 66:

$$f(\mathbf{x}) = 0.2x_3 - 0.8x_1$$

such that

$$x_2 - \exp(x_1) \geq 0$$
$$x_3 - \exp(x_2) \geq 0$$
$$0 \leq x_1 \leq 100$$
$$0 \leq x_2 \leq 100$$
$$0 \leq x_3 \leq 10$$

$$\mathbf{x}^0 = \begin{bmatrix} 0 \\ 1.05 \\ 2.9 \end{bmatrix} \quad \mathbf{x}^* = \begin{bmatrix} 0.1841264879 \\ 1.202167873 \\ 3.327322322 \end{bmatrix} \quad \lambda_{max}^* = 0.096, \ \lambda_{min}^* = 0.096$$

10. Hock & Schittkowski Problem 104:

$$f(\mathbf{x}) = 0.4x_1^{0.67}x_7^{-0.67} + 0.4x_2^{0.67}x_8^{-0.67} + 10 - x_1 - x_2$$

such that

$$1 - 0.0588x_5x_7 - 0.1x_1 \geq 0$$
$$1 - 0.0588x_6x_8 - 0.1x_1 - 0.1x_2 \geq 0$$
$$1 - 4x_3x_5^{-1} - 2x_3^{-0.71}x_5^{-1} - 0.0588x_3^{-1.3}x_7 \geq 0$$
$$1 - 4x_4x_6^{-1} - 2x_4^{-0.71}x_6^{-1} - 0.0588x_4^{-1.3}x_8 \geq 0$$
$$1 \leq f(\mathbf{x}) \leq 4.2$$
$$1 \leq x_i \leq 10, \ i = 1, \ldots, 8$$

$$\mathbf{x}^0 = \begin{bmatrix} 6, \ 3, \ 0.4, \ 0.2, \ 6, \ 6, \ 1, \ 0.5 \end{bmatrix}^{\mathrm{T}}$$
$$\mathbf{x}^* = [6.465114, \ 2.232709, \ 0.6673975, \ 0.5957564,$$
$$5.932676, \ 5.527235, \ 1.013322, \ 0.4006682]^{\mathrm{T}}$$
$$\lambda_{max}^* = 1.87, \ \lambda_{min}^* = 0.043$$

1.7   Exercises

1.7.1 Sketch the graphical solution to the following problem:

$$\min_{\mathbf{x}} f(\mathbf{x}) = (x_1 - 2)^2 + (x_2 - 2)^2$$
$$\text{such that } x_1 + 2x_2 = 4; \ x_1 \geq 0; \ x_2 \geq 0.$$

In particular indicate the feasible region:

$$F = \{(x_1, x_2) | x_1 + 2x_2 = 4; x_1 \geq 0; x_2 \geq 0\}$$

and the solution point $\mathbf{x}^*$.

1.7.2 Show that x^2 is a convex function.

1.7.3 Show that the sum of convex functions is also convex.

1.7.4 Determine the gradient vector and Hessian matrix of the Rosenbrock function given in Section 1.6.1.

1.7.5 Write the quadratic function $f(\mathbf{x}) = x_1^2 + 2x_1x_2 + 3x_2^2$ in the standard matrix-vector notation. Is $f(\mathbf{x})$ positive-definite?

1.7.6 Write each of the following objective functions in standard form:

$$f(\mathbf{x}) = \tfrac{1}{2}\mathbf{x}^T\mathbf{A}\mathbf{x} + \mathbf{b}^T\mathbf{x} + c.$$

(i) $f(\mathbf{x}) = x_1^2 + 2x_1x_2 + 4x_1x_3 + 3x_2^2 + 2x_2x_3 + 5x_3^2 + 4x_1 - 2x_2 + 3x_3$.

(ii) $f(\mathbf{x}) = 5x_1^2 + 12x_1x_2 - 16x_1x_3 + 10x_2^2 - 26x_2x_3 + 17x_3^2 - 2x_1 - 4x_2 - 6x_3$.

(iii) $f(\mathbf{x}) = x_1^2 - 4x_1x_2 + 6x_1x_3 + 5x_2^2 - 10x_2x_3 + 8x_3^2$.

1.7.7 Determine the definiteness of the following quadratic form:

$$f(\mathbf{x}) = x_1^2 - 4x_1x_2 + 6x_1x_3 + 5x_2^2 - 10x_2x_3 + 8x_3^2. \tag{1.29}$$

1.7.8 Approximate the Rosenbrock function given in Section 1.6.1 using a first order Taylor series expansion around $\mathbf{x}^0$. Compute the accuracy of the approximation at $\mathbf{x} = \mathbf{x}^0 + \Delta\mathbf{x}$, with $\Delta\mathbf{x} = [0, \ 1.0]^T$.

1.7.9 Approximate the Rosenbrock function given in Section 1.6.1 using a second order Taylor series expansion around $\mathbf{x}^0$. Compute the accuracy of the approximation at $\mathbf{x} = \mathbf{x}^0 + \Delta\mathbf{x}$, with $\Delta\mathbf{x} = [0, \ 1.0]^{\mathrm{T}}$.

1.7.10 Compute the directional derivatives for the Rosenbrock function given in Section 1.6.1 at $\mathbf{x}^0$ along the following three directions

$$\mathbf{u}^1 = [1, \ 0],$$
$$\mathbf{u}^2 = [0, \ 1],$$
$$\mathbf{u}^3 = [\frac{1}{\sqrt{2}}, \ \frac{1}{\sqrt{2}}].$$

Compare the first two computed directional derivatives to the components of the gradient vector. What conclusions can you draw.

1.7.11 Clearly state which of the directions computed in Exercise 1.7.10 are descent directions, i.e. directions along which the function will decrease for small positive steps along the direction.

1.7.12 Propose a descent direction of unit length that would result in the largest directional derivative magnitude at $\mathbf{x}^0$.

1.7.13 Compute the eigenvalues and eigenvectors for the computed $\mathbf{A}$ matrices in Exercise 1.7.6.

1.7.14 Determine whether the $\mathbf{A}$ matrices computed in Exercise 1.7.6 are positive-definite, negative-definite or indefinite.

1.7.15 Compare the associated eigenvalue for each computed eigenvector $\mathbf{u}^i$ in Exercise 1.7.13 against the second derivative of the univariate function $f(\lambda) = (\mathbf{x}^0 + \lambda\mathbf{u}^i)^{\mathrm{T}}\mathbf{A}(\mathbf{x}^0 + \lambda\mathbf{u}^i)$ and draw concrete conclusions.

1.7.16 Consider the following constrained optimization problem

$$\min_{\mathbf{x}} f(\mathbf{x}) = x_1^4 - 6x_1^3 + 9x_1^2 + x_2^4 - 0.5x_2^3 + 0.0625x_2^2,$$
such that
$$x_1 \geq 0,$$
$$x_2 \leq 0.$$

Utilizing only transformation of variables reformulate the constrained minimization problem as an unconstrained minimization problem.

1.7.17 Given the function $f(\mathbf{x})$ and the non-linear variable scaling $\mathbf{z}(\mathbf{x}) = \mathbf{G}(\mathbf{x})$ that transforms the domain, $\mathbf{x} \in \mathcal{R}^n$, to the domain, $\mathbf{z} \in \mathcal{R}^n$, with inverse relation $\mathbf{x}(\mathbf{z}) = \mathbf{G}^{-1}(\mathbf{z})$ transforming $\mathbf{z}$ back to the $\mathbf{x}$ domain. By substituting $\mathbf{x}(\mathbf{z})$ into $f(\mathbf{x})$ we obtain $f(\mathbf{z})$. Utilize the chain rule to derive the expression for computing $\nabla_{\mathbf{z}} f(\mathbf{z})$.

1.7.18 Perform the first five Newton steps in solving the *Rosenbrock function* listed in Section 1.6.1.

1.7.19 Perform the first five Newton steps in solving the *Quadratic function* listed in Section 1.6.1.

1.7.20 Perform the first five Newton steps in solving the *Freudenstein and Roth function* listed in Section 1.6.1.

1.7.21 Perform the first five Newton steps in solving the *Powell's badly scaled function* listed in Section 1.6.1.

1.7.22 Perform the first five Newton steps in solving the *Brown's badly scaled function* listed in Section 1.6.1.

1.7.23 Perform the first five Newton steps in solving the *Beale's function* listed in Section 1.6.1.

Chapter 2

LINE SEARCH DESCENT METHODS FOR UNCONSTRAINED MINIMIZATION

2.1 General line search descent algorithm for unconstrained minimization

Over the last 40 years many powerful *direct search algorithms* have been developed for the unconstrained minimization of general functions. These algorithms require an initial estimate to the optimum point, denoted by $\mathbf{x}^0$. With this estimate as starting point, the algorithm generates a sequence of estimates $\mathbf{x}^0$, $\mathbf{x}^1$, $\mathbf{x}^2$, ..., by successively searching *directly* from each point in a direction of *descent* to determine the next point. The process is terminated if either no further progress is made, or if a point $\mathbf{x}^k$ is reached (for smooth functions) at which the first necessary condition in (1.24), i.e. $\nabla f(\mathbf{x}) = \mathbf{0}$ is sufficiently accurately satisfied, in which case $\mathbf{x}^* \cong \mathbf{x}^k$. It is usually, although not always, required that the function value at the new iterate $\mathbf{x}^{i+1}$ be lower than that at $\mathbf{x}^i$.

© Springer International Publishing AG, part of Springer Nature 2018
J.A. Snyman and D.N. Wilke, *Practical Mathematical Optimization,*
Springer Optimization and Its Applications 133,
https://doi.org/10.1007/978-3-319-77586-9_2

An important sub-class of direct search methods, specifically suitable for smooth functions, are the so-called *line search* descent methods. Basic to these methods is the selection of a descent direction $\mathbf{u}^{i+1}$ at each iterate $\mathbf{x}^i$ that ensures descent at $\mathbf{x}^i$ in the direction $\mathbf{u}^{i+1}$, i.e. it is required that the directional derivative in the direction $\mathbf{u}^{i+1}$ be negative:

$$\left. \frac{df(\mathbf{x}^i)}{d\lambda} \right|_{\mathbf{u}^{i+1}} = \boldsymbol{\nabla}^T f(\mathbf{x}^i)\mathbf{u}^{i+1} < 0. \tag{2.1}$$

The general structure of such descent methods is given below.

2.1.1   General structure of a line search descent method

1. Given starting point $\mathbf{x}^0$ and positive tolerances ε_1, ε_2 and ε_3, set $i = 1$.

2. Select a descent direction $\mathbf{u}^i$ (see descent condition (2.1)).

3. Perform a *one-dimensional line search* in direction $\mathbf{u}^i$: i.e.

$$\min_{\lambda} F(\lambda) = \min_{\lambda} f(\mathbf{x}^{i-1} + \lambda \mathbf{u}^i)$$

   to give minimizer λ_i.

4. Set $\mathbf{x}^i = \mathbf{x}^{i-1} + \lambda_i \mathbf{u}^i$.

5. Test for convergence:

   if $\|\mathbf{x}^i - \mathbf{x}^{i-1}\| < \varepsilon_1$, or $\|\boldsymbol{\nabla}f(\mathbf{x}^i)\| < \varepsilon_2$, or $|f(\mathbf{x}^i) - f(\mathbf{x}^{i-1})| < \varepsilon_3$, then stop and $\mathbf{x}^* \cong \mathbf{x}^i$,

   else go to Step 6.

6. Set $i = i + 1$ and go to Step 2.

In testing for termination in step 5, a combination of the stated termination criteria may be used, i.e. instead of *or*, *and* may be specified. The structure of the above descent algorithm is depicted in Figure 2.1. Different descent methods, within the above sub-class, differ according to the way in which the descent directions $\mathbf{u}^i$ are chosen. Another important consideration is the method by means of which the one-dimensional line search is performed.

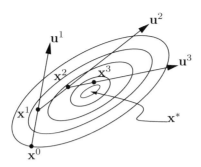

Figure 2.1: Sequence of line search descent directions and steps

2.2 One-dimensional line search

Clearly, in implementing descent algorithms of the above type, the one-dimensional minimization problem:

$$\min_{\lambda} F(\lambda), \ \lambda \in \mathbb{R} \tag{2.2}$$

is an important sub-problem. Here the minimizer is denoted by λ^*, i.e.

$$F(\lambda^*) = \min_{\lambda} F(\lambda).$$

Many one-dimensional minimization techniques have been proposed and developed over the years. These methods differ according to whether they are to be applied to smooth functions or poorly conditioned functions. For smooth functions *interpolation methods*, such as the quadratic interpolation method of Powell (1964) and the cubic interpolation algorithm of Davidon (1959), are the most efficient and accurate methods. For poorly conditioned functions, *bracketing methods*, such as the Fibonacci search method (Kiefer 1957), which is optimal with respect to the number of function evaluations required for a prescribed accuracy, and the golden section method (Walsh 1975), which is near optimal but much simpler and easier to implement, are preferred. Here *Powell's quadratic interpolation* method and the *golden section* method, are respectively presented as representative of the two different approaches that may be adopted to one-dimensional minimization.

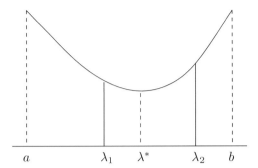

Figure 2.2: Unimodal function $F(\lambda)$ over interval $[a, b]$

2.2.1 Golden section method

It is assumed that $F(\lambda)$ is *unimodal* over the interval $[a, b]$, i.e. that it has a minimum λ^* within the interval and that $F(\lambda)$ is strictly descending for $\lambda < \lambda^*$ and strictly ascending for $\lambda > \lambda^*$, as shown in Figure 2.2.

Note that if $F(\lambda)$ is unimodal over $[a, b]$ with λ^* in $[a, b]$, then to determine a sub-unimodal interval, at least *two* evaluations of $F(\lambda)$ in $[a, b]$ must be made as indicated in Figure 2.2.

If $F(\lambda_2) > F(\lambda_1) \Rightarrow$ new unimodal interval $= [a, \lambda_2]$, and set $b = \lambda_2$ and select new λ_2; otherwise new unimodal interval $= [\lambda_1, b]$ and set $a = \lambda_1$ and select new λ_1.

Thus, the unimodal interval may successively be reduced by inspecting values of $F(\lambda_1)$ and $F(\lambda_2)$ at interior points λ_1 and λ_2.

The question arises: How can λ_1 and λ_2 be chosen in the most economic manner, i.e. such that a least number of function evaluations are required for a prescribed accuracy (i.e. for a specified uncertainty interval)? The most economic method is the Fibonacci search method. It is however a complicated method. A near optimum and more straightforward method is the golden section method. This method is a limiting form of the Fibonacci search method. Use is made of the golden ratio r when selecting the values for λ_1 and λ_2 within the unimodal interval. The value of r corresponds to the positive root of the quadratic equation: $r^2 + r - 1 = 0$, thus $r = \frac{\sqrt{5}-1}{2} = 0.618034$.

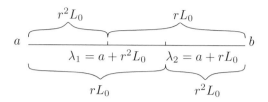

Figure 2.3: Selection of interior points λ_1 and λ_2 for golden section search

The details of the selection procedure are as follows. Given initial unimodal interval $[a, b]$ of length L_0, then choose interior points λ_1 and λ_2 as shown in Figure 2.3.

Then, if $F(\lambda_1) > F(\lambda_2) \Rightarrow$ new $[a, b] = [\lambda_1, b]$ with new interval length $L_1 = rL_0$, and

if $F(\lambda_2) > F(\lambda_1) \Rightarrow$ new $[a, b] = [a, \lambda_2]$ also with $L_1 = rL_0$.

The detailed formal algorithm is stated below.

2.2.1.1   Basic golden section algorithm

Given interval $[a, b]$ and prescribed accuracy ε; then set $i = 0$; $L_0 = b - a$, and perform the following steps:

1. Set $\lambda_1 = a + r^2 L_0$; $\lambda_2 = a + rL_0$.

2. Compute $F(\lambda_1)$ and $F(\lambda_2)$; set $i = i + 1$.

3. *If* $F(\lambda_1) > F(\lambda_2)$ *then*

   set $a = \lambda_1$; $\lambda_1 = \lambda_2$; $L_i = (b - a)$; and $\lambda_2 = a + rL_i$,

   else

   set $b = \lambda_2$; $\lambda_2 = \lambda_1$; $L_i = (b - a)$; and $\lambda_1 = a + r^2 L_i$.

4. *If* $L_i < \varepsilon$ *then*

   set $\lambda^* = \dfrac{b + a}{2}$; compute $F(\lambda^*)$ and stop,

   else go to Step 2.

Note that only one function evaluation is required in each iteration in Step 2 after the first one, as one of the interior points is inherited from the former iteration.

2.2.2 Powell's quadratic interpolation algorithm

In Powell's method successive quadratic interpolation curves are fitted to function data giving a sequence of approximations of the minimum point λ^*.

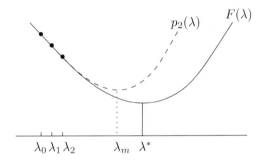

Figure 2.4: Approximate minimum λ_m via quadratic curve fitting

With reference to Figure 2.4, the basic idea is the following. Given three data points $\{(\lambda_i, F(\lambda_i)),\ i = 1, 2, 3\}$, then the interpolating quadratic polynomial through these points $p_2(\lambda)$ is given by

$$p_2(\lambda) = F(\lambda_0) + F[\lambda_0, \lambda_1](\lambda - \lambda_0) + F[\lambda_0, \lambda_1, \lambda_2](\lambda - \lambda_0)(\lambda - \lambda_1) \quad (2.3)$$

where $F[\ ,\ ]$ and $F[\ ,\ ,\ ]$ respectively denote the first order and second order divided differences.

The turning point of $p_2(\lambda)$ occurs where the slope is zero, i.e. where

$$\frac{dp_2}{d\lambda} = F[\lambda_0, \lambda_1] + 2\lambda F[\lambda_0, \lambda_1, \lambda_2] - F[\lambda_0, \lambda_1, \lambda_2](\lambda_0 + \lambda_1) = 0$$

which gives the turning point λ_m as

$$\lambda_m = \frac{F[\lambda_0, \lambda_1, \lambda_2](\lambda_0 + \lambda_1) - F[\lambda_0, \lambda_1]}{2F[\lambda_0, \lambda_1, \lambda_2]} \cong \lambda^* \quad (2.4)$$

with the further condition that for a minimum the second derivative must be non-negative, i.e. $F[\lambda_0, \lambda_1, \lambda_2] > 0$.

The detailed formal algorithm is as follows.

2.2.2.1 Powell's interpolation algorithm

Given starting point λ_0, step size h, tolerance ε and maximum step size H; perform following steps:

1. Compute $F(\lambda_0)$ and $F(\lambda_0 + h)$.

2. *If* $F(\lambda_0) < F(\lambda_0 + h)$ evaluate $F(\lambda_0 - h)$,

   else evaluate $F(\lambda_0 + 2h)$. (The three initial values of λ so chosen constitute the initial set $(\lambda_0, \lambda_1, \lambda_2)$ with corresponding function values $F(\lambda_i)$, $i = 0, 1, 2$.)

3. Compute turning point λ_m by formula (2.4) and test for minimum or maximum.

4. *If* λ_m a minimum point *and* $|\lambda_m - \lambda_n| > H$, where λ_n is the nearest point to λ_m, then discard the point furthest from λ_m and take a step of size H from the point with lowest value in direction of descent, and go to Step 3;

   if λ_m a maximum point, then discard point nearest λ_m and take a step of size H from the point with lowest value in the direction of descent and go to Step 3;

   else continue.

5. *If* $|\lambda_m - \lambda_n| < \varepsilon$ then $F(\lambda^*) \cong \min[F(\lambda_m), F(\lambda_n)]$ and stop,

   else continue.

6. Discard point with highest F value and replace it by λ_m; go to Step 3

Note: It is always safer to compute the next turning point by interpolation rather than by extrapolation. Therefore in Step 6: if the maximum value of F corresponds to a point which lies alone on one side of λ_m, then rather discard the point with highest value on the other side of λ_m.

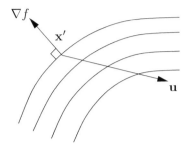

Figure 2.5: Search direction **u** relative to gradient vector at $\mathbf{x}'$

2.3 First order line search descent methods

Line search descent methods (see Section 2.1.1), that use the gradient vector $\boldsymbol{\nabla} f(\mathbf{x})$ to determine the search direction for each iteration, are called *first order methods* because they employ *first order* partial derivatives of $f(\mathbf{x})$ to compute the search direction at the current iterate. The simplest and most famous of these methods is the *method of steepest descent*, first proposed by Cauchy in 1847.

2.3.1 The method of steepest descent

In this method the direction of steepest descent is used as the search direction in the line search descent algorithm given in section 2.1.1. The expression for the direction of steepest descent is derived below.

2.3.1.1 The direction of steepest descent

At $\mathbf{x}'$ we seek a unit vector **u** (here understood as a vector of length one), such that for $F(\lambda) = f(\mathbf{x}' + \lambda\mathbf{u})$, the directional derivative

$$\frac{df(\mathbf{x}')}{d\lambda}\bigg|_{\mathbf{u}} = \frac{dF(0)}{d\lambda} = \boldsymbol{\nabla}^T f(\mathbf{x}')\mathbf{u}$$

assumes a minimum value with respect to all possible choices for **u** at $\mathbf{x}'$ (see Figure 2.5).

By Schwartz's inequality:

$$\boldsymbol{\nabla}^T f(\mathbf{x}')\mathbf{u} \geq -\|\boldsymbol{\nabla} f(\mathbf{x}')\|\|\mathbf{u}\| = -\|\boldsymbol{\nabla} f(\mathbf{x}')\| = \text{ least value.}$$

Clearly for the particular choice $\mathbf{u} = \dfrac{-\boldsymbol{\nabla} f(\mathbf{x}')}{\|\boldsymbol{\nabla} f(\mathbf{x}')\|}$ the directional derivative at $\mathbf{x}'$ is given by

$$\frac{dF(0)}{d\lambda} = -\boldsymbol{\nabla}^T f(\mathbf{x}')\frac{\boldsymbol{\nabla} f(\mathbf{x}')}{\|\boldsymbol{\nabla} f(\mathbf{x}')\|} = -\|\boldsymbol{\nabla} f(\mathbf{x}')\| = \text{ least value.}$$

Thus this particular choice for the unit vector corresponds to the direction of steepest descent.

The search direction

$$\mathbf{u} = \frac{-\boldsymbol{\nabla} f(\mathbf{x})}{\|\boldsymbol{\nabla} f(\mathbf{x})\|} \tag{2.5}$$

is called the *normalized steepest descent direction* at $\mathbf{x}$.

2.3.1.2 Steepest descent algorithm

Given $\mathbf{x}^0$, do for iteration $i = 1, 2, \ldots$ until convergence:

1. set $\mathbf{u}^i = \dfrac{-\boldsymbol{\nabla} f(\mathbf{x}^{i-1})}{\|\boldsymbol{\nabla} f(\mathbf{x}^{i-1})\|}$

2. set $\mathbf{x}^i = \mathbf{x}^{i-1} + \lambda_i \mathbf{u}^i$ where λ_i is such that

$$F(\lambda_i) = f(\mathbf{x}^{i-1} + \lambda_i \mathbf{u}^i) = \min_\lambda f(\mathbf{x}^{i-1} + \lambda \mathbf{u}^i) \text{ (line search).}$$

2.3.1.3 Characteristic property

Successive steepest descent search directions can be shown to be *orthogonal*. Consider the line search through $\mathbf{x}^{i-1}$ in the direction $\mathbf{u}^i$ to give $\mathbf{x}^i$. The condition for a minimum at λ_i, i.e. for optimal descent, is

$$\frac{df(\mathbf{x}^{i-1} + \lambda_i \mathbf{u}^i)}{d\lambda}\bigg|_{\mathbf{u}^i} = \frac{dF(\lambda_i)}{d\lambda}\bigg|_{\mathbf{u}^i} = \boldsymbol{\nabla}^T f(\mathbf{x}^i)\mathbf{u}^i = 0$$

and with $\mathbf{u}^{i+1} = -\dfrac{\boldsymbol{\nabla} f(\mathbf{x}^i)}{\|\boldsymbol{\nabla} f(\mathbf{x}^i)\|}$ it follows that $\mathbf{u}^{i+1^T}\mathbf{u}^i = 0$ as shown in Figure 2.6.

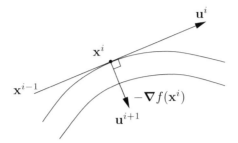

Figure 2.6: Orthogonality of successive steepest descent search directions

2.3.1.4 Convergence criteria

In practice the algorithm is terminated if some convergence criterion is satisfied. Usually termination is enforced at iteration i if one, or a combination, of the following criteria is met:

(i) $\|\mathbf{x}^i - \mathbf{x}^{i-1}\| < \varepsilon_1$

(ii) $\|\nabla f(\mathbf{x}^i)\| < \varepsilon_2$

(iii) $|f(\mathbf{x}^i) - f(\mathbf{x}^{i-1})| < \varepsilon_3$.

where ε_1, ε_2 and ε_3 are prescribed small positive tolerances.

2.3.1.5 Conditions for sufficient improvement

The associated computational cost of performing the exact line searches for the steepest descent method, and indeed for the first order line search descent methods in general, has driven the development of conditions that economically indicate sufficient improvement along a search direction (Armijo (1966); Wolfe (1969, 1971)). The aim of these conditions are to ensure that the step sizes λ_i are neither too large so as to diverge, or too small so as to make insufficient progress along a search direction $\mathbf{u}^i$.

The following four conditions have successfully been employed in the past to update step sizes:

1. Improvement:

$$f(\mathbf{x}^{i-1} + \lambda_i \mathbf{u}^i) \leq f(\mathbf{x}^{i-1}),$$

2. Armijo:

$$f(\mathbf{x}^{i-1} + \lambda_i \mathbf{u}^i) \leq f(\mathbf{x}^{i-1}) + c_1 \lambda_i \mathbf{u}^{i^T} \nabla f(\mathbf{x}^{i-1}),$$

3. Curvature:

$$c_2 \mathbf{u}^{i^T} \nabla f(\mathbf{x}^{i-1}) \leq \mathbf{u}^{i^T} \nabla f(\mathbf{x}^{i-1} + \lambda_i \mathbf{u}^i),$$

4. Strong curvature:

$$|\mathbf{u}^{i^T} \nabla f(\mathbf{x}^{i-1} + \lambda_i \mathbf{u}^i)| \leq c_3 |\mathbf{u}^{i^T} \nabla f(\mathbf{x}^{i-1})|,$$

with c_1, c_2 and c_3 required to be selected as non-negative parameters. These parameters control the degree to which the conditions are enforced. The second and third conditions are collectively referred to as the Wolfe conditions, while the second and fourth conditions designate the strong Wolfe conditions.

Although these conditions do not bespeak a line search strategy per se, they do act as additional tests to indicate whether a proposed step length signifies sufficient improvement. Hence, they can be used in termination strategies for first order line searches.

2.3.1.6 Gradients by finite differences

As analytical sensitivities are not always accessible for computation, it would be convenient to compute sensitivities numerically using finite differences. Often the components of the gradient vector may be approximated by forward finite differences:

$$\frac{\partial f(\mathbf{x})}{\partial x_j} \cong \frac{\Delta f(\mathbf{x})}{\delta_j} = \frac{f(\mathbf{x} + \boldsymbol{\delta}_j) - f(\mathbf{x})}{\delta_j} \tag{2.6}$$

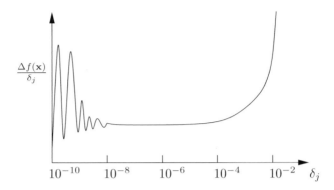

Figure 2.7: Sensitivity of finite difference approximation to δ_j

where $\boldsymbol{\delta}_j = [0, 0, \ldots \delta_j, 0, \ldots, 0]^T$, $\delta_j > 0$ in the j-th position.

Usually $\delta_j \equiv \delta$ for all $j = 1, 2, \ldots, n$. A typically choice is $\delta = 10^{-6}$. If however "*numerical noise*" is present in the computation of $f(\mathbf{x})$, special care should be taken in selecting δ_j. This may require doing some numerical experiments such as, for example, determining the sensitivity of approximation (2.6) to the value of δ_j, for each j. Typically the sensitivity graph obtained is as depicted in Figure 2.7, and for the implementation of the optimization algorithm a value for δ_j should be chosen which corresponds to a point on the plateau as shown in Figure 2.7. Better approximations, at of course greater computational expense, may be obtained through the use of central finite differences.

In elaborating on the nature of Figure 2.7, consider the conventional finite difference approximation schemes for $f'(x)$ that can be derived from the one-dimensional Taylor series expansion around x,

$$f(x + \delta) = f(x) + \delta f'(x) + \delta^2 \frac{f''(x)}{2} + \text{higher order terms}, \qquad (2.7)$$

by making appropriate choices for δ. Clearly the forward difference approximation is of order $O(\delta)$, while the central difference scheme, recovered by subtracting the backward difference expansion from the forward difference expansion is of order $O(\delta^2)$ results in the improved accuracy of the latter method.

An additional error is present in the numerical finite difference computations, namely the subtraction error as a result of finite precision arithmetic. This is a direct result of subtracting two finite values $f(\mathbf{x} - \boldsymbol{\delta}_j)$ and $f(\mathbf{x})$ to recover a small difference. The smaller the step size, the larger the subtraction error. Consequently some numerical experimentation requiring function evaluations are usually required to find a suitable step size.

The complex-step method, originally proposed by Lyness and Moler (1967) and popularized by Squire and Trapp (1998) and Martins et al. (2001), avoids the subtraction between two finite values when computing derivatives, and thus it has no subtraction error. The implication is that there is no increase in the sensitivity error as the step size is reduced. The advantage of the complex-step method is that extremely small step sizes δ can be taken, e.g. $\delta = 10^{-20}$ is not uncommon in order to reduce the Taylor series truncation error to be negligible. Thus numerical sensitivities can be computed that is as accurate as numerically evaluated analytical sensitivities.

The complex-step method is based on the complex Taylor series expansion of an analytic function $f(x)$ using a complex step $i\delta$,

$$f(x + i\delta) = f(x) + i\delta f'(x) - \delta^2 \frac{f''(x)}{2} + \text{higher order terms.} \qquad (2.8)$$

By neglecting the higher order terms and equating the imaginary parts (Im) on both sides of (2.8), the complex-step derivative approximation,

$$f'(x) \approx \frac{Im[f(x + i\delta)]}{\delta},$$

is obtained as a second order accurate derivative approximation to $f'(x)$. It is important to note that the complex-step method requires double the amount of computer memory as complex numbers require a real and imaginary part to be stored. In addition, some modifications to software may be required to correctly compute the complex derivatives as outlined by Martins et al. (2003).

Since directional derivatives are computed by projecting the gradient vector, $\nabla f(\mathbf{x})$, onto a chosen search direction $\mathbf{u}$, they can be computed efficiently using finite difference strategies. The directional derivative can be computed by taking a single finite difference step along a search direction $\mathbf{u}$, that is, evaluating f at $\mathbf{x} + i\delta_u \mathbf{u}$. Hence, one complex step is sufficient to compute the directional derivative as opposed to

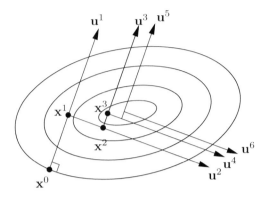

Figure 2.8: Orthogonal zigzagging behaviour of the steepest descent method

computing the gradient vector $\nabla f(\mathbf{x})$ and then projecting it onto the search direction $\mathbf{u}$, i.e. by computing the product $\mathbf{u}^T \nabla f(\mathbf{x})$.

2.3.2   Conjugate gradient methods

In spite of its local optimal descent property, the method of steepest descent often performs poorly, following a zigzagging path of ever decreasing steps. This results in slow convergence and becomes extreme when the problem is poorly scaled, i.e. when the contours are extremely elongated. This poor performance is mainly due to the fact that the method enforces successive orthogonal search directions (see Section 2.3.1.3) as shown in Figure 2.8. Although, from a theoretical point of view, the method can be proved to be convergent, in practice the method may not effectively converge within a finite number of steps. Depending on the starting point, this poor convergence also occurs when applying the method to positive-definite quadratic functions.

There is, however, a class of first order line search descent methods, known as *conjugate gradient methods*, for which it can be proved that *whatever the scaling, a method from this class will converge exactly in a finite number of iterations when applied to a positive-definite quadratic function*, i.e. to a function of the form

$$f(\mathbf{x}) = \tfrac{1}{2}\mathbf{x}^T \mathbf{A}\mathbf{x} + \mathbf{b}^T\mathbf{x} + c \tag{2.9}$$

where $c \in \mathbb{R}$, $\mathbf{b}$ is a real n-vector and $\mathbf{A}$ is a positive-definite $n \times n$ real symmetric matrix. Methods that have this property of *quadratic termination* are highly rated, because they are expected to also perform well on other non-quadratic functions in the neighbourhood of a local minimum. This is so, because by the Taylor expansion (1.21), it can be seen that many general differentiable functions approximate the form (2.9) near a local minimum.

2.3.2.1 Mutually conjugate directions

Two vectors $\mathbf{u}$, $\mathbf{v} \neq \mathbf{0}$ are defined to be *orthogonal* if the scalar product $\mathbf{u}^T\mathbf{v} = (\mathbf{u}, \mathbf{v}) = 0$. The concept of *mutual conjugacy* may be defined in a similar manner. Two vectors $\mathbf{u}$, $\mathbf{v} \neq \mathbf{0}$, are defined to be *mutually conjugate* with respect to the matrix $\mathbf{A}$ in (2.9) if $\mathbf{u}^T\mathbf{A}\mathbf{v} = (\mathbf{u}, \mathbf{A}\mathbf{v}) = 0$. Note that $\mathbf{A}$ is a positive-definite symmetric matrix.

It can also be shown (see Theorem 5.5.1 in Chapter 5) that if the set of vectors $\mathbf{u}^i$, $i = 1, 2, \ldots, n$ are *mutually conjugate*, then they form a *basis* in $\mathbb{R}^n$, i.e. any $\mathbf{x} \in \mathbb{R}^n$ may be expressed as

$$\mathbf{x} = \sum_{i=1}^{n} \tau_i \mathbf{u}^i \qquad (2.10)$$

where

$$\tau_i = \frac{(\mathbf{u}^i, \mathbf{A}\mathbf{x})}{(\mathbf{u}^i, \mathbf{A}\mathbf{u}^i)}. \qquad (2.11)$$

2.3.2.2 Convergence theorem for mutually conjugate directions

Suppose $\mathbf{u}^i$, $i = 1, 2, \ldots, n$ are mutually conjugate with respect to positive-definite $\mathbf{A}$, then the optimal line search descent method in Section 2.1.1, using $\mathbf{u}^i$ as search directions, converges to the unique minimum $\mathbf{x}^*$ of $f(\mathbf{x}) = \frac{1}{2}\mathbf{x}^T\mathbf{A}\mathbf{x} + \mathbf{b}^T\mathbf{x} + c$ in less than or equal to n steps.

Proof:

If $\mathbf{x}^0$ the starting point, then after i iterations:

$$\mathbf{x}^i = \mathbf{x}^{i-1} + \lambda_i \mathbf{u}^i = \mathbf{x}^{i-2} + \lambda_{i-1} \mathbf{u}^{i-1} + \lambda_i \mathbf{u}^i = \dots$$

$$= \mathbf{x}^0 + \sum_{k=1}^{i} \lambda_k \mathbf{u}^k. \tag{2.12}$$

The condition for *optimal descent* at iteration i is

$$\frac{dF(\lambda_i)}{d\lambda} = \left. \frac{df(\mathbf{x}^{i-1} + \lambda_i \mathbf{u}^i)}{d\lambda} \right|_{\mathbf{u}^i} = [\mathbf{u}^i, \nabla f(\mathbf{x}^{i-1} + \lambda_i \mathbf{u}^i)] = 0$$

$$= [\mathbf{u}^i, \nabla f(\mathbf{x}^i)] = 0$$

i.e.

$$0 = (\mathbf{u}^i, \mathbf{A}\mathbf{x}^i + \mathbf{b})$$

$$= \left(\mathbf{u}^i, \mathbf{A}\left(\mathbf{x}^0 + \sum_{k=1}^{i} \lambda_k \mathbf{u}^k \right) + \mathbf{b} \right) = (\mathbf{u}^i, \mathbf{A}\mathbf{x}^0 + \mathbf{b}) + \lambda_i(\mathbf{u}^i, \mathbf{A}\mathbf{u}^i)$$

because $\mathbf{u}^i$, $i = 1, 2, \dots, n$ are mutually conjugate, and thus

$$\lambda_i = -(\mathbf{u}^i, \mathbf{A}\mathbf{x}^0 + \mathbf{b})/(\mathbf{u}^i, \mathbf{A}\mathbf{u}^i). \tag{2.13}$$

Substituting (2.13) into (2.12) above gives

$$\mathbf{x}^n = \mathbf{x}^0 + \sum_{i=1}^{n} \lambda_i \mathbf{u}^i = \mathbf{x}^0 - \sum_{i=1}^{n} \frac{(\mathbf{u}^i, \mathbf{A}\mathbf{x}^0 + \mathbf{b})\mathbf{u}^i}{(\mathbf{u}^i, \mathbf{A}\mathbf{u}^i)}$$

$$= \mathbf{x}^0 - \sum_{i=1}^{n} \frac{(\mathbf{u}^i, \mathbf{A}\mathbf{x}^0)\mathbf{u}^i}{(\mathbf{u}^i, \mathbf{A}\mathbf{u}^i)} - \sum_{i=1}^{n} \frac{(\mathbf{u}^i, \mathbf{A}(\mathbf{A}^{-1}\mathbf{b}))\mathbf{u}^i}{(\mathbf{u}^i, \mathbf{A}\mathbf{u}^i)}.$$

Now by utilizing (2.10) and (2.11) it follows that

$$\mathbf{x}^n = \mathbf{x}^0 - \mathbf{x}^0 - \mathbf{A}^{-1}\mathbf{b} = -\mathbf{A}^{-1}\mathbf{b} = \mathbf{x}^*.$$

The implication of the above theorem for the case $n = 2$ and where mutually conjugate line search directions $\mathbf{u}^1$ and $\mathbf{u}^2$ are used, is depicted in Figure 2.9.

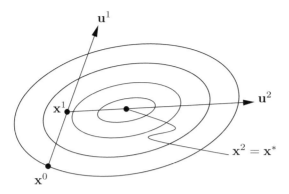

Figure 2.9: Quadratic termination of the conjugate gradient method in two steps for the case $n = 2$

2.3.2.3 Determination of mutually conjugate directions

How can mutually conjugate search directions be found? One way is to determine all the eigenvectors $\mathbf{u}^i$, $i = 1, 2, \ldots, n$ of $\mathbf{A}$. For $\mathbf{A}$ positive-definite, all the eigenvectors are mutually orthogonal and since $\mathbf{A}\mathbf{u}^i = \mu_i \mathbf{u}^i$ where μ_i is the associated eigenvalue, it follows directly that for all $i \neq j$ that $(\mathbf{u}^i, \mathbf{A}\mathbf{u}^j) = (\mathbf{u}^i, \mu_j \mathbf{u}^j) = \mu_j(\mathbf{u}^i, \mathbf{u}^j) = 0$, i.e. the eigenvectors are mutually conjugate with respect to $\mathbf{A}$. It is, however, not very practical to determine mutually conjugate directions by finding all the eigenvectors of $\mathbf{A}$, since the latter task in itself represents a computational problem of magnitude equal to that of solving the original unconstrained optimization problem via any other numerical algorithm. An easier method for obtaining mutually conjugate directions, is by means of the Fletcher-Reeves formulæ (Fletcher and Reeves 1964).

2.3.2.4 The Fletcher-Reeves directions

The *Fletcher-Reeves directions* $\mathbf{u}^i$, $i = 1, 2, \ldots, n$, that are listed below, can be shown (see Theorem 5.5.3 in Chapter 5) to be mutually conjugate with respect to the matrix $\mathbf{A}$ in the expression for the quadratic function in (2.9) (for which $\nabla f(\mathbf{x}) = \mathbf{A}\mathbf{x} + \mathbf{b}$). The explicit directions are:

$$\mathbf{u}^1 = -\nabla f(\mathbf{x}^0)$$

and for $i = 1, 2, \ldots, n - 1$

$$\mathbf{u}^{i+1} = -\nabla f(\mathbf{x}^i) + \beta_i \mathbf{u}^i \tag{2.14}$$

where $\mathbf{x}^i = \mathbf{x}^{i-1} + \lambda_i \mathbf{u}^i$, and λ_i corresponds to the optimal descent step in iteration i, and

$$\beta_i = \frac{\|\nabla f(\mathbf{x}^i)\|^2}{\|\nabla f(\mathbf{x}^{i-1})\|^2}. \tag{2.15}$$

The *Polak-Ribiere* directions are obtained if, instead of using (2.15), β_i is computed using

$$\beta_i = \frac{(\nabla f(\mathbf{x}^i) - \nabla f(\mathbf{x}^{i-1}))^T \nabla f(\mathbf{x}^i)}{\|\nabla f(\mathbf{x}^{i-1})\|^2}. \tag{2.16}$$

If $f(\mathbf{x})$ is quadratic it can be shown (Fletcher 1987) that (2.16) is equivalent to (2.15).

2.3.2.5 Formal Fletcher-Reeves conjugate gradient algorithm for general functions

Given $\mathbf{x}^0$ perform the following steps:

1. Compute $\nabla f(\mathbf{x}^0)$ and set $\mathbf{u}^1 = -\nabla f(\mathbf{x}^0)$.

2. For $i = 1, 2, \ldots, n$ do:

    2.1 set $\mathbf{x}^i = \mathbf{x}^{i-1} + \lambda_i \mathbf{u}^i$ where λ_i such that

    $$f(\mathbf{x}^{i-1} + \lambda_i \mathbf{u}^i) = \min_{\lambda} f(\mathbf{x}^{i-1} + \lambda \mathbf{u}^i) \text{ (line search)},$$

    2.2 compute $\nabla f(\mathbf{x}^i)$,

    2.3 *if* convergence criteria satisfied, then stop and $\mathbf{x}^* \cong \mathbf{x}^i$, *else* go to Step 2.4.

    2.4 *if* $1 \leq i \leq n - 1$, $\mathbf{u}^{i+1} = -\nabla f(\mathbf{x}^i) + \beta_i \mathbf{u}^i$ with β_i given by (2.15).

3. Set $\mathbf{x}^0 = \mathbf{x}^n$ and go to Step 2 (restart).

If β_i is computed by (2.16) instead of (2.15) the method is known as the *Polak-Ribiere* method.

2.3.2.6 Simple illustrative example

Apply the Fletcher-Reeves method to minimize

$$f(\mathbf{x}) = \tfrac{1}{2}x_1^2 + x_1 x_2 + x_2^2$$

with $\mathbf{x}^0 = [10, -5]^T$.

Solution:

Iteration 1:

$$\nabla f(\mathbf{x}) = \begin{bmatrix} x_1 + x_2 \\ x_1 + 2x_2 \end{bmatrix} \text{ and therefore } \mathbf{u}^1 = -\nabla f(\mathbf{x}^0) = \begin{bmatrix} -5 \\ 0 \end{bmatrix}.$$

$$\mathbf{x}^1 = \mathbf{x}^0 + \lambda \mathbf{u}^1 = \begin{bmatrix} 10 - 5\lambda \\ -5 \end{bmatrix} \text{ and}$$

$$F(\lambda) = f(\mathbf{x}^0 + \lambda \mathbf{u}^1) = \tfrac{1}{2}(10 - 5\lambda)^2 + (10 - 5\lambda)(-5) + 25.$$

For optimal descent

$$\tfrac{dF}{d\lambda}(\lambda) = \left.\tfrac{df}{d\lambda}\right|_{\mathbf{u}^1} = -5(10 - 5\lambda) + 25 = 0 \text{ (line search)}.$$

This gives $\lambda_1 = 1$, $\mathbf{x}^1 = \begin{bmatrix} 5 \\ -5 \end{bmatrix}$ and $\nabla f(\mathbf{x}^1) = \begin{bmatrix} 0 \\ -5 \end{bmatrix}$.

Iteration 2:

$$\mathbf{u}^2 = -\nabla f(\mathbf{x}^1) + \frac{\|\nabla f(\mathbf{x}^1)\|^2}{\|\nabla f(\mathbf{x}^0)\|^2} \mathbf{u}^1 = -\begin{bmatrix} 0 \\ -5 \end{bmatrix} + \tfrac{25}{25}\begin{bmatrix} -5 \\ 0 \end{bmatrix} = \begin{bmatrix} -5 \\ 5 \end{bmatrix}.$$

$$\mathbf{x}^2 = \mathbf{x}^1 + \lambda \mathbf{u}^2 = \begin{bmatrix} 5 \\ -5 \end{bmatrix} + \lambda \begin{bmatrix} -5 \\ 5 \end{bmatrix} = \begin{bmatrix} 5(1 - \lambda) \\ -5(1 - \lambda) \end{bmatrix} \text{ and}$$

$$F(\lambda) = f(\mathbf{x}^1 + \lambda \mathbf{u}^2) = \tfrac{1}{2}[25(1 - \lambda)^2 - 50(1 - \lambda)^2 + 50(1 - \lambda)^2].$$

Again for optimal descent

$$\tfrac{dF}{d\lambda}(\lambda) = \left.\tfrac{df}{d\lambda}\right|_{\mathbf{u}^2} = -25(1 - \lambda) = 0 \text{ (line search)}.$$

This gives $\lambda_2 = 1$, $\mathbf{x}^2 = \begin{bmatrix} 0 \\ 0 \end{bmatrix}$ and $\nabla f(\mathbf{x}^2) = \begin{bmatrix} 0 \\ 0 \end{bmatrix}$. Therefore stop.

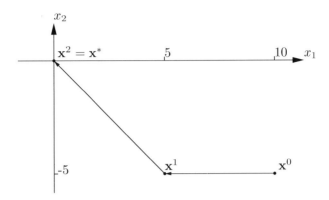

Figure 2.10: Convergence of Fletcher-Reeves method for illustrative example

The two iteration steps are shown in Figure 2.10.

2.4   Second order line search descent methods

These methods are based on Newton's method (see Section 1.5.4.1) for solving $\nabla f(\mathbf{x}) = \mathbf{0}$ iteratively: Given $\mathbf{x}^0$, then

$$\mathbf{x}^i = \mathbf{x}^{i-1} - \mathbf{H}^{-1}(\mathbf{x}^{i-1})\nabla f(\mathbf{x}^{i-1}), \ i = 1, 2, \ldots \qquad (2.17)$$

As stated in Chapter 1, the main characteristics of this method are:

1. In the neighbourhood of the solution it may converge very fast. In fact, it has the very desirous property of being quadratically convergent if it converges. Unfortunately convergence is not guaranteed and it may sometimes diverge, even from close to the solution.

2. The implementation of the method requires that $\mathbf{H}(\mathbf{x})$ be evaluated at each step.

3. To obtain the Newton step, $\boldsymbol{\Delta} = \mathbf{x}^i - \mathbf{x}^{i-1}$ it is also necessary to solve a $n \times n$ linear system $\mathbf{H}(\mathbf{x})\boldsymbol{\Delta} = -\nabla f(\mathbf{x})$ at each itera-

tion. This is computationally very expensive for large n, since an order n^3 multiplication operations are required to solve the system numerically.

2.4.1 Modified Newton's method

To avoid the problem of convergence (point 1. above), the computed Newton step $\mathbf{\Delta}$ is rather used as a search direction in the general line search descent algorithm given in Section 2.1.1. Thus at iteration i: select $\mathbf{u}^i = \mathbf{\Delta} = -\mathbf{H}^{-1}(\mathbf{x}^{i-1})\nabla f(\mathbf{x}^{i-1})$, and minimize in that direction to obtain a λ_i such that

$$f(\mathbf{x}^{i-1} + \lambda_i \mathbf{u}^i) = \min_{\lambda} f(\mathbf{x}^{i-1} + \lambda \mathbf{u}^i)$$

and then set $\mathbf{x}^i = \mathbf{x}^{i-1} + \lambda_i \mathbf{u}^i$.

2.4.2 Quasi-Newton methods

To avoid the above mentioned computational problems (2. and 3.), methods have been developed in which approximations of $\mathbf{H}^{-1}$ are applied at each iteration. Starting with an approximation $\mathbf{G}_0$ to $\mathbf{H}^{-1}$ for the first iteration, the approximation is updated after each line search. An example of such a method is the Davidon-Fletcher-Powell (DFP) method.

2.4.2.1 DFP quasi-Newton method

The structure of this (rank-1 update) method (Fletcher 1987) is as follows.

1. Choose $\mathbf{x}^0$ and set $\mathbf{G}_0 = \mathbf{I}$.

2. Do for iteration $i = 1, 2, \ldots, n$:

   2.1 set $\mathbf{x}^i = \mathbf{x}^{i-1} + \lambda_i \mathbf{u}^i$, where $\mathbf{u}^i = -\mathbf{G}_{i-1}\nabla f(\mathbf{x}^{i-1})$ and λ_i is such that $f(\mathbf{x}^{i-1} + \lambda_i \mathbf{u}^i) = \min_{\lambda} f(\mathbf{x}^{i-1} + \lambda \mathbf{u}^i)$, $\lambda_i \geq 0$ (line search),

2.2 if stopping criteria satisfied then stop, $\mathbf{x}^* \cong \mathbf{x}^i$,

2.3 set $\mathbf{v}^i = \lambda_i \mathbf{u}^i$ and

set $\mathbf{y}^i = \boldsymbol{\nabla} f(\mathbf{x}^i) - \boldsymbol{\nabla} f(\mathbf{x}^{i-1})$,

2.4 set

$$\mathbf{G}_i = \mathbf{G}_{i-1} + \mathbf{A}_i + \mathbf{B}_i \text{ (rank 1-update)} \qquad (2.18)$$

where $\mathbf{A}_i = \dfrac{\mathbf{v}^i \mathbf{v}^{iT}}{\mathbf{v}^{iT} \mathbf{y}^i}$, $\mathbf{B}_i = \dfrac{-\mathbf{G}_{i-1} \mathbf{y}^i (\mathbf{G}_{i-1} \mathbf{y}^i)^T}{\mathbf{y}^{iT} \mathbf{G}_{i-1} \mathbf{y}^i}$.

3. Set $\mathbf{x}^0 = \mathbf{x}^n$; $\mathbf{G}_0 = \mathbf{G}_n$ (or $\mathbf{G}_0 = \mathbf{I}$), and go to Step 2 (restart).

2.4.2.2   Characteristics of DFP method

1. The method does not require the evaluation of $\mathbf{H}$ or the explicit solution of a linear system.

2. If $\mathbf{G}_{i-1}$ is positive-definite then so is $\mathbf{G}_i$ (see Theorem 5.6.1).

3. If $\mathbf{G}_i$ is positive-definite then descent is ensured at $\mathbf{x}^i$ because

$$\frac{df(\mathbf{x}^i)}{d\lambda}\bigg|_{\mathbf{u}^{i+1}} = \boldsymbol{\nabla}^T f(\mathbf{x}^i) \mathbf{u}^{i+1}$$

$$= -\boldsymbol{\nabla}^T f(\mathbf{x}^i) \mathbf{G}_i \boldsymbol{\nabla} f(\mathbf{x}^i) < 0, \text{ for all } \boldsymbol{\nabla} f(\mathbf{x}) \neq \mathbf{0}.$$

4. The directions $\mathbf{u}^i$, $i = 1, 2, \ldots, n$ are mutually conjugate for a quadratic function with $\mathbf{A}$ positive-definite (see Theorem 5.6.2). The method therefore possesses the desirable property of *quadratic termination* (see Section 2.3.2).

5. For quadratic functions: $\mathbf{G}_n = \mathbf{A}^{-1}$ (see again Theorem 5.6.2).

2.4.2.3   The BFGS method

The state-of-the-art quasi-Newton method is the Broyden-Fletcher-Goldfarb-Shanno (BFGS) method developed during the early 1970s (see Fletcher 1987). This method uses a more complicated rank-2 update formula for $\mathbf{H}^{-1}$. For this method the update formula to be used in Step 2.4 of the algorithm given in Section 2.4.2.1 becomes

$$\mathbf{G}_i = \mathbf{G}_{i-1} + \left[1 + \frac{\mathbf{y}^{iT}\mathbf{G}_{i-1}\mathbf{y}^i}{\mathbf{v}^{iT}\mathbf{y}^i}\right]\left[\frac{\mathbf{v}^i\mathbf{v}^{iT}}{\mathbf{v}^{iT}\mathbf{y}^i}\right]$$
$$- \left[\frac{\mathbf{v}^i\mathbf{y}^{iT}\mathbf{G}_{i-1} + \mathbf{G}_{i-1}\mathbf{y}^i\mathbf{v}^{iT}}{\mathbf{v}^{iT}\mathbf{y}^i}\right]. \tag{2.19}$$

2.5 Zero order methods and computer optimization subroutines

This chapter would not be complete without mentioning something about the large number of so-called *zero order* methods that have been developed. These methods are called such because they neither use first order nor second order derivative information, but only function values, i.e. only zero order derivative information.

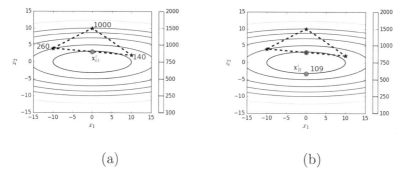

(a)   (b)

Figure 2.11: Example of some of the *downhill simplex method* steps which includes (a) computing the centroid, $\mathbf{x}_O^i$, using all vertices excluding the worst vertex, i.e. vertex with highest function value, followed by (b) reflecting the worst vertex to compute the reflection point $\mathbf{x}_R^i$

Zero order methods are of the earliest methods and many of them are based on rough and ready ideas with few theoretical background. Although these ad hoc methods are, as one may expect, much slower and computationally much more expensive than the higher order methods, they are usually reliable and easy to program. One of the most successful of these methods is the *downhill simplex method* of Nelder

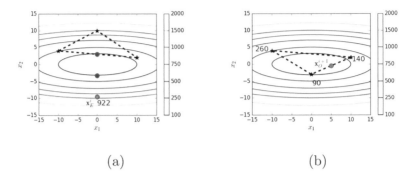

(a)   (b)

Figure 2.12: Example of some of the *downhill simplex method* steps. Since the reflection point computed in Figure 2.11 had the lowest function value an (a) expansion point, $\mathbf{x}_E^i$, is computed and (b) the simplex updated, the vertices sorted and the next centroid, $\mathbf{x}_O^{i+1}$, computed

and Mead (1965), also known as *Nelder-Mead*. This method should not be confused with Dantzig's simplex method for linear programming.

The *downhill simplex method* first initializes a simplex with $n+1$ vertices in n-dimensional space, as depicted in Figure 2.11(a) for $n = 2$. After initialization, the vertices are ordered according to their function values from low to high. Confirming that the convergence criteria have not been met allows for the first n vertices to be used to compute the centroid as indicated by $\mathbf{x}_O^i$ in Figure 2.11(a). This is followed by the first simplex operator, namely reflection of the worst point $(n+1)^{\text{th}}$ over the centroid to obtain the reflection point $\mathbf{x}_R^i$ as illustrated in Figure 2.11(b). Based on the function value of the reflection point, the next operator is determined as listed in Algorithm 2.5. In this example, the function value at the reflection point is lower than the best computed point so far, which requires us to compute an expansion point, $\mathbf{x}_E^i$, following Step 4 as listed and depicted in Figure 2.12(a). Since the function value at the expansion point is higher than the function value at the reflection point, the reflection point is kept by replacing the $(n + 1)^{\text{th}}$ vertex with it. Again the points are ordered and convergence of the algorithm checked. Thereafter the centroid for the new simplex, $\mathbf{x}_O^{i+1}$, is computed as depicted in Figure 2.12(b) in accordance with Step 2 of the listed algorithm. The general framework of the *downhill simplex method* that includes contraction and shrinking, in addition to reflection and expansion already explained, is given in Algorithm 2.5.

Algorithm 2.5 Nelder-Mead derivative free heuristic search method

---

Initialization: Select real constants $\rho_R > 0$, $\rho_E > 1$, $0 \leq \rho_C \leq 0.5$ and $0 \leq \rho_S \leq 1$. Select the maximum number of iterations i_{max} and convergence tolerance $\epsilon > 0$. Set $i := 1$, and initialize the initial simplex by randomly generating $n + 1$ vectors $\mathbf{x}_1^i, \ldots, \mathbf{x}_{n+1}^i$ in the n-dimensional design space and evaluate the function at the $n+1$ simplex vertices. Perform the following steps:

1. **Order:** Order the vertices according to their function values:

$$f(\mathbf{x}_1^i) \leq f(\mathbf{x}_2^i) \leq \cdots \leq f(\mathbf{x}_{n+1}^i)$$

If $\|\mathbf{x}_1^i - \mathbf{x}_{n+1}^i\| \leq \epsilon$ then $\mathbf{x}^* := \mathbf{x}_1^i$ and stop, else go to Step 2.

2. **Centroid:** Compute the centroid $\mathbf{x}_O^i$ of the first n vertices, i.e. the $(n+1)^{\text{th}}$ vertex is excluded:

$$\mathbf{x}_O^i = \frac{1}{n} \sum_{j=1}^{n} \mathbf{x}_j^i.$$

3. **Reflection:** Compute the reflected point:

$$\mathbf{x}_R^i = \mathbf{x}_O^i + \rho_R(\mathbf{x}_O^i - \mathbf{x}_{n+1}^i).$$

If, $f(\mathbf{x}_O^i) \leq f(\mathbf{x}_R^i) < f(\mathbf{x}_n^i)$ then set $\mathbf{x}_{n+1}^i := \mathbf{x}_R^i$, set $i := i+1$ and go to Step 1, else if $f(\mathbf{x}_R^i) < f(\mathbf{x}_O^i)$ go to Step 4, otherwise go to Step 5.

4. **Expansion:** Compute the expansion point:

$$\mathbf{x}_E^i = \mathbf{x}_O^i + \rho_E(\mathbf{x}_O^i - \mathbf{x}_{n+1}^i).$$

If $f(\mathbf{x}_E^i) < f(\mathbf{x}_R^i)$ then set $\mathbf{x}_{n+1}^i := \mathbf{x}_E^i$ else set $\mathbf{x}_{n+1}^i := \mathbf{x}_R^i$. Set $i := i+1$ and go to Step 1.

5. **Contraction:** Compute the contraction point:

$$\mathbf{x}_C^i = \mathbf{x}_O^i + \rho_C(\mathbf{x}_O^i - \mathbf{x}_{n+1}^i).$$

If $f(\mathbf{x}_C^i) < f(\mathbf{x}_{n+1}^i)$ then set $\mathbf{x}_{n+1}^i := \mathbf{x}_C^i$, set $i := i+1$ and go to Step 1, otherwise go to Step 6.

6. **Shrinkage:** Shrink all the points except the best point:

$$\mathbf{x}_j^i = \mathbf{x}_1^i + \rho_S(\mathbf{x}_j^i - \mathbf{x}_1^i), \quad j = 2, \ldots, n+1,$$

set $i := i+1$ and go to Step 1.

Another very powerful and popular method that only uses function values is the multi-variable method of Powell (1964). This method generates mutually conjugate directions by performing sequences of line searches in which only function evaluations are used. For this method Theorem 2.3.2.2 applies and the method therefore possesses the property of quadratic termination. The basic procedure of *Powell's method* is outlined in Algorithm 2.6.

Algorithm 2.6 Powell's derivative free conjugate direction method

---

Initialization: For an n-dimensional problem, choose a starting point $\mathbf{x}^0$ and n search directions $\mathbf{u}^k$, $k = 1, \ldots, n$. The initial directions are usually chosen as the Cartesian coordinate directions $\mathbf{u}^1 = \mathbf{e}^1, \ldots, \mathbf{u}^n = \mathbf{e}^n$. Select $\epsilon > 0$ and set $l := 1$. Perform the following steps:

1. **Minimize in n search directions:** Set $\mathbf{x}^1 := \mathbf{x}^0$. For each $k = 1, \ldots, n$, find λ_k such that

$$f(\mathbf{x}^k + \lambda_k \mathbf{u}^k) = \min_\lambda f(\mathbf{x}^k + \lambda \mathbf{u}^k)$$

   and update $\mathbf{x}^{k+1} := \mathbf{x}^k + \lambda_k \mathbf{u}^k$.

2. **Update n search directions:** For each $k = 1, \ldots, n - 1$, set $\mathbf{u}^k := \mathbf{u}^{k+1}$ and finally set $\mathbf{u}^n := \mathbf{x}^{n+1} - \mathbf{x}^1$. Check for convergence: $\|\mathbf{x}^{n+1} - \mathbf{x}^1\| \leq \epsilon$, if satisfied then set $\mathbf{x}^* := \mathbf{x}^{n+1}$ and stop, else set $l := l + 1$, $\mathbf{x}^0 := \mathbf{x}^{n+1}$ and go to Step 3.

3. If l is divisible by $n + 1$ then reinitialize the search directions to the Cartesian coordinate directions $\mathbf{u}^1 = \mathbf{e}^1, \ldots, \mathbf{u}^n = \mathbf{e}^n$. Go to Step 1.

Amongst the more recently proposed and modern zero order methods, the method of *simulated annealing* and the so-called *genetic algorithms* (GA's) are the most prominent (see for example, Haftka and Gürdal 1992). Other zero order methods include *Differential Evolution* proposed by Storn and Price (1997), *Basin Hopping* proposed by Wales and Doye (1997) and the contribution by Kennedy and Eberhart (1995) the so-called *Particle Swarm Optimization Algorithm*. A strongly interacting dynamic particle swarm variant of this method has been proposed by Kok and Snyman (2008).

Computer programs are commercially available for all the unconstrained optimization methods presented in this chapter. Most of the algorithms may, for example, be found in the *Matlab Optimization Toolbox* and in the *IMSL* and *NAG* mathematical subroutine libraries. In Chapter 9 we cover most of the algorithms using the `scipy.optimize` module in *Python*, which includes *Nelder-Mead*, *Powell's method* and *Basin Hopping*.

2.6 Exercises

2.6.1 Apply the golden section method and Powell's quadratic interpolation method to the problems below. Compare their respective performances with regard to the number of function evaluations required to attain the prescribed accuracies.

(i) minimize $F(\lambda) = \lambda^2 + 2e^{-\lambda}$ over $[0, 2]$ with $\varepsilon = 0.01$.

(ii) maximize $F(\lambda) = \lambda \cos(\lambda)$ over $[0, \pi/2]$ with $\varepsilon = 0.001$.

(iii) minimize $F(\lambda) = 4(\lambda - 7)/(\lambda^2 + \lambda - 2)$ over $[-1.9; 0.9]$ using up to 10 function evaluations.

(iv) minimize $F(\lambda) = \lambda^4 - 20\lambda^3 + 0.1\lambda$ over $[0; 20]$ with $\varepsilon = 10^{-5}$.

2.6.2 Plot the functions and derivatives for the problems given in (i)–(iv). Indicate the $[0, \lambda]$ interval that satisfies each of the four conditions for sufficient improvement in Section 2.3.1.5, taking a step from $\lambda = 0$ and using $c_1 = 0.5$, $c_2 = 0.5$ and $c_3 = 0.5$ for conditions for 2, 3 and 4 respectively.

2.6.3 Reconsidering Exercise 2.6.2, would the interval increase or decrease if c_1 is increased?

2.6.4 Reconsidering Exercise 2.6.2, would the interval increase or decrease if c_2 is increased?

2.6.5 Reconsidering Exercise 2.6.2, would the interval increase or decrease if c_3 is increased?

2.6.6 Apply one step of Powell's method to the problems listed in (i)–(iv) and determine whether the Wolfe conditions are satisfied using $c_1 = 0.2$ and $c_2 = 0.8$.

2.6.7 For the Quadratic function in Section 1.6.1, how many iterations would Newton's method, and the BFGS using exact line
searches, respectively require to solve the problem?

2.6.8 Compute the gradient vector about $\mathbf{x}^0$ using the forward, backward and central difference schemes by guessing appropriate
step sizes for the Rosenbrock function listed in Section 1.6.1.

2.6.9 Derive the central finite difference formula and show that it is
second order accurate.

2.6.10 Show that the eigenvectors of the Hessian matrix are orthogonal?

2.6.11 The steepest descent method with exact line searches is expected
to converge in less than n, n or more than n iterations when
optimizing a

  (i) general quadratic function of dimension n,

  (ii) general non-linear function of dimension n, and

  (iii) spherical quadratic function of dimension n.

2.6.12 A conjugate gradient method with exact line searches is expected
to converge in less than n, n or more than n iterations when
optimizing a

  (i) general quadratic function of dimension n,

  (ii) general non-linear function of dimension n, and

  (iii) spherical quadratic function of dimension n.

2.6.13 A Quasi-Newton method with exact line searches is expected
to converge in less than n, n or more than n iterations when
optimizing a

  (i) general quadratic function of dimension n,

  (ii) general non-linear function of dimension n, and

  (iii) spherical quadratic function of dimension n.

2.6.14 The modified Newton's method with exact line searches is
expected to converge in less than n, n or more than n iterations when optimizing a

    (i) general quadratic function of dimension n,

   (ii) general non-linear function of dimension n, and

  (iii) spherical quadratic function of dimension n.

Chapter 3

STANDARD METHODS FOR CONSTRAINED OPTIMIZATION

3.1 Penalty function methods for constrained minimization

Consider the general constrained optimization problem:

$$
\begin{aligned}
\underset{\mathbf{x}}{\text{minimize}} \quad & f(\mathbf{x}) \\
\text{such that} \quad & g_j(\mathbf{x}) \le 0 \quad j = 1, 2, \ldots, m \\
& h_j(\mathbf{x}) = 0 \quad j = 1, 2, \ldots, r.
\end{aligned}
\tag{3.1}
$$

The most simple and straightforward approach to handling constrained problems of the above form is to apply a suitable unconstrained optimization algorithm to *a penalty function formulation of* constrained problem (3.1).

3.1.1 The penalty function formulation

A penalty function formulation of the general constrained problem (3.1) is

$$
\underset{\mathbf{x}}{\text{minimize}} \, P(\mathbf{x})
$$

© Springer International Publishing AG, part of Springer Nature 2018
J.A. Snyman and D.N. Wilke, *Practical Mathematical Optimization,*
Springer Optimization and Its Applications 133,
https://doi.org/10.1007/978-3-319-77586-9_3

where

$$P(\mathbf{x}, \boldsymbol{\rho}, \boldsymbol{\beta}) = f(\mathbf{x}) + \sum_{j=1}^{r} \rho_j h_j^2(\mathbf{x}) + \sum_{j=1}^{m} \beta_j g_j^2(\mathbf{x}) \qquad (3.2)$$

and where the components of the penalty parameter vectors $\boldsymbol{\rho}$ and $\boldsymbol{\beta}$ are given by

$$\rho_j \gg 0; \; \beta_j = \begin{cases} 0 & \text{if } g_j(\mathbf{x}) \leq 0 \\ \mu_j \gg 0 & \text{if } g_j(\mathbf{x}) > 0. \end{cases}$$

The latter parameters, ρ_j and β_j, are called *penalty parameters*, and $P(\mathbf{x}, \boldsymbol{\rho}, \boldsymbol{\beta})$ the *penalty function*. The solution to this unconstrained minimization problem is denoted by $\mathbf{x}^*(\boldsymbol{\rho}, \boldsymbol{\beta})$, where $\boldsymbol{\rho}$ and $\boldsymbol{\beta}$ denote the respective vectors of penalty parameters.

Often $\rho_j \equiv$ constant $\equiv \rho$, for all j, and also $\mu_j \equiv \rho$ for all j such that $g_j(\mathbf{x}) > 0$. Thus P in (3.2) is denoted by $P(\mathbf{x}, \rho)$ and the corresponding minimum by $\mathbf{x}^*(\rho)$. It can be shown that under normal continuity conditions the $\lim_{\rho \to \infty} \mathbf{x}^*(\rho) = \mathbf{x}^*$. Typically the overall penalty parameter ρ is set at $\rho = 10^4$ if the constraints functions are normalized in some sense.

3.1.2   Illustrative examples

Consider the following two one-dimensional constrained optimization problems:

(a) $\min f(x)$
   such that $h(x) = x - a = 0,$
        then $P(x) = f(x) + \rho(x - a)^2;$

and

(b) $\min f(x)$
   such that $g(x) = x - b \leq 0,$
        then $P(x) = f(x) + \beta(x - b)^2.$

The penalty function solutions to these two problems are as depicted in Figures 3.1 (a) and (b). The penalty function method falls in the class of *external methods* because it converges externally from the infeasible region.

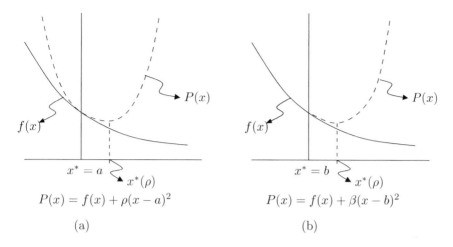

$$P(x) = f(x) + \rho(x - a)^2$$

(a)

$$P(x) = f(x) + \beta(x - b)^2$$

(b)

Figure 3.1: Behaviour of penalty function for one-dimensional (a) equality constrained, and (b) inequality constrained minimization problems

3.1.3 Sequential unconstrained minimization technique (SUMT)

Unfortunately the penalty function method becomes unstable and inefficient for very large ρ if high accuracy is required. This is because rounding errors result in the computation of unreliable descent directions. If second order unconstrained minimization methods are used for the minimization of $P(\mathbf{x}, \rho)$, then the associated Hessian matrices become ill-conditioned and again the method is inclined to break down. A remedy to this situation is to apply the penalty function method to a sequence of sub-problems, starting with moderate penalty parameter values, and successively increasing their values for the sub-problems. The details of this approach (SUMT) is as follows.

SUMT algorithm:

1. Choose tolerances ε_1 and ε_2, starting point $\mathbf{x}^0$ and initial overall penalty parameter value ρ_0, and set $k = 0$.

2. Minimize $P(\mathbf{x}, \rho_k)$ by any unconstrained optimization algorithm to give $\mathbf{x}^*(\rho_k)$.

3. *If (for $k > 0$) the convergence criteria are satisfied: STOP*

   i.e. stop if $\|\mathbf{x}^*(\rho_k) - \mathbf{x}^*(\rho_{k-1})\| < \varepsilon_1$

   and/or $|P(\mathbf{x}^*(\rho_{k-1}) - P(\mathbf{x}^*(\rho_k))| < \varepsilon_2$,

   else

   set $\rho_{k+1} = c\rho_k$, $c > 1$ and $\mathbf{x}^0 = \mathbf{x}^*(\rho_k)$,

   set $k = k + 1$ and go to Step 2.

Typically choose $\rho_0 = 1$ and $c = 10$.

3.1.4   Simple example

Consider the constrained problem:

$$\min f(\mathbf{x}) = \tfrac{1}{3}(x_1 + 1)^3 + x_2$$

such that

$$1 - x_1 \leq 0; \quad -x_2 \leq 0.$$

The problem is depicted in Figure 3.2.

Define the penalty function:

$$P(\mathbf{x}, \rho) = \tfrac{1}{3}(x_1 + 1)^3 + x_2 + \rho(1 - x_1)^2 + \rho x_2^2.$$

(Of course the ρ only comes into play if the corresponding constraint is violated.) The penalty function solution may now be obtained analytically as follows.

The first order necessary conditions for an unconstrained minimum of P are

$$\frac{\partial P}{\partial x_1} = (x_1 + 1)^2 - 2\rho(1 - x_1) = 0 \tag{3.3}$$

$$\frac{\partial P}{\partial x_2} = 1 + 2\rho x_2 = 0. \tag{3.4}$$

From (3.4): $x_2^*(\rho) = -\tfrac{1}{2\rho}$

and from (3.3): $x_1^2 + 2(1 + \rho)x_1 + 1 - 2\rho = 0$.

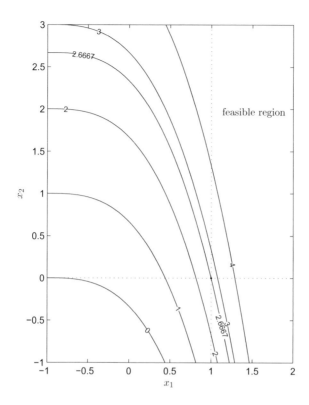

Figure 3.2: Contour representation of objective function and feasible region for simple example

Solving the quadratic equation and taking the positive root gives

$$x_1^*(\rho) = -(1+\rho) + (1+\rho)\left(1 + \frac{(2\rho - 1)}{(1+\rho)^2}\right)^{1/2}.$$

Clearly $\lim\limits_{\rho \to \infty} x_2^*(\rho) = 0$ and $\lim\limits_{\rho \to \infty} x_1^*(\rho) = 1$ (where use has been made of the expansion $(1+\varepsilon)^{1/2} = 1 + \frac{1}{2}\varepsilon + \dots$) giving $\mathbf{x}^* = [1, 0]^T$, as one expects from Figure 3.2. Here the solution to the penalty function formulated problem has been obtained analytically via the first order necessary conditions. In general, of course, the solution is obtained numerically by applying a suitable unconstrained minimization algorithm.

3.2 Classical methods for constrained optimization problems

3.2.1 Equality constrained problems and the Lagrangian function

Consider the equality constrained problem:

$$\begin{aligned} \text{minimize} \quad & f(\mathbf{x}) \\ \text{such that} \quad & h_j(\mathbf{x}) = 0, \ j = 1, 2, \ldots, r < n. \end{aligned} \tag{3.5}$$

In 1760 Lagrange transformed this constrained problem to an unconstrained problem via the introduction of so-called *Lagrange multipliers* λ_j, $j = 1, 2, \ldots, r$ in the formulation of the *Lagrangian function*:

$$L(\mathbf{x}, \boldsymbol{\lambda}) = f(\mathbf{x}) + \sum_{j=1}^{r} \lambda_j h_j(\mathbf{x}) = f(\mathbf{x}) + \boldsymbol{\lambda}^T \mathbf{h}(\mathbf{x}). \tag{3.6}$$

The necessary conditions for a constrained minimum of the above equality constrained problem may be stated in terms of the Lagrangian function and the Lagrange multipliers.

3.2.1.1 Necessary conditions for an equality constrained minimum

Let the functions f and $h_j \in C^1$ then, on the assumption that the $n \times r$ *Jacobian* matrix

$$\frac{\partial \mathbf{h}(\mathbf{x}^*)}{\partial \mathbf{x}} = [\boldsymbol{\nabla} h_1(\mathbf{x}^*), \boldsymbol{\nabla} h_2(\mathbf{x}^*), \ldots]$$

is of rank r, the *necessary conditions* for $\mathbf{x}^*$ to be a constrained internal *local minimum* of the equality constrained problem (3.5) is that $\mathbf{x}^*$ corresponds to a stationary point $(\mathbf{x}^*, \boldsymbol{\lambda}^*)$ of the Lagrangian function, i.e. that a vector $\boldsymbol{\lambda}^*$ exists such that

$$\begin{aligned} \frac{\partial L}{\partial x_i}(\mathbf{x}^*, \boldsymbol{\lambda}^*) &= 0, \ i = 1, 2, \ldots, n \\ \\ \frac{\partial L}{\partial \lambda_j}(\mathbf{x}^*, \boldsymbol{\lambda}^*) &= 0, \ j = 1, 2, \ldots, r. \end{aligned} \tag{3.7}$$

For a formal proof of the above, see Theorem 5.2.1.

3.2.1.2 The Lagrangian method

Note that necessary conditions (3.7) represent $n+r$ equations in the $n+r$ unknowns $x_1^*, x_2^*, \ldots, x_n^*, \lambda_1^*, \ldots, \lambda_r^*$. The solutions to these, in general non-linear equations, therefore give candidate solutions $\mathbf{x}^*$ to problem (3.5). This indirect approach to solving the constrained problem is illustrated in solving the following simple example problem.

3.2.1.3 Example

$$\begin{aligned}
\text{minimize} \quad & f(\mathbf{x}) = (x_1 - 2)^2 + (x_2 - 2)^2 \\
\text{such that} \quad & h(\mathbf{x}) = x_1 + x_2 - 6 = 0.
\end{aligned}$$

First formulate the Lagrangian:

$$L(\mathbf{x}, \lambda) = (x_1 - 2)^2 + (x_2 - 2)^2 + \lambda(x_1 + x_2 - 6).$$

By the Theorem in Section 3.2.1.1 the necessary conditions for a constrained minimum are

$$\begin{aligned}
\frac{\partial L}{\partial x_1} &= 2(x_1 - 2) + \lambda = 0 \\
\frac{\partial L}{\partial x_2} &= 2(x_2 - 2) + \lambda = 0 \\
\frac{\partial L}{\partial \lambda} &= x_1 + x_2 - 6 = 0.
\end{aligned}$$

Solving these equations gives a candidate point: $x_1^* = 3$, $x_2^* = 3$, $\lambda^* = -2$ with $f(\mathbf{x}^*) = 2$. This solution is depicted in Figure 3.3.

3.2.1.4 Sufficient conditions

In general the necessary conditions (3.7) are not sufficient to imply a constrained local minimum at $\mathbf{x}^*$. A more general treatment of the sufficiency conditions is however very complicated and will not be discussed

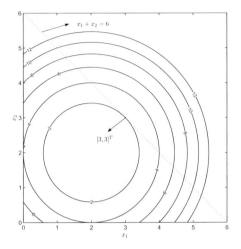

Figure 3.3: Graphical solution to example problem 3.2.1.3

here. It can, however, be shown that if over the whole domain of inter-
est, $f(\mathbf{x})$ is convex, and $h_j(\mathbf{x})$ convex or concave depending on whether
the corresponding Lagrange multiplier λ_j^* is positive or negative, then
conditions (3.7) indeed constitute sufficiency conditions. More generally,
this is also true if the Hessian of the Lagrange function with respect to
$\mathbf{x}$ is positive-definite at $(\mathbf{x}^*, \boldsymbol{\lambda}^*)$. In these cases the local constrained
minimum is unique and represents the global minimum.

3.2.1.5   Saddle point of the Lagrangian function

Assume

(i) $f(\mathbf{x})$ has a constrained minimum at $\mathbf{x}^*$ (with associated $\boldsymbol{\lambda}^*$) and

(ii) that if $\boldsymbol{\lambda}$ is chosen in the neighbourhood of $\boldsymbol{\lambda}^*$, then $L(\mathbf{x}, \boldsymbol{\lambda})$ has a
    local minimum with respect to $\mathbf{x}$ in the neighbourhood of $\mathbf{x}^*$.

The latter assumption can be expected to be true if the Hessian matrix
of L with respect to $\mathbf{x}$ at $(\mathbf{x}^*, \boldsymbol{\lambda}^*)$ is positive-definite.

It *can be shown* (see Theorem 5.2.2) that if (i) and (ii) applies then
$L(\mathbf{x}, \boldsymbol{\lambda})$ has a saddle point at $(\mathbf{x}^*, \boldsymbol{\lambda}^*)$. Indeed it is a degenerate saddle

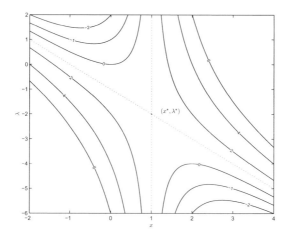

Figure 3.4: Saddle point of Lagrangian function $L = x^2 + \lambda(x - 1)$

point since

$$L(\mathbf{x}, \boldsymbol{\lambda}^*) \geq L(\mathbf{x}^*, \boldsymbol{\lambda}^*) = L(\mathbf{x}^*, \boldsymbol{\lambda}).$$

Consider the example:

$$\text{minimize } f(x) = x^2 \text{ such that } h(x) = x - 1 = 0.$$

With $L = x^2 + \lambda(x - 1)$ it follows directly that $x^* = 1$ and $\lambda^* = -2$.

Since along the straight line asymptotes through (x^*, λ^*), $\Delta L = 0$ for changes Δx and $\Delta \lambda$, it follows that

$$[\Delta x \ \Delta \lambda] \begin{bmatrix} 2 & 1 \\ 1 & 0 \end{bmatrix} \begin{bmatrix} \Delta x \\ \Delta \lambda \end{bmatrix} = 0, \text{ or}$$

$$\Delta x(\Delta x + \Delta \lambda) = 0.$$

The asymptotes therefore are the lines through (x^*, λ^*) with $\Delta x = 0$, and $\frac{\Delta \lambda}{\Delta x} = -1$ respectively, as shown in Figure 3.4, i.e. the lines $x = 1$ and $\lambda = -x - 1$.

In general, if it can be shown that candidate point $(\mathbf{x}^*, \boldsymbol{\lambda}^*)$ is a saddle point of L, then the Hessian of the Lagrangian with respect to $\mathbf{x}$, $\mathbf{H}_L$, at the saddle point is positive-definite for a constrained minimum.

3.2.1.6   Special case: quadratic function with linear equality constraints

From a theoretical point of view, an important application of the Lagrangian method is to the minimization of the positive-definite quadratic function:

$$f(\mathbf{x}) = \tfrac{1}{2}\mathbf{x}^T \mathbf{A}\mathbf{x} + \mathbf{b}^T \mathbf{x} + c \qquad (3.8)$$

subject to the linear constraints

$$\mathbf{C}\mathbf{x} = \mathbf{d}.$$

Here $\mathbf{A}$ is a $n \times n$ positive-definite matrix and $\mathbf{C}$ a $r \times n$ constraint matrix, $r < n$, $\mathbf{b}$ is a n-vector and $\mathbf{d}$ a r-vector.

In this case the Lagrangian is

$$L(\mathbf{x}, \boldsymbol{\lambda}) = \tfrac{1}{2}\mathbf{x}^T \mathbf{A}\mathbf{x} + \mathbf{b}^T \mathbf{x} + c + \boldsymbol{\lambda}^T (\mathbf{C}\mathbf{x} - \mathbf{d})$$

and the necessary conditions (3.7) for a constrained minimum at $\mathbf{x}^*$ is the existence of a vector $\boldsymbol{\lambda}^*$ such that

$$\begin{aligned}
\nabla_{\mathbf{x}} L(\mathbf{x}^*, \boldsymbol{\lambda}^*) &= \mathbf{A}\mathbf{x}^* + \mathbf{b} + \mathbf{C}^T \boldsymbol{\lambda}^* = 0 \\
\nabla_{\boldsymbol{\lambda}} L(\mathbf{x}^*, \boldsymbol{\lambda}^*) &= \mathbf{C}\mathbf{x}^* - \mathbf{d} = 0
\end{aligned}$$

i.e.

$$\begin{bmatrix} \mathbf{A} & \mathbf{C}^T \\ \mathbf{C} & \mathbf{0} \end{bmatrix} \begin{bmatrix} \mathbf{x}^* \\ \boldsymbol{\lambda}^* \end{bmatrix} = \begin{bmatrix} -\mathbf{b} \\ \mathbf{d} \end{bmatrix}. \qquad (3.9)$$

The solution to this linear system is given by

$$\begin{bmatrix} \mathbf{x}^* \\ \boldsymbol{\lambda}^* \end{bmatrix} = \mathbf{M}^{-1} \begin{bmatrix} -\mathbf{b} \\ \mathbf{d} \end{bmatrix} \quad \text{where } \mathbf{M} = \begin{bmatrix} \mathbf{A} & \mathbf{C}^T \\ \mathbf{C} & \mathbf{0} \end{bmatrix}.$$

3.2.1.7   Inequality constraints as equality constraints

Consider the more general problem:

$$\begin{aligned}
\text{minimize} \quad & f(\mathbf{x}) \\
\text{such that} \quad & g_j(\mathbf{x}) \le 0, \ j = 1, 2, \ldots, m \\
& h_j(\mathbf{x}) = 0, \ j = 1, 2, \ldots, r.
\end{aligned} \qquad (3.10)$$

The inequality constraints may be transformed to equality constraints by the introduction of so-called *auxiliary variables* θ_j, $j = 1, 2, \ldots, m$:

$$g_j(\mathbf{x}) + \theta_j^2 = 0.$$

Since $g_j(\mathbf{x}) = -\theta_j^2 \le 0$ for all j, the inequality constraints are automatically satisfied.

The Lagrangian method for equality constrained problems may now be applied, where

$$L(\mathbf{x}, \boldsymbol{\theta}, \boldsymbol{\lambda}, \boldsymbol{\mu}) = f(\mathbf{x}) + \sum_{j=1}^{m} \lambda_j (g_j(\mathbf{x}) + \theta_j^2) + \sum_{j=1}^{r} \mu_j h_j(\mathbf{x}) \qquad (3.11)$$

and λ_j and μ_j denote the respective Lagrange multipliers.

From (3.7) the associated necessary conditions for a minimum at $\mathbf{x}$ are

$$\frac{\partial L}{\partial x_i} = \frac{\partial f(\mathbf{x})}{\partial x_i} + \sum_{j=1}^{m} \lambda_j \frac{\partial g_j(\mathbf{x})}{\partial x_i} + \sum_{j=1}^{r} \mu_j \frac{\partial h_j(\mathbf{x})}{\partial x_i} = 0, \ i = 1, 2, \ldots, n$$

$$\frac{\partial L}{\partial \theta_j} = 2\lambda_j \theta_j = 0, \ j = 1, 2, \ldots, m$$

$$\frac{\partial L}{\partial \lambda_j} = g_j(\mathbf{x}) + \theta_j^2 = 0, \ j = 1, 2, \ldots, m \qquad (3.12)$$

$$\frac{\partial L}{\partial \mu_j} = h_j(\mathbf{x}) = 0, \ j = 1, 2, \ldots, r.$$

The above system (3.12) represents a system of $n + 2m + r$ simultaneous non-linear equations in the $n + 2m + r$ unknowns $\mathbf{x}$, $\boldsymbol{\theta}$, $\boldsymbol{\lambda}$ and $\boldsymbol{\mu}$. Obtaining the solutions to system (3.12) yields candidate solutions $\mathbf{x}^*$ to the general optimization problem (3.10). The application of this approach is demonstrated in the following example.

3.2.1.8 Example

Minimize $f(\mathbf{x}) = 2x_1^2 - 3x_2^2 - 2x_1$

such that $x_1^2 + x_2^2 \le 1$ by making use of auxiliary variables.

Introduce auxiliary variable θ such that

$$x_1^2 + x_2^2 - 1 + \theta^2 = 0$$

then

$$L(\mathbf{x}, \theta, \lambda) = 2x_1^2 - 3x_2^2 - 2x_1 + \lambda(x_1^2 + x_2^2 - 1 + \theta^2).$$

The necessary conditions at the minimum are

$$\frac{\partial L}{\partial x_1} = 4x_1 - 2 + 2\lambda x_1 = 0 \qquad (3.13)$$

$$\frac{\partial L}{\partial x_2} = -6x_2 + 2\lambda x_2 = 0 \qquad (3.14)$$

$$\frac{\partial L}{\partial \theta} = 2\lambda\theta = 0 \qquad (3.15)$$

$$\frac{\partial L}{\partial \lambda} = x_1^2 + x_2^2 - 1 + \theta^2 = 0. \qquad (3.16)$$

As first choice in (3.15), select $\lambda = 0$ which gives $x_1 = 1/2$, $x_2 = 0$, and $\theta^2 = 3/4$.

Since $\theta^2 > 0$ the problem is unconstrained at this point. Further since $\mathbf{H} = \begin{bmatrix} 4 & 0 \\ 0 & -6 \end{bmatrix}$ is non-definite the candidate point $\mathbf{x}^0$ corresponds to a saddle point where $f(\mathbf{x}^0) = -0.5$.

Select as second choice in (3.15),

$$\theta = 0 \text{ which gives } x_1^2 + x_2^2 - 1 = 0 \qquad (3.17)$$

i.e. the constraint is active. From (3.14) it follows that for $x_2 \neq 0$, $\lambda = 3$, and substituting into (3.13) gives $x_1 = 1/5$ and from (3.17): $x_2 = \pm\sqrt{24}/5 = \pm 0.978$ which give the two possibilities: $\mathbf{x}^* = \left(\frac{1}{5}, \frac{\sqrt{24}}{5}\right)$, $f(\mathbf{x}^*) = -3.189$ and $\mathbf{x}^* = \left(\frac{1}{5}, \frac{-\sqrt{24}}{5}\right)$, with $f(\mathbf{x}^*) = -3.189$.

Also the choice $\theta = 0$ with $x_2 = 0$, gives $x_1 = \pm 1$. These two points correspond to maxima with respective function values $f = 4$ and $f = 0$.

3.2.1.9   Directions of asymptotes at a saddle point $\mathbf{x}^0$

If, in the example above, the direction of an asymptote at saddle point $\mathbf{x}^0$ is denoted by the unit vector $\mathbf{u} = [u_1, u_2]^T$. Then for a displacement

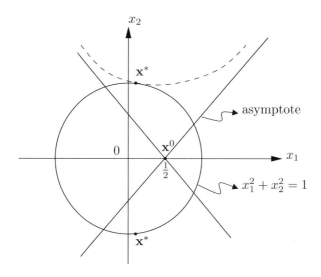

Figure 3.5: Directions of asymptotes at a saddle point $\mathbf{x}^0$

$\Delta\mathbf{x} = \mathbf{u}$ along the asymptote the change in the function value is $\Delta f = 0$. It follows from the Taylor expansion that

$$\Delta f = f(\mathbf{x}^0 + \Delta\mathbf{x}) - f(\mathbf{x}^0) = \mathbf{u}^T\nabla f(\mathbf{x}^0) + \tfrac{1}{2}\mathbf{u}^T\mathbf{H}\mathbf{u} = 0$$

for step $\Delta\mathbf{x} = \mathbf{u}$ at saddle point $\mathbf{x}^0$. Since $\mathbf{H} = \begin{bmatrix} 4 & 0 \\ 0 & -6 \end{bmatrix}$ and $\nabla f(\mathbf{x}^0) = \mathbf{0}$, it follows that $2u_1^2 - 3u_2^2 = 0$ and also, since $\|\mathbf{u}\| = 1$: $u_1^2 + u_2^2 = 1$.

Solving for u_1 and u_2 in the above gives

$$u_1 = \pm\sqrt{\tfrac{3}{5}}; \ u_2 = \pm\sqrt{\tfrac{2}{5}}$$

which, taken in combinations, correspond to the directions of the four asymptotes at the saddle point $\mathbf{x}^0$, as shown in Figure 3.5.

3.2.2 Classical approach to optimization with inequality constraints: the KKT conditions

Consider the *primal problem* (PP):

$$\begin{array}{ll} \text{minimize} & f(\mathbf{x}) \\ \text{such that} & g_j(\mathbf{x}) \leq 0, \ j = 1, 2, \ldots, m. \end{array} \qquad (3.18)$$

Define again the Lagrangian:

$$L(\mathbf{x}, \boldsymbol{\lambda}) = f(\mathbf{x}) + \sum_{j=1}^{m} \lambda_j g_j(\mathbf{x}). \qquad (3.19)$$

Karush (1939) and Kuhn and Tucker (1951) independently derived the necessary conditions that must be satisfied at the solution $\mathbf{x}^*$ of the primary problem (3.18). These conditions are generally known as the *KKT conditions* which are expressed in terms of the Lagrangian $L(\mathbf{x}, \boldsymbol{\lambda})$.

3.2.2.1 The KKT necessary conditions for an inequality constrained minimum

Let the functions f and $g_j \in C^1$, and assume the existence of Lagrange multipliers $\boldsymbol{\lambda}^*$, then at the point $\mathbf{x}^*$, corresponding to the solution of the primal problem (3.18), the following conditions must be satisfied:

$$\begin{array}{rcl} \dfrac{\partial f}{\partial x_i}(\mathbf{x}^*) + \displaystyle\sum_{j=1}^{m} \lambda_j^* \dfrac{\partial g_j}{\partial x_i}(\mathbf{x}^*) & = & 0, \ i = 1, 2, \ldots, n \\[2mm] g_j(\mathbf{x}^*) & \leq & 0, \ j = 1, 2, \ldots, m \\[2mm] \lambda_j^* g_j(\mathbf{x}^*) & = & 0, \ j = 1, 2, \ldots, m \\[2mm] \lambda_j^* & \geq & 0, \ j = 1, 2, \ldots, m. \end{array} \qquad (3.20)$$

For a formal proof of the above see Theorem 5.3.1. It can be shown, that the KKT conditions also constitute *sufficient* conditions (those that imply that) for $\mathbf{x}^*$ to be a constrained minimum, if $f(\mathbf{x})$ and the $g_j(\mathbf{x})$ are all convex functions.

Let $\mathbf{x}^*$ be a solution to problem (3.18), and suppose that the KKT conditions (3.20) are satisfied. If now $g_k(\mathbf{x}^*) = 0$ for some $k \in \{1, 2, \ldots, m\}$, then the corresponding inequality constraint k is said to be *active* and *binding* at $\mathbf{x}^*$, if the corresponding Lagrange multiplier $\lambda_k^* \geq 0$. It is strongly active if $\lambda_k^* > 0$, and weakly active if $\lambda_k^* = 0$. However, if for some candidate KKT point $\bar{\mathbf{x}}$, $g_k(\bar{\mathbf{x}}) = 0$ for some k, and all the KKT conditions are satisfied except that the corresponding Lagrange multiplier $\bar{\lambda}_k < 0$, then the inequality constraint k is said to be inactive, and must be deleted from the set of active constraints at $\bar{\mathbf{x}}$.

3.2.2.2 Constraint qualification

It can be shown that the existence of $\boldsymbol{\lambda}^*$ is guaranteed if the so-called *constraint qualification* is satisfied at $\mathbf{x}^*$, i.e. if a vector $\mathbf{h} \in \mathbb{R}^n$ exists such that for each active constraint j at $\mathbf{x}^*$

$$\nabla^T g_j(\mathbf{x}^*)\mathbf{h} < 0 \qquad (3.21)$$

then $\boldsymbol{\lambda}^*$ exists.

The constraint qualification (3.21) is always satisfied

(i) if all the constraints are convex and at least one $\mathbf{x}$ exists within the feasible region, or

(ii) if the rank of the Jacobian of all active and binding constraints at $\mathbf{x}^*$ is maximal, or

(iii) if all the constraints are linear.

3.2.2.3 Illustrative example

Minimize $f(\mathbf{x}) = (x_1 - 2)^2 + x_2^2$

such that $x_1 \geq 0$, $x_2 \geq 0$, $(1 - x_1)^3 \geq x_2$.

A minimum is attained at the point $x_1^* = 1$, $x_2^* = 0$ where

$$g_1(\mathbf{x}^*) = -x_1^* < 0, \; g_2(\mathbf{x}^*) = -x_2^* = 0 \text{ and } g_3(\mathbf{x}^*) = x_2^* - (1 - x_1^*)^3 = 0$$

and therefore g_2 and g_3 are active at $\mathbf{x}^*$.

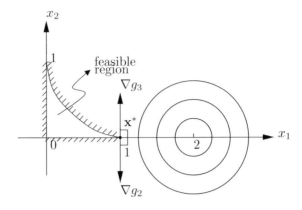

Figure 3.6: Failure of constraint qualification

At this point

$$\nabla g_2 = \begin{bmatrix} 0 \\ -1 \end{bmatrix}, \ \nabla g_3 = \begin{bmatrix} 0 \\ 1 \end{bmatrix}$$

giving $\nabla g_2 = -\nabla g_3$ at $\mathbf{x}^*$ as shown in Figure 3.6.

Thus no vector $\mathbf{h}$ exists that satisfies the constraint qualification (3.21). Therefore the application of KKT theory to this problem will break down since no $\boldsymbol{\lambda}^*$ exists.

Also note that the Jacobian of the active constraints at $\mathbf{x}^* = [1, 0]^T$:

$$\frac{\partial \mathbf{g}}{\partial \mathbf{x}}(\mathbf{x}^*) = \begin{bmatrix} \dfrac{\partial g_2}{\partial x_1} & \dfrac{\partial g_3}{\partial x_1} \\ \dfrac{\partial g_2}{\partial x_2} & \dfrac{\partial g_3}{\partial x_2} \end{bmatrix} = \begin{bmatrix} 0 & 0 \\ -1 & 1 \end{bmatrix}$$

is of rank $1 < 2$, and therefore not maximal which also indicates that $\boldsymbol{\lambda}^*$ does not exist, and therefore the problem cannot be analysed via the KKT conditions.

3.2.2.4   Simple example of application of KKT conditions

Minimize $f(\mathbf{x}) = (x_1 - 1)^2 + (x_2 - 2)^2$

such that $x_1 \geq 0$, $x_2 \geq 0$, $x_1 + x_2 \leq 2$ and $x_2 - x_1 = 1$.

In standard form the constraints are:

$$-x_1 \leq 0, \ -x_2 \leq 0, \ x_1 + x_2 - 2 \leq 0, \ \text{and} \ x_2 - x_1 - 1 = 0$$

and the Lagrangian, now including the equality constraint with Lagrange multiplier μ as well:

$$L(\mathbf{x}, \boldsymbol{\lambda}, \mu) = (x_1 - 1)^2 + (x_2 - 2)^2 + \lambda_3(x_1 + x_2 - 2) - \lambda_1 x_1 - \lambda_2 x_2 + \mu(x_2 - x_1 - 1).$$

The KKT conditions are

$$2(x_1 - 1) + \lambda_3 - \lambda_1 - \mu = 0; \ \ 2(x_2 - 2) + \lambda_3 - \lambda_2 + \mu = 0$$
$$-x_1 \leq 0; \ -x_2 \leq 0; \ x_1 + x_2 - 2 \leq 0; \ x_2 - x_1 - 1 = 0$$
$$\lambda_3(x_1 + x_2 - 2) = 0; \ \ \lambda_1 x_1 = 0; \ \ \lambda_2 x_2 = 0$$
$$\lambda_1 \geq 0; \ \lambda_2 \geq 0; \ \lambda_3 \geq 0.$$

In general, the approach to the solution is combinatorial. Try different possibilities and test for contradictions. One possible choice is $\lambda_3 \neq 0$ that implies $x_1 + x_2 - 2 = 0$. This together with the equality constraint gives

$$x_1^* = \tfrac{1}{2}, \ x_2^* = \tfrac{3}{2}, \ \lambda_3^* = 1, \ \lambda_1^* = \lambda_2^* = 0, \ \mu^* = 0 \text{ and } f(\mathbf{x}^*) = \tfrac{1}{2}.$$

This candidate solution satisfies all the KKT conditions and is indeed the optimum solution. Why? The graphical solution is depicted in Figure 3.7.

3.3 Saddle point theory and duality

3.3.1 Saddle point theorem

If the point $(\mathbf{x}^*, \boldsymbol{\lambda}^*)$, with $\boldsymbol{\lambda}^* \geq \mathbf{0}$, is a saddle point of the Lagrangian associated with the primal problem (3.18), then $\mathbf{x}^*$ is a solution of the primal problem. For a proof of this statement see Theorem 5.4.2 in Chapter 5.

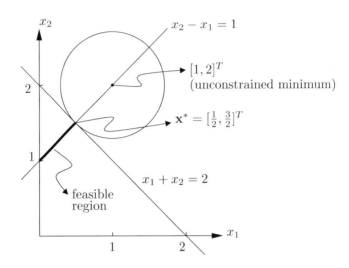

Figure 3.7: Graphical solution to example problem in 3.2.2.4

3.3.2   Duality

Define the *dual function*:

$$h(\boldsymbol{\lambda}) = \min_{\mathbf{x}} L(\mathbf{x}, \boldsymbol{\lambda}). \tag{3.22}$$

Note that the minimizer $\mathbf{x}^*(\boldsymbol{\lambda})$ does not necessarily satisfy $\mathbf{g}(\mathbf{x}) \leq \mathbf{0}$, and indeed the minimum may not even exist for all $\boldsymbol{\lambda}$.

Defining the set

$$D = \{\boldsymbol{\lambda} | h(\boldsymbol{\lambda}) \ \exists \ \text{and} \ \boldsymbol{\lambda} \geq \mathbf{0}\} \tag{3.23}$$

allows for the formulation of the *dual problem* (DP):

$$\underset{\boldsymbol{\lambda} \in D}{\text{maximize}} \, h(\boldsymbol{\lambda}) \tag{3.24}$$

which is equivalent to $\left\{ \underset{\boldsymbol{\lambda} \in D}{\max} (\underset{\mathbf{x}}{\min} \, L(\mathbf{x}, \boldsymbol{\lambda})) \right\}$.

3.3.3   Duality theorem

The point $(\mathbf{x}^*, \boldsymbol{\lambda}^*)$, with $\boldsymbol{\lambda}^* \geq \mathbf{0}$, is a saddle point of the Lagrangian function of the primal problem (PP), defined by (3.18), *if and only if*:

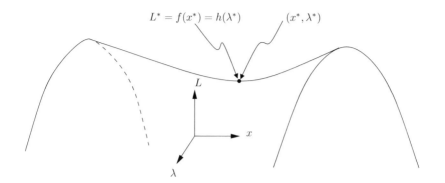

Figure 3.8: Schematic representation of saddle point solution to PP

(i) $\mathbf{x}^*$ is a solution to the primal problem (PP),

(ii) $\boldsymbol{\lambda}^*$ is a solution to the dual problem (DP), and

(iii) $f(\mathbf{x}^*) = h(\boldsymbol{\lambda}^*)$.

A schematic representation of this theorem is given in Figure 3.8 and a formal proof is given in listed Theorem 5.4.4 in Chapter 5.

3.3.3.1   Practical significance of the duality theorem

The implication of the Duality Theorem is that the PP may be solved by carrying out the following steps:

1. If possible, solve the DP separately to give $\boldsymbol{\lambda}^* \geq \mathbf{0}$, i.e. solve an essentially unconstrained problem.

2. With $\boldsymbol{\lambda}^*$ known solve the unconstrained problem: $\min_{\mathbf{x}} L(\mathbf{x}, \boldsymbol{\lambda}^*)$ to give $\mathbf{x}^* = \mathbf{x}^*(\boldsymbol{\lambda}^*)$.

3. Test whether $(\mathbf{x}^*, \boldsymbol{\lambda}^*)$ satisfy the KKT conditions.

3.3.3.2   Example of the application of duality

Consider the problem:
minimize $f(\mathbf{x}) = x_1^2 + 2x_2^2$ such that $x_1 + x_2 \geq 1$.

Here the Lagrangian is

$$L(\mathbf{x}, \lambda) = x_1^2 + 2x_2^2 + \lambda(1 - x_1 - x_2).$$

For a given λ the necessary conditions for $\min_{\mathbf{x}} L(\mathbf{x}, \lambda)$ at $\mathbf{x}^*(\lambda)$ are

$$\frac{\partial L}{\partial x_1} = 2x_1 - \lambda = 0 \quad \Rightarrow \quad x_1 = x_1^*(\lambda) = \tfrac{\lambda}{2}$$

$$\frac{\partial L}{\partial x_2} = 4x_2 - \lambda = 0 \quad \Rightarrow \quad x_2 = x_2^*(\lambda) = \tfrac{\lambda}{4}.$$

Note that $\mathbf{x}^*(\lambda)$ is a minimum since the Hessian of the Lagrangian with respect to $\mathbf{x}$, $\mathbf{H}_L = \begin{bmatrix} 2 & 0 \\ 0 & 4 \end{bmatrix}$ is positive-definite.

Substituting the minimizing values (i.t.o. λ) into L gives

$$L(\mathbf{x}^*(\lambda), \lambda) = h(\lambda) = \left(\tfrac{\lambda}{2}\right)^2 + 2\left(\tfrac{\lambda}{4}\right)^2 + \lambda\left(1 - \tfrac{\lambda}{2} - \tfrac{\lambda}{4}\right)$$

i.e. the dual function is $h(\lambda) = -\tfrac{3}{8}\lambda^2 + \lambda$.

Now solve the DP: $\max_{\lambda} h(\lambda)$.

The necessary condition for a for maximum is $\frac{dh}{d\lambda} = 0$, i.e.

$$-\tfrac{3}{4}\lambda + 1 = 0 \Rightarrow \lambda^* = \tfrac{4}{3} > 0 \text{ (a maximum since } \tfrac{d^2h}{d\lambda^2} = -\tfrac{3}{4} < 0).$$

Substituting λ^* in $\mathbf{x}^*(\lambda)$ above gives

$$x_1^* = \tfrac{2}{3}; \ x_2^* = \tfrac{1}{3} \Rightarrow f(\mathbf{x}^*) = \tfrac{2}{3} = h(\lambda^*).$$

and since $(\mathbf{x}^*, \lambda^*)$, with $\lambda^* = \tfrac{4}{3} > 0$, clearly satisfies the KKT conditions, it indeed represents the optimum solution.

The dual approach is an important method in some structural optimization problems (Fleury 1979). It is also employed in the development of the *augmented Lagrange multiplier method* to be discussed later.

3.4 Quadratic programming

The problem of minimizing a positive-definite quadratic function subject to linear constraints, dealt with in Section 3.2.1.6, is a special case of

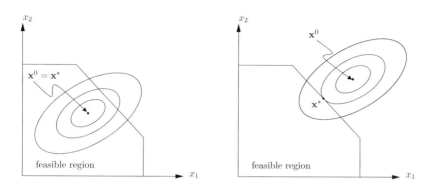

Figure 3.9: The solution of the QP problem: may be an interior point or lie on the boundary of the feasible region

a *quadratic programming* (QP) problem. Consider now a more general case of a QP problem with inequality constraints:

$$\text{minimize } f(\mathbf{x}) = \tfrac{1}{2}\mathbf{x}^T \mathbf{A}\mathbf{x} + \mathbf{b}^T\mathbf{x} + c$$

subject to

$$\mathbf{C}\mathbf{x} \le \mathbf{d} \tag{3.25}$$

where $\mathbf{C}$ is a $m \times n$ matrix and $\mathbf{d}$ is a m-vector.

The solution point may be an interior point or may lie on the boundary of the feasible region as shown for the two-dimensional case in Figure 3.9.

If the solution point is an interior point then no constraints are active, and $\mathbf{x}^* = \mathbf{x}^0 = -\mathbf{A}^{-1}\mathbf{b}$ as shown in the figure.

The QP problem is often an important sub-problem to be solved when applying modern methods to more general problems (see the discussion of the SQP method later).

3.4.1 Active set of constraints

It is clear (see Section 3.2.1.6) that if the set of constraints active at $\mathbf{x}^*$ is known, then the problem is greatly simplified. Suppose the active set at $\mathbf{x}^*$ is known, i.e. $\mathbf{c}^j\mathbf{x}^* = d_j$ for some $j \in \{1, 2, \ldots, m\}$, where $\mathbf{c}^j$ here

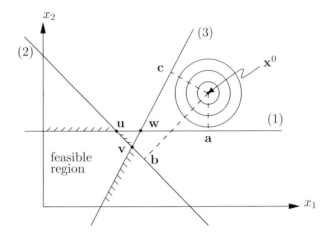

Figure 3.10: Graphical illustration of the method of Theil and Van de Panne

denotes the $1 \times n$ matrix corresponding to the jth row of $\mathbf{C}$. Represent this active set in matrix form by $\mathbf{C}'\mathbf{x} = \mathbf{d}'$. The solution $\mathbf{x}^*$ is then obtained by minimizing $f(\mathbf{x})$ over the set $\{\mathbf{x}|\mathbf{C}'\mathbf{x} = \mathbf{d}'\}$. Applying the appropriate Lagrange theory (Section 3.2.1.6), the solution is obtained by solving the linear system:

$$\begin{bmatrix} \mathbf{A} & \mathbf{C}'^T \\ \mathbf{C}' & \mathbf{0} \end{bmatrix} \begin{bmatrix} \mathbf{x}^* \\ \boldsymbol{\lambda}^* \end{bmatrix} = \begin{bmatrix} -\mathbf{b}^* \\ \mathbf{d}' \end{bmatrix}. \tag{3.26}$$

In solving the QP the major task therefore lies in the *identification of the active set* of constraints.

3.4.2   The method of Theil and Van de Panne

The method of Theil and van de Panne (1961) is a straightforward method for identifying the active set. A description of the method, after Wismer and Chattergy (1978), for the problem graphically depicted in Figure 3.10 is now given.

Let $V[\mathbf{x}]$ denote the set of constraints violated at $\mathbf{x}$. Select as initial candidate solution the unconstrained minimum $\mathbf{x}^0 = \mathbf{A}^{-1}\mathbf{b}$. Clearly for the example sketched in Figure 3.10, $V[\mathbf{x}^0] = \{1, 2, 3\}$. Therefore $\mathbf{x}^0$

is not the solution. Now consider, as active constraint (set S_1), each constraint in $V[\mathbf{x}^0]$ separately, and let $\mathbf{x}[S_1]$ denote the corresponding minimizer:

$$S_1 = \{1\} \Rightarrow \quad \mathbf{x}[S_1] = \mathbf{a} \Rightarrow V[\mathbf{a}] = \{2, 3\} \neq \phi \text{ (not empty)}$$
$$S_1 = \{2\} \Rightarrow \quad \mathbf{x}[S_1] = \mathbf{b} \Rightarrow V[\mathbf{b}] = \{3\} \neq \phi$$
$$S_1 = \{3\} \Rightarrow \quad \mathbf{x}[S_1] = \mathbf{c} \Rightarrow V[\mathbf{c}] = \{1, 2\} \neq \phi.$$

Since all the solutions with a single active constraint violate one or more constraints, the next step is to consider different combinations S_2 of two simultaneously active constraints from $V[\mathbf{x}^0]$:

$$S_2 = \{1, 2\} \Rightarrow \quad \mathbf{x}[S_2] = \mathbf{u} \Rightarrow V[\mathbf{u}] = \phi$$
$$S_2 = \{1, 3\} \Rightarrow \quad \mathbf{x}[S_2] = \mathbf{w} \Rightarrow V[\mathbf{w}] = \{2\}$$
$$S_2 = \{2, 3\} \Rightarrow \quad \mathbf{x}[S_2] = \mathbf{v} \Rightarrow V[\mathbf{v}] = \phi.$$

Since both $V[\mathbf{u}]$ and $V[\mathbf{v}]$ are empty, $\mathbf{u}$ and $\mathbf{v}$ are both candidate solutions. Apply the KKT conditions to both $\mathbf{u}$ and $\mathbf{v}$ separately, to determine which point is the optimum one. Assume it can be shown, from the solution of (3.26) that for $\mathbf{u}$: $\lambda_1 < 0$ (which indeed is apparently so from the geometry of the problem sketched in Figure 3.10), then it follows that $\mathbf{u}$ is non-optimal. hand, assume it can be shown that for $\mathbf{v}$: $\lambda_2 > 0$ and $\lambda_3 > 0$ (which is evidently the case in the figure), then $\mathbf{v}$ is optimum, i.e. $\mathbf{x}^* = \mathbf{v}$.

3.4.2.1 Explicit example

Solve by means of the method of *Theil and van de Panne* the following QP problem:

$$\text{minimize } f(\mathbf{x}) = \tfrac{1}{2}x_1^2 - x_1 x_2 + x_2^2 - 2x_1 + x_2$$

such that

$$x_1 \geq 0; \ x_2 \geq 0; \ x_1 + x_2 \leq 3; \ 2x_1 - x_2 \leq 4.$$

In matrix form $f(\mathbf{x})$ is given by $f(\mathbf{x}) = \tfrac{1}{2}\mathbf{x}^T \mathbf{A} \mathbf{x} + \mathbf{b}^T \mathbf{x}$, with $\mathbf{A} = \begin{bmatrix} 1 & -1 \\ -1 & 2 \end{bmatrix}$ and $\mathbf{b} = [-2, 1]^T$.

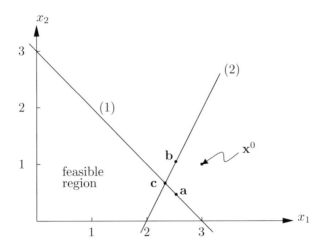

Figure 3.11: Graphical solution to explicit example 3.4.2.1

The unconstrained solution is $\mathbf{x}^0 = \mathbf{A}^{-1}\mathbf{b} = [3, 1]^T$. Clearly $V[\mathbf{x}^0] = \{1, 2\}$ and therefore $\mathbf{x}^0$ is not the solution. Continuing, the method yields:

$$
\begin{aligned}
S_1 = \{1\} &\Rightarrow \quad \mathbf{x}[S_1] = \mathbf{a} \Rightarrow V[\mathbf{a}] = \{2\} \neq \phi \\
S_1 = \{2\} &\Rightarrow \quad \mathbf{x}[S_1] = \mathbf{b} \Rightarrow V[\mathbf{b}] = \{1\} \neq \phi \\
S_2 = \{1, 2\} &\Rightarrow \quad \mathbf{x}[S_2] = \mathbf{c} \Rightarrow V[\mathbf{c}] = \phi, \text{ where } \mathbf{c} = \left[\tfrac{7}{3}, \tfrac{2}{3}\right]^T.
\end{aligned}
$$

Applying the KKT conditions to $\mathbf{c}$ establishes the optimality of $\mathbf{c}$ since $\lambda_1 = \lambda_2 = \frac{1}{9} > 0$.

3.5 Modern methods for constrained optimization

The most established gradient-based methods for constrained optimization are

(i) gradient projection methods (Rosen 1960, 1961),

(ii) augmented Lagrangian multiplier methods (see Haftka and Gürdal 1992), and

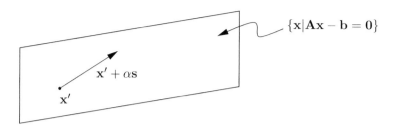

Figure 3.12: Schematic representation of the subspace $\{\mathbf{x}|\mathbf{Ax} - \mathbf{b} = \mathbf{0}\}$

(iii) successive or sequential quadratic programming (SQP) methods (see Bazaraa et al. 1993).

All these methods are largely based on the theory already presented here. SQP methods are currently considered to represent the state-of-the-art gradient-based approach to the solution of constrained optimization problems.

3.5.1 The gradient projection method

3.5.1.1 Equality constrained problems

The gradient projection method is due to Rosen (1960, 1961). Consider the linear equality constrained problem:

$$\begin{aligned} \text{minimize} \quad & f(\mathbf{x}) \\ \text{such that} \quad & \mathbf{Ax} - \mathbf{b} = \mathbf{0} \end{aligned} \qquad (3.27)$$

where $\mathbf{A}$ is a $r \times n$ matrix, $r < n$ and $\mathbf{b}$ a r-vector.

The gradient projection method for solving (3.27) is based on the following argument.

Assume that $\mathbf{x}'$ is feasible, i.e. $\mathbf{Ax}' - \mathbf{b} = \mathbf{0}$ in Figure 3.12. A direction $\mathbf{s}$, ($\|\mathbf{s}\| = 1$) is sought such that a step $\alpha \mathbf{s}$ ($\alpha > 0$) from $\mathbf{x}'$ in the direction $\mathbf{s}$ also gives a feasible point, i.e. $\mathbf{A}(\mathbf{x}' + \alpha \mathbf{s}) - \mathbf{b} = \mathbf{0}$. This condition reduces to

$$\mathbf{As} = \mathbf{0}. \qquad (3.28)$$

Also since $\|\mathbf{s}\| = 1$, it also follows that

$$1 - \mathbf{s}^T \mathbf{s} = 0. \tag{3.29}$$

It is now required that $\mathbf{s}$ be chosen such that it corresponds to the direction which gives the steepest descent at $\mathbf{x}'$ subject to satisfying the constraints specified by (3.28) and (3.29). This requirement is equivalent to determining a $\mathbf{s}$ such that the directional derivative at $\mathbf{x}'$:

$$R(\mathbf{s}) = \left.\frac{dF(0)}{d\alpha}\right|_{\mathbf{s}} = \mathbf{\nabla}^T f(\mathbf{x}')\mathbf{s} \tag{3.30}$$

is minimized with respect to $\mathbf{s}$, where $F(\alpha) = f(\mathbf{x}' + \alpha\mathbf{s})$.

Applying the classical Lagrange theory for minimizing a function subject to equality constraints, requires the formulation of the following Lagrangian function:

$$L(\mathbf{s}, \boldsymbol{\lambda}, \lambda_0) = \mathbf{\nabla}^T f(\mathbf{x}')\mathbf{s} + \boldsymbol{\lambda}^T \mathbf{A}\mathbf{s} + \lambda_0(1 - \mathbf{s}^T\mathbf{s}) \tag{3.31}$$

where the variables $\mathbf{s} = [s_1, s_2, \ldots, s_n]^T$ correspond to the direction cosines of $\mathbf{s}$. The Lagrangian necessary conditions for the constrained minimum are

$$\mathbf{\nabla}_{\mathbf{s}}L = \mathbf{\nabla}f(\mathbf{x}') + \mathbf{A}^T\boldsymbol{\lambda} - 2\lambda_0\mathbf{s} = \mathbf{0} \tag{3.32}$$

$$\mathbf{\nabla}_{\boldsymbol{\lambda}}L = \mathbf{A}\mathbf{s} = \mathbf{0} \tag{3.33}$$

$$\mathbf{\nabla}_{\lambda_0}L = (1 - \mathbf{s}^T\mathbf{s}) = 0. \tag{3.34}$$

Equation (3.32) yields

$$\mathbf{s} = \frac{1}{2\lambda_0}(\mathbf{\nabla}f(\mathbf{x}') + \mathbf{A}^T\boldsymbol{\lambda}). \tag{3.35}$$

Substituting (3.35) into (3.34) gives

$$1 = \frac{1}{4\lambda_0^2}(\mathbf{\nabla}f(\mathbf{x}') + \mathbf{A}^T\boldsymbol{\lambda})^T(\mathbf{\nabla}f(\mathbf{x}') + \mathbf{A}^T\boldsymbol{\lambda})$$

and thus

$$\lambda_0 = \pm\frac{1}{2}\|\mathbf{\nabla}f(\mathbf{x}') + \mathbf{A}^T\boldsymbol{\lambda}\|. \tag{3.36}$$

Substituting (3.36) into (3.35) gives $\mathbf{s} = \pm(\mathbf{\nabla}f(\mathbf{x}') + \mathbf{A}^T\boldsymbol{\lambda})/\|\mathbf{\nabla}f(\mathbf{x}') + \mathbf{A}^T\boldsymbol{\lambda}\|$.

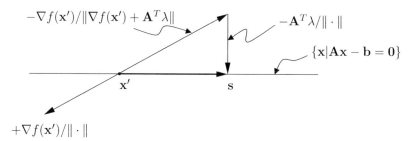

Figure 3.13: Schematic representation of projected direction of steepest descent **s**

For maximum descent, choose the negative sign as shown in Figure 3.13. This also ensures that the Hessian of the Lagrangian with respect to **s** is positive-definite (sufficiency condition) which ensures that $R(\mathbf{s})$ in (3.30) indeed assumes a minimum value with respect to **s**. Thus the constrained (projected) direction of steepest descent is chosen as

$$\mathbf{s} = -(\boldsymbol{\nabla} f(\mathbf{x}') + \mathbf{A}^T\boldsymbol{\lambda})/\|\boldsymbol{\nabla} f(\mathbf{x}') + \mathbf{A}^T\boldsymbol{\lambda}\|. \tag{3.37}$$

It remains to solve for $\boldsymbol{\lambda}$. Equations (3.33) and (3.37) imply $\mathbf{A}(\boldsymbol{\nabla} f(\mathbf{x}') + \mathbf{A}^T\boldsymbol{\lambda}) = \mathbf{0}$. Thus if $\mathbf{s} \neq \mathbf{0}$, then $\mathbf{A}\mathbf{A}^T\boldsymbol{\lambda} = -\mathbf{A}\boldsymbol{\nabla} f(\mathbf{x}')$ with solution

$$\boldsymbol{\lambda} = -(\mathbf{A}\mathbf{A}^T)^{-1}\mathbf{A}\boldsymbol{\nabla} f(\mathbf{x}'). \tag{3.38}$$

The direction **s**, called the *gradient projection* direction, is therefore finally given by

$$\mathbf{s} = -(\mathbf{I} - \mathbf{A}^T(\mathbf{A}\mathbf{A}^T)^{-1}\mathbf{A})\boldsymbol{\nabla} f(\mathbf{x}')/\|\boldsymbol{\nabla} f(\mathbf{x}') + \mathbf{A}^T\boldsymbol{\lambda}\|. \tag{3.39}$$

A *projection matrix* is defined by

$$\mathbf{P} = (\mathbf{I} - \mathbf{A}^T(\mathbf{A}\mathbf{A}^T)^{-1}\mathbf{A}). \tag{3.40}$$

The un-normalized gradient projection search vector **u**, that is used in practice, is then simply

$$\mathbf{u} = -\mathbf{P}\boldsymbol{\nabla} f(\mathbf{x}'). \tag{3.41}$$

3.5.1.2   Extension to non-linear constraints

Consider the more general problem:

$$\text{minimize} \quad f(\mathbf{x})$$
$$\text{such that} \quad h_i(\mathbf{x}) = 0, \ i = 1, 2, \ldots, r, \quad (3.42)$$

or in vector form $\mathbf{h}(\mathbf{x}) = \mathbf{0}$, where the constraints may be non-linear.

Linearize the constraint functions $h_i(\mathbf{x})$ at the feasible point $\mathbf{x}'$, by the truncated Taylor expansions:

$$h_i(\mathbf{x}) = h_i(\mathbf{x}' + (\mathbf{x} - \mathbf{x}')) \cong h_i(\mathbf{x}') + \boldsymbol{\nabla}^T h_i(\mathbf{x}')(\mathbf{x} - \mathbf{x}')$$

which allows for the following approximations of the constraints:

$$\boldsymbol{\nabla}^T h_i(\mathbf{x}')(\mathbf{x} - \mathbf{x}') = 0, \ i = 1, 2, \ldots, r \quad (3.43)$$

in the neighbourhood of $\mathbf{x}'$, since the $h_i(\mathbf{x}') = 0$. This set of linearized constraints may be written in matrix form as

$$\left[\frac{\partial \mathbf{h}(\mathbf{x}')}{\partial \mathbf{x}}\right]^T \mathbf{x} - \mathbf{b} = \mathbf{0}$$

where

$$\mathbf{b} = \left[\frac{\partial \mathbf{h}(\mathbf{x}')}{\partial \mathbf{x}}\right]^T \mathbf{x}'. \quad (3.44)$$

The *linearized problem* at $\mathbf{x}'$ therefore becomes

$$\text{minimize } f(\mathbf{x}) \text{ such that } \mathbf{A}\mathbf{x} - \mathbf{b} = \mathbf{0} \quad (3.45)$$

where $\mathbf{A} = \left[\frac{\partial \mathbf{h}(\mathbf{x}')}{\partial \mathbf{x}}\right]^T$ and $\mathbf{b} = \left[\frac{\partial \mathbf{h}(\mathbf{x}')}{\partial \mathbf{x}}\right]^T \mathbf{x}'$.

Since the problem is now linearized, the computation of the gradient projection direction at $\mathbf{x}'$ is identical to that before, i.e.:

$\mathbf{u} = \mathbf{P}(\mathbf{x}')\boldsymbol{\nabla} f(\mathbf{x}')$, but the projection matrix $\mathbf{P}(\mathbf{x}') = (\mathbf{I} - \mathbf{A}^T(\mathbf{A}\mathbf{A}^T)^{-1}\mathbf{A})$ is now dependent on $\mathbf{x}'$, since $\mathbf{A}$ is given by $\left[\frac{\partial \mathbf{h}(\mathbf{x}')}{\partial \mathbf{x}}\right]^T$.

For an initial feasible point $\mathbf{x}^0 (= \mathbf{x}')$, a new point in the gradient projection direction of descent $\mathbf{u}^1 = -\mathbf{P}(\mathbf{x}^0)\boldsymbol{\nabla} f(\mathbf{x}^0)$, is $\overline{\mathbf{x}}^1 = \mathbf{x}^0 + \alpha_1 \mathbf{u}^1$ for step size $\alpha_1 > 0$ as shown in Figure 3.14.

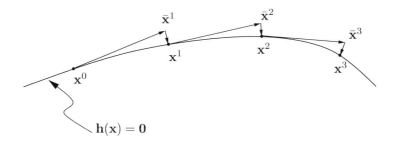

Figure 3.14: Schematic representation of correction steps when applying the gradient projection method to nonlinear constrained problems

In general, $\mathbf{h}(\bar{\mathbf{x}}^1) \neq 0$, and a correction step must be calculated: $\bar{\mathbf{x}}^1 \rightarrow \mathbf{x}^1$. How is this correction step computed? Clearly (see Figure 3.14), the step should be such that its projection at $\mathbf{x}^1$ is zero, i.e. $\mathbf{P}(\mathbf{x}^1)(\mathbf{x}^1 - \bar{\mathbf{x}}^1) = 0$ and also $\mathbf{h}(\mathbf{x}^1) = 0$. These two conditions imply that

$$(\mathbf{I} - \mathbf{A}^T(\mathbf{AA}^T)^{-1}\mathbf{A})(\mathbf{x}^1 - \bar{\mathbf{x}}^1) = 0$$

with $\mathbf{A}$ evaluated at $\mathbf{x}^1$, which gives

$$\mathbf{x}^1 \cong \bar{\mathbf{x}}^1 - \mathbf{A}^T(\mathbf{AA}^T)^{-1}\mathbf{h}(\bar{\mathbf{x}}^1) \tag{3.46}$$

as the correction step, where use was made of the expression $\mathbf{h}(\mathbf{x}) \cong \mathbf{Ax} - \mathbf{b}$ for both $\mathbf{h}(\bar{\mathbf{x}}^1)$ and $\mathbf{h}(\mathbf{x}^1)$ and of the fact that $\mathbf{h}(\mathbf{x}^1) = 0$.

Since the correction step is based on an approximation it may have to be applied repeatedly until $\mathbf{h}(\mathbf{x}^1)$ is sufficiently small. Having found a satisfactory $\mathbf{x}^1$, the procedure is repeated successively for $k = 2, 3, \ldots$ to give $\mathbf{x}^2$, $\mathbf{x}^3, \ldots$, until $\mathbf{P}(\mathbf{x}^k)\nabla f(\mathbf{x}^k) \cong 0$.

3.5.1.3   Example problem

$$\begin{aligned}
\text{Minimize} \quad & f(\mathbf{x}) = x_1^2 + x_2^2 + x_3^2 \\
\text{such that} \quad & h(\mathbf{x}) = x_1 + x_2 + x_3 = 1
\end{aligned}$$

with initial feasible point $\mathbf{x}^0 = [1, 0, 0]^T$.

First evaluate the projection matrix $\mathbf{P} = \mathbf{I} - \mathbf{A}^T(\mathbf{AA}^T)^{-1}\mathbf{A}$.

Here $\mathbf{A} = [1\ 1\ 1]$, $\mathbf{A}\mathbf{A}^T = 3$ and $(\mathbf{A}\mathbf{A}^T)^{-1} = \frac{1}{3}$, thus giving

$$
\mathbf{P} = \begin{bmatrix} 1 & 0 & 0 \\ 0 & 1 & 0 \\ 0 & 0 & 1 \end{bmatrix} - \frac{1}{3} \begin{bmatrix} 1 & 1 & 1 \\ 1 & 1 & 1 \\ 1 & 1 & 1 \end{bmatrix} = \frac{1}{3} \begin{bmatrix} 2 & -1 & -1 \\ -1 & 2 & -1 \\ -1 & -1 & 2 \end{bmatrix}
$$

The gradient vector is $\nabla f(\mathbf{x}) = \begin{bmatrix} 2x_1 \\ 2x_2 \\ 2x_3 \end{bmatrix}$ giving at $\mathbf{x}^0$, $\nabla f(\mathbf{x}) = \begin{bmatrix} 2 \\ 0 \\ 0 \end{bmatrix}$.

The search at direction at $\mathbf{x}^0$ is therefore given by

$$
-\mathbf{P}\nabla f(\mathbf{x}^0) = -\frac{1}{3} \begin{bmatrix} 2 & -1 & -1 \\ -1 & 2 & -1 \\ -1 & -1 & 2 \end{bmatrix} \begin{bmatrix} 2 \\ 0 \\ 0 \end{bmatrix} = -\frac{1}{3} \begin{bmatrix} 4 \\ -2 \\ -2 \end{bmatrix} = \frac{2}{3} \begin{bmatrix} -2 \\ 1 \\ 1 \end{bmatrix}
$$

or more conveniently, for this example, choose the search direction simply as $\mathbf{u} = \begin{bmatrix} -2 \\ 1 \\ 1 \end{bmatrix}$.

For a suitable value of λ the next point is given by

$$
\mathbf{x}^1 = \mathbf{x}^0 + \lambda \begin{bmatrix} -2 \\ 1 \\ 1 \end{bmatrix} = \begin{bmatrix} 1 - 2\lambda \\ \lambda \\ \lambda \end{bmatrix}.
$$

Substituting the above in $f(\mathbf{x}^1) = f(\mathbf{x}^0 + \lambda\mathbf{u}) = F(\lambda) = (1 - 2\lambda)^2 + \lambda^2 + \lambda^2$, it follows that for optimal descent in the direction $\mathbf{u}$:

$$
\frac{dF}{d\lambda} = -2(1 - 2\lambda)2 + 2\lambda + 2\lambda = 0 \Rightarrow \lambda = \frac{1}{3}
$$

which gives
$$
\mathbf{x}^1 = \left[\tfrac{1}{3}, \tfrac{1}{3}, \tfrac{1}{3}\right]^T, \quad \nabla f_1(\mathbf{x}^1) = \left[\tfrac{2}{3}, \tfrac{2}{3}, \tfrac{2}{3}\right]^T
$$
and

$$
\mathbf{P}\nabla f(\mathbf{x}^1) = \frac{1}{3} \begin{bmatrix} 2 & -1 & -1 \\ -1 & 2 & -1 \\ -1 & -1 & 2 \end{bmatrix} \frac{2}{3} \begin{bmatrix} 1 \\ 1 \\ 1 \end{bmatrix} = \frac{2}{9} \begin{bmatrix} 0 \\ 0 \\ 0 \end{bmatrix} = \mathbf{0}.
$$

Since the projection of the gradient vector at $\mathbf{x}^1$ is zero, it is the optimum point.

3.5.1.4   Extension to linear inequality constraints

Consider the case of linear inequality constraints:

$$\mathbf{Ax} - \mathbf{b} \leq \mathbf{0} \qquad\qquad (3.47)$$

where $\mathbf{A}$ is a $m \times n$ matrix and $\mathbf{b}$ a m-vector, i.e. the individual constraints are of the form

$$g_j(\mathbf{x}) = \mathbf{a}^j\mathbf{x} - b_j \leq 0, \ j = 1, 2, \ldots, m \ ,$$

where $\mathbf{a}^j$ denotes the $1 \times n$ matrix corresponding to the jth row of $\mathbf{A}$. Suppose at the current point $\mathbf{x}^k$, $r(\leq m)$ constraints are enforced, i.e. are active. Then a set of equality constraints, corresponding to the active constraints in (3.47) apply at $\mathbf{x}^k$, i.e. $\mathbf{A}_r\mathbf{x}^k - \mathbf{b}^r = \mathbf{0}$, where $\mathbf{A}_r$ and $\mathbf{b}^r$ correspond to the set of r active constraints in (3.47).

Now apply the gradient projection method as depicted in Figure 3.15 where the recursion is as follows:

$$\mathbf{u}^{k+1} = -\mathbf{P}(\mathbf{x}^k)\nabla f(\mathbf{x}^k) \text{ and } \mathbf{x}^{k+1} = \mathbf{x}^k + \alpha_{k+1}\mathbf{u}^{k+1}$$

where

$$f(\mathbf{x}^k + \alpha_{k+1}\mathbf{u}^{k+1}) = \min_{\alpha} f(\mathbf{x}^k + \alpha\mathbf{u}^{k+1}). \qquad (3.48)$$

Two possibilities may arise:

(i) No additional constraint is encountered along $\mathbf{u}^{k+1}$ before $\mathbf{x}^{k+1} = \mathbf{x}^k + \alpha_{k+1}\mathbf{u}^{k+1}$. Test whether $\mathbf{P}\nabla f(\mathbf{x}^{k+1}) = \mathbf{0}$. If so then $\mathbf{x}^* = \mathbf{x}^{k+1}$, otherwise set $\mathbf{u}^{k+2} = -\mathbf{P}(\mathbf{x}^{k+1})\nabla f(\mathbf{x}^{k+1})$ and continue.

(ii) If an additional constraint is encountered before $\mathbf{x}^{k+1} = \mathbf{x}^k + \alpha_{k+1}\mathbf{u}^{k+1}$ at $\overline{\mathbf{x}}^{k+1}$, (see Figure 3.15), then set $\mathbf{x}^{k+1} = \overline{\mathbf{x}}^{k+1}$ and add new constraint to active set, with associated matrix $\mathbf{A}_{r+1}$. Compute new $\mathbf{P}$ and set $\mathbf{u}^{k+2} = -\mathbf{P}(\mathbf{x}^{k+1})\nabla f(\mathbf{x}^{k+1})$. Continue this process until for some active set at $\mathbf{x}^p$, $\mathbf{P}\nabla f(\mathbf{x}^p) = \mathbf{0}$.

How do we know that if $\mathbf{P}\nabla f(\mathbf{x}^k) = \mathbf{0}$ occurs, that all the identified constraints are active? The answer is given by the following argument.

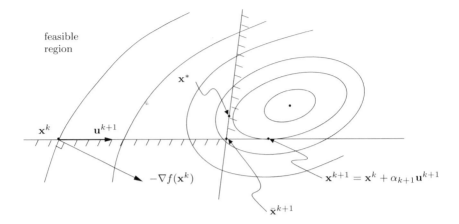

Figure 3.15: Representation of the gradient projection method for linear inequality constraints

If $\mathbf{P}\nabla f(\mathbf{x}^k) = \mathbf{0}$ it implies that $(\mathbf{I} - \mathbf{A}^T(\mathbf{A}\mathbf{A}^T)^{-1}\mathbf{A})\nabla f(\mathbf{x}^k) = \mathbf{0}$ which, using expression (3.38), is equivalent to $\nabla f(\mathbf{x}^k) + \mathbf{A}^T\boldsymbol{\lambda} = \mathbf{0}$, i.e. $\nabla f(\mathbf{x}^k) + \sum \lambda_i \nabla g_i(\mathbf{x}^k) = \mathbf{0}$ for all $i \in I_a$ = set of all active constraints. This expression is nothing else but the KKT conditions for the optimum at $\mathbf{x}^k$, provided that $\lambda_i \geq 0$ for all $i \in I_a$. Now, if $\mathbf{P}\nabla f(\mathbf{x}^k) = \mathbf{0}$ occurs, then if $\lambda_i < 0$ for some i, remove the corresponding constraint from the active set. In practice remove the constraint with the most negative multiplier, compute the new projection matrix $\mathbf{P}$, and continue.

3.5.2 Multiplier methods

These methods combine the classical Lagrangian method with the penalty function approach. In the Lagrangian approach, the minimum point of the constrained problem coincides with a stationary point $(\mathbf{x}^*, \boldsymbol{\lambda}^*)$ of the Lagrangian function which, in general, is *difficult to determine analytically*. On the other hand, in the penalty function approach, the constrained minimum approximately coincides with the minimum of the penalty function. If, however, high accuracy is required, the problem *becomes ill-conditioned*.

In the *multiplier methods* (see Bertsekas 1976) both approaches are combined to give an unconstrained problem which is not ill-conditioned.

As an introduction to the multiplier method consider the equality constrained problem:

$$\begin{aligned} \text{minimize} \quad & f(\mathbf{x}) \\ \text{such that} \quad & h_j(\mathbf{x}) = 0, \ j = 1, 2, \dots, r. \end{aligned} \tag{3.49}$$

The *augmented Lagrange function* $\mathcal{L}$ is introduced as

$$\mathcal{L}(\mathbf{x}, \boldsymbol{\lambda}, \rho) = f(\mathbf{x}) + \sum_{j=1}^{r} \lambda_j h_j(\mathbf{x}) + \rho \sum_{j=1}^{r} h_j^2(\mathbf{x}). \tag{3.50}$$

If all the multipliers λ_j are chosen equal to zero, $\mathcal{L}$ becomes the usual external penalty function. On the other hand, if all the stationary values λ_j^* are available, then *it can be shown* (Fletcher 1987) that for any positive value of ρ, the minimization of $\mathcal{L}(\mathbf{x}, \boldsymbol{\lambda}^*, \rho)$ with respect to $\mathbf{x}$ gives the solution $\mathbf{x}^*$ to problem (3.49). This result is not surprising, since it can be shown that the classical Lagrangian function $L(\mathbf{x}, \boldsymbol{\lambda})$ has a saddle point at $(\mathbf{x}^*, \boldsymbol{\lambda}^*)$.

The multiplier methods are based on the use of approximations of the Lagrange multipliers. If $\boldsymbol{\lambda}^k$ is a good approximation to $\boldsymbol{\lambda}^*$, then it is possible to approach the optimum through the unconstrained minimization of $\mathcal{L}(\mathbf{x}, \boldsymbol{\lambda}^k, \rho)$ without using large values of ρ. The value of ρ must only be sufficiently large to ensure that $\mathcal{L}$ has a local minimum point with respect to $\mathbf{x}$ rather than simply a stationary point at the optimum.

How is the approximation to the Lagrange multiplier vector $\boldsymbol{\lambda}^k$ obtained? To answer this question, compare the stationary conditions with respect to $\mathbf{x}$ for $\mathcal{L}$ (the augmented Lagrangian) with those for L (the classical Lagrangian) at $\mathbf{x}^*$.

For $\mathcal{L}$:

$$\frac{\partial \mathcal{L}}{\partial x_i} = \frac{\partial f}{\partial x_i} + \sum_{j=1}^{r} (\lambda_j^k + 2\rho h_j) \frac{\partial h_j}{\partial x_j} = 0, \ i = 1, 2, \dots, n. \tag{3.51}$$

For L:

$$\frac{\partial L}{\partial x_i} = \frac{\partial f}{\partial x_i} + \sum_{j=1}^{r} \lambda_j^* \frac{\partial h_j}{\partial x_j} = 0, \ i = 1, 2, \dots, n. \tag{3.52}$$

The comparison clearly indicates that as the minimum point of $\mathcal{L}$ tends to $\mathbf{x}^*$ that

$$\lambda_j^k + 2\rho h_j \to \lambda_j^*. \tag{3.53}$$

This observation prompted Hestenes (1969) to suggest the following scheme for approximating $\boldsymbol{\lambda}^*$. For a given approximation $\boldsymbol{\lambda}^k$, $k = 1, 2, \ldots$, minimize $\mathcal{L}(\mathbf{x}, \boldsymbol{\lambda}^k, \rho)$ by some standard unconstrained minimization technique to give a minimum $\mathbf{x}^{*k}$. The components of the new estimate to $\boldsymbol{\lambda}^*$, as suggested by (3.53), is then given by

$$\lambda_j^* \cong \lambda_j^{k+1} = \lambda_j^k + 2\rho h_j(\mathbf{x}^{*k}). \tag{3.54}$$

The value of the penalty parameter $\rho (= \rho_k)$ may also be adjusted iteratively.

3.5.2.1   Example

Minimize $f(\mathbf{x}) = x_1^2 + 10x_2^2$ such that $h(\mathbf{x}) = x_1 + x_2 - 4 = 0$.

Here $\mathcal{L} = x_1^2 + 10x_2^2 + \lambda(x_1 + x_2 - 4) + \rho(x_1 + x_2 - 4)^2$.

The first order necessary conditions for a constrained minimum with respect to $\mathbf{x}$ for any given values of λ and ρ are

$$\begin{aligned}
2x_1 + \lambda + 2\rho(x_1 + x_2 - 4) &= 0 \\
20x_2 + \lambda + 2\rho(x_1 + x_2 - 4) &= 0
\end{aligned}$$

from which it follows that

$$x_1 = 10x_2 = \frac{-5\lambda + 40\rho}{10 + 11\rho}.$$

Taking $\lambda^1 = 0$ and $\rho_1 = 1$ gives $\mathbf{x}^{*1} = [1.905, 0.1905]^T$ and $h(\mathbf{x}^{*1}) = -1.905$.

Using the approximation scheme (3.54) for λ^* gives $\lambda^2 = 0 + 2(1)(-1.905) = -3.81$.

Now repeat the minimization of $\mathcal{L}$ with $\lambda^2 = -3.81$ and $\rho_2 = 10$. This gives $\mathbf{x}^{*2} = [3.492, 0.3492]^T$ and $h(\mathbf{x}^{*2}) = -0.1587$, resulting in the new approximation: $\lambda^3 = -3.81 + 2(10)(-0.1587) = -6.984$.

Using λ^3 in the next iteration with $\rho_3 = 10$ gives $\mathbf{x}^{*3} = [3.624, 0.3624]^T$, $h(\mathbf{x}^{*3}) = 0.0136$, which shows good convergence to the exact solution $[3.6363, 0.3636]^T$ without the need for increasing ρ.

3.5.2.2 The multiplier method extended to inequality constraints

The multiplier method can be extended to apply to inequality constraints (Fletcher 1975).

Consider the problem:

$$\begin{aligned} \text{minimize} \quad & f(\mathbf{x}) \\ \text{such that} \quad & g_j(\mathbf{x}) \leq 0, \ j = 1, 2, \ldots, m. \end{aligned} \tag{3.55}$$

Here the augmented Lagrangian function is

$$\mathcal{L}(\mathbf{x}, \boldsymbol{\lambda}, \rho) = f(\mathbf{x}) + \rho \sum_{j=1}^{m} \left\langle \frac{\lambda_j}{2\rho} + g_j(\mathbf{x}) \right\rangle^2 \tag{3.56}$$

where $\langle a \rangle = \max(a, 0)$.

In this case, the stationary conditions at $\mathbf{x}^*$ for the augmented and classical Lagrangians are respectively

$$\frac{\partial \mathcal{L}}{\partial x_i} = \frac{\partial f}{\partial x_i} + 2\rho \sum_{j=1}^{m} \left\langle \frac{\lambda_j}{2\rho} + g_j \right\rangle \frac{\partial g_j}{\partial x_i} = 0, \ i = 1, 2, \ldots, n \tag{3.57}$$

and

$$\frac{\partial L}{\partial x_i} = \frac{\partial f}{\partial x_i} + \sum_{j=1}^{m} \lambda_j^* \frac{\partial g_j}{\partial x_i} = 0, \ i = 1, 2, \ldots, n. \tag{3.58}$$

The latter classical KKT conditions require in addition that $\lambda_j^* g_j(\mathbf{x}^*) = 0$, $j = 1, 2, \ldots, m$.

A comparison of conditions (3.57) and (3.58) leads to the following iterative approximation scheme for $\boldsymbol{\lambda}^*$:

$$\lambda_j^* \cong \lambda_j^{k+1} = \langle \lambda_j^k + 2\rho_k g_j(\mathbf{x}^{*k}) \rangle \tag{3.59}$$

where $\mathbf{x}^{*k}$ minimizes $\mathcal{L}(\mathbf{x}, \boldsymbol{\lambda}^k, \rho_k)$.

3.5.2.3 Example problem with inequality constraint

Minimize $f(\mathbf{x}) = x_1^2 + 10x_2^2$ such that $g(\mathbf{x}) = 4 - x_1 - x_2 \le 0$.

Here $\mathcal{L}(\mathbf{x}, \lambda, \rho) = x_1^2 + 10x_2^2 + \rho \left\langle \dfrac{\lambda}{2\rho} + (4 - x_1 - x_2) \right\rangle^2$.

Now perform the *unconstrained* minimization of $\mathcal{L}$ with respect to $\mathbf{x}$ for given λ and ρ. This is usually done by some standard method such as Fletcher-Reeves or BFGS but here, because of the simplicity of the illustrative example, the necessary conditions for a minimum are used, namely:

$$\frac{\partial \mathcal{L}}{\partial x_1} = 2x_1 - \langle \lambda + 2\rho(4 - x_1 - x_2) \rangle = 0 \qquad (3.60)$$

$$\frac{\partial \mathcal{L}}{\partial x_2} = 20x_2 - \langle \lambda + 2\rho(4 - x_1 - x_2) \rangle = 0. \qquad (3.61)$$

Clearly

$$x_1 = 10x_2 \qquad (3.62)$$

provided that $\langle\ \rangle$ is nonzero. Substituting $x_2 = 0.1x_1$ into (3.60) gives

$$x_1 = \frac{5\lambda + 40\rho}{10 + 11\rho}. \qquad (3.63)$$

Successive iterations can now be performed as shown below.

Iteration 1: Use $\lambda^1 = 0$ and $\rho_1 = 1$, then (3.63) and (3.62) imply

$$x_1^{*1} = 1.9048 \text{ and } x_2^{*1} = 0.19048.$$

The new value λ^2 for the Lagrange multiplier determined via (3.59) is $\lambda^2 = \langle \lambda^1 + 2\rho_1(4 - x_1^{*1} - x_2^{*1}) \rangle = 3.8095$.

Iteration 2: Now with $\lambda^2 = 3.8095$ choose $\rho_2 = 10$ which gives

$$x_1^{*2} = 3.4921,\ x_2^{*2} = 0.34921 \text{ and } \lambda^3 = \langle \lambda^2 + 2\rho_2(4 - x_1^{*2} - x_2^{*2}) \rangle = 6.9842.$$

Iteration 3: Using the current value λ^3 for the Lagrange multiplier and $\rho = 10$ for the penalty, the iterations proceed as follows:

Iteration	x_1^{*k}	x_2^{*k}		λ^{k+1}
3	3.6243	0.36243	$\rightarrow$	7.2488
4	3.6354	0.36354	$\rightarrow$	7.2707
5	3.6363	0.36363	$\rightarrow$	7.2725.

Thus the iterative procedure converges satisfactorily to the numerical approximate solution: $\mathbf{x}^* = [3.6363, 0.36363]^T$.

3.5.3 Sequential quadratic programming (SQP)

The SQP method is based on the application of Newton's method to determine $\mathbf{x}^*$ and $\boldsymbol{\lambda}^*$ from the KKT conditions of the constrained optimization problem. It can be shown (see Bazaraa et al. 1993) that the determination of the Newton step is equivalent to the solution of a quadratic programming (QP) problem.

Consider the general problem:

$$\begin{array}{ll} \text{minimize} & f(\mathbf{x}) \\ \text{such that} & g_j(\mathbf{x}) \le 0; \;\; j = 1, 2, \ldots, m \\ & h_j(\mathbf{x}) = 0; \;\; j = 1, 2, \ldots, r. \end{array} \qquad (3.64)$$

It *can be shown* (Bazaraa et al. 1993) that given estimates $(\mathbf{x}^k, \boldsymbol{\lambda}^k, \boldsymbol{\mu}^k)$, $k = 0, 1, \ldots$, of the solution and the respective Lagrange multipliers values, with $\boldsymbol{\lambda}^k \ge \mathbf{0}$, then the Newton step $\mathbf{s}$ of iteration $k+1$, such that $\mathbf{x}^{k+1} = \mathbf{x}^k + \mathbf{s}$ is given by the solution to the following *k-th QP problem*:

QP-k $(\mathbf{x}^k, \boldsymbol{\lambda}^k, \boldsymbol{\mu}^k)$: Minimize with respect to $\mathbf{s}$:

$$F(\mathbf{s}) = f(\mathbf{x}^k) + \nabla^T f(\mathbf{x}^k)\mathbf{s} + \tfrac{1}{2}\mathbf{s}^T \mathbf{H}_L(\mathbf{x}^k)\mathbf{s}$$

such that

$$\mathbf{g}(\mathbf{x}^k) + \left[\frac{\partial \mathbf{g}(\mathbf{x}^k)}{\partial \mathbf{x}}\right]^T \mathbf{s} \le \mathbf{0} \qquad (3.65)$$

and

$$\mathbf{h}(\mathbf{x}^k) + \left[\frac{\partial \mathbf{h}(\mathbf{x}^k)}{\partial \mathbf{x}}\right]^T \mathbf{s} = \mathbf{0}$$

and where $\mathbf{g} = [g_1, g_2, \ldots, g_m]^T$, $\mathbf{h} = [h_1, h_2, \ldots, h_r]^T$ and the Hessian of the classical Lagrangian with respect to $\mathbf{x}$ is

$$\mathbf{H}_L(\mathbf{x}^k) = \boldsymbol{\nabla}^2 f(\mathbf{x}^k) + \sum_{j=1}^{m} \lambda_j^k \boldsymbol{\nabla}^2 g_j(\mathbf{x}^k) + \sum_{j=1}^{r} \mu_j^k \boldsymbol{\nabla}^2 h_j(\mathbf{x}^k).$$

Note that the solution of QP-k does not only yield step $\mathbf{s}$, but also the Lagrange multipliers $\boldsymbol{\lambda}^{k+1}$ and $\boldsymbol{\mu}^{k+1}$ via the solution of equation (3.26) in Section 3.4. Thus with $\mathbf{x}^{k+1} = \mathbf{x}^k + \mathbf{s}$ we may construct the next QP problem: QP-$k+1$.

The solution of successive QP problems is continued until $\mathbf{s} = \mathbf{0}$. It can be shown that if this occurs, then the KKT conditions of the original problem (3.64) are satisfied.

In practice, since convergence from a point far from the solution is not guaranteed, the full step $\mathbf{s}$ is usually not taken. To improve convergence, $\mathbf{s}$ is rather used as a search direction in performing a line search minimization of a so-called merit or descent function. A popular choice for the merit function is (Bazaraa et al. 1993):

$$F_E(\mathbf{x}) = f(\mathbf{x}) + \gamma \left(\sum_{i=1}^{m} \max\{0, g_i(\mathbf{x})\} + \sum_{i=1}^{r} |h_i(\mathbf{x})| \right) \qquad (3.66)$$

where $\gamma \geq \max\{\lambda_1, \lambda_2, \ldots, \lambda_m, |\mu_1|, \ldots, |\mu_r|\}$.

Note that it is not advisable here to use curve fitting techniques for the line search since the function is non-differentiable. More stable methods, such as the golden section method, are therefore preferred.

The great advantage of the SQP approach, above the classical Newton method, is that it allows for a systematic and natural way of selecting the active set of constraints and in addition, through the use of the merit function, the convergence process may be controlled.

3.5.3.1   Example

Solve the problem below by means of the basic SQP method, using $\mathbf{x}^0 = [0, 1]^T$ and $\boldsymbol{\lambda}^0 = \mathbf{0}$.

Minimize $f(\mathbf{x}) = 2x_1^2 + 2x_2^2 - 2x_1x_2 - 4x_1 - 6x_2$

such that

$$
\begin{aligned}
g_1(\mathbf{x}) &= 2x_1^2 - x_2 \leq 0 \\
g_2(\mathbf{x}) &= x_1 + 5x_2 - 5 \leq 0 \\
g_3(\mathbf{x}) &= -x_1 \leq 0 \\
g_4(\mathbf{x}) &= -x_2 \leq 0.
\end{aligned}
$$

Since $\boldsymbol{\nabla} f(\mathbf{x}) = \begin{bmatrix} 4x_1 - 2x_2 - 4 \\ 4x_2 - 2x_1 - 6 \end{bmatrix}$ it follows that $\boldsymbol{\nabla} f(\mathbf{x}^0) = \begin{bmatrix} -6 \\ -2 \end{bmatrix}$.

Thus with $\boldsymbol{\lambda}^0 = \mathbf{0}$, $\mathbf{H}_L = \begin{bmatrix} 4 & -2 \\ -2 & 4 \end{bmatrix}$ and it follows from (3.65) that the starting quadratic programming problem is:

QP-0: Minimize with respect to $\mathbf{s}$:

$$
F(\mathbf{s}) = -4 - 6s_1 - 2s_2 + \tfrac{1}{2}(4s_1^2 + 4s_2^2 - 4s_1s_2)
$$

such that

$$
\begin{aligned}
-1 - s_2 \leq 0, && s_1 + 5s_2 \leq 0 \\
-s_1 \leq 0, \text{ and } && -1 - s_2 \leq 0
\end{aligned}
$$

where $\mathbf{s} = [s_1, s_2]^T$.

The solution to this QP can be obtained via the method described in Section 3.4, which firstly shows that only the second constraint is active, and then obtains the solution by solving the corresponding equation (3.26) giving $\mathbf{s} = [1.1290, -0.2258]^T$ and $\boldsymbol{\lambda}^1 = [0, 1.0322, 0, 0]^T$ and therefore $\mathbf{x}^1 = \mathbf{x}^0 + \mathbf{s} = [1.1290, 0.7742]^T$ which completes the *first iteration*.

The next quadratic program QP-1 can now be constructed. It is left to the reader to perform the further iterations. The method, because it is basically a Newton method, converges rapidly to the optimum $\mathbf{x}^* = [0.6589, 0.8682]^T$.

3.6   Exercises

3.6.1 Solve for all the points satisfying the KKT conditions for the problem:

$$\text{minimize} \quad f(\mathbf{x}) = (x_1 - 4)^2 + (x_2 - 6)^2$$
$$\text{such that}$$
$$g_1(\mathbf{x}) = x_1 + x_2 \leq 12,$$
$$g_2(\mathbf{x}) = x_1 \leq 6,$$
$$g_3(\mathbf{x}) = -x_1 \leq 0,$$
$$g_4(\mathbf{x}) = -x_2 \leq 0.$$

3.6.2 Sketch the solution(s) to Exercise 3.6.1 and identify the optimum.

3.6.3 Given the problem

$$\text{minimize} \quad f(\mathbf{x}) = x_1^3 - 6x_1^2 + 11x_1 + x_3$$
$$\text{such that}$$
$$g_1(\mathbf{x}) = x_1^2 + x_2^2 - x_3^2 \leq 0,$$
$$g_2(\mathbf{x}) = 4 - x_1^2 - x_2^2 - x_3^2 \leq 0,$$
$$g_3(\mathbf{x}) = x_3 - 5 \leq 0,$$
$$g_4(\mathbf{x}) = x_1 \geq 0,$$
$$g_5(\mathbf{x}) = x_2 \geq 0,$$
$$g_6(\mathbf{x}) = x_3 \geq 0.$$

Formulate an approximate linear programming problem about the point $\mathbf{x} = [1,\ 1,\ 1]^{\text{T}}$.

3.6.4 For the problem given in Exercise 3.6.3, formulate an approximate quadratic programming problem about the point $\mathbf{x} = [1,\ 1,\ 1]$.

3.6.5 Transform the problem in Exercise 3.6.3 to an unconstrained optimization problem using a penalty function formulation.

3.6.6 Consider the following primal problem:

$$\text{minimize} \quad (x - 1)^2,$$
$$\text{such that}$$
$$2x - 1 = 0.$$

Give the Lagrangian dual problem.

3.6.7 Solve the primal and dual problems in Exercise 3.6.6 and compare their respective solutions with each other.

3.6.8 Consider the problem:

$$\text{minimize} \quad f(\mathbf{x}) = (x_1 - 2)^2 + (x_2 - 2)^2,$$
$$\text{such that}$$
$$h(\mathbf{x}) = x_1 + x_2 - 6 = p,$$

for which the optimum can be expressed as $f^*(\mathbf{x}^*(p)) = f^*(p)$ with optimal Lagrange multiplier $\lambda^*(p)$. Confirm numerically that $f^*(p) \approx f^*(0) - \lambda^*(0)p$ for $p = 0.1$.

3.6.9 Consider the problem:

$$\text{minimize} \quad f(\mathbf{x}) = 2x_1^2 - 3x_2^2 - 2x_1,$$
$$\text{such that}$$
$$g(\mathbf{x}) = x_1^2 + x_2^2 - 1 \leq q,$$

for which the optimum can be expressed as $f^*(\mathbf{x}^*(q)) = f^*(q)$ with optimal Lagrange multiplier $\lambda^*(q)$. Confirm numerically that $f^*(q) \approx f^*(0) - \lambda^*(0)q$ for $q = -0.1$.

3.6.10 How does the scaling the objective function influence the value of the Lagrange multiplier, and inversely how does the scaling of the constraint function influence the value of the Lagrange multiplier.

3.6.11 Consider the constrained maximization problem:

$$f(\mathbf{x}) = \mathbf{x}^T \mathbf{A} \mathbf{x}, \quad s.t. \quad \mathbf{x}^T \mathbf{x} = 1,$$

for the symmetric matrix $\mathbf{A}$, would a bounded solution exist with the constraint $\mathbf{x}^T \mathbf{x} = 1$ omitted in the formulation.

3.6.12 For symmetric matrix $\mathbf{A}$, show that the solution $\mathbf{x}^*$ to the constrained maximization problem:

$$f(\mathbf{x}) = \mathbf{x}^T \mathbf{A} \mathbf{x}, \quad s.t. \quad \mathbf{x}^T \mathbf{x} = 1,$$

is the unit eigenvector corresponding to the maximum eigenvalue of $\mathbf{A}$.

3.6.13 Consider the symmetric matrix $\mathbf{A}$ with known unit eigenvector $\mathbf{u}^1$ corresponding to the maximum eigenvalue of $\mathbf{A}$. Formulate a constrained optimization problem that would solve for the unit eigenvector associated with the second highest eigenvalue of $\mathbf{A}$.

Chapter 4

BASIC EXAMPLE PROBLEMS

An extensive set of worked-out example optimization problems are presented in this chapter. They demonstrate the application of the basic concepts and methods introduced and developed in the previous three chapters. The reader is encouraged to attempt each problem separately before consulting the given detailed solution. This set of example problems is not only convenient for students to test their understanding of basic mathematical optimization, but also provides models for easily formulating additional optimization exercises.

4.1 Introductory examples

Problem 4.1.1

Sketch the geometrical solution to the optimization problem:

$$\text{minimize } f(\mathbf{x}) = 2x_2 - x_1$$
$$\text{subject to } g_1(\mathbf{x}) = x_1^2 + 4x_2^2 - 16 \le 0,$$
$$g_2(\mathbf{x}) = (x_1 - 3)^2 + (x_2 - 3)^2 - 9 \le 0$$
$$\text{and } x_1 \ge 0 \text{ and } x_2 \ge 0.$$

© Springer International Publishing AG, part of Springer Nature 2018
J.A. Snyman and D.N. Wilke, *Practical Mathematical Optimization,*
Springer Optimization and Its Applications 133,
https://doi.org/10.1007/978-3-319-77586-9_4

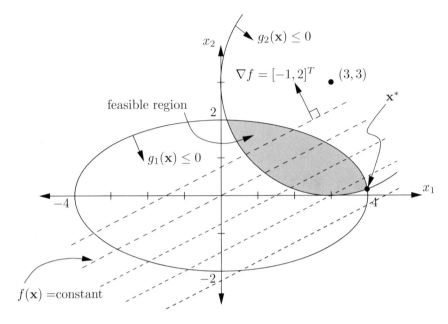

Figure 4.1: Solution to problem 4.1.1

In particular sketch the *contours of the objective function* and the *constraint curves*. Indicate the feasible region and the position of the *optimum* $\mathbf{x}^*$ and the active constraint(s).

The solution to this problem is indicated in Figure 4.1.

Problem 4.1.2

Consider the function $f(\mathbf{x}) = 100(x_2 - x_1^2)^2 + (1 - x_1)^2$.

(i) Compute the gradient vector and the Hessian matrix.

(ii) Let $\mathbf{x}^* = [1, 1]^T$. Show that $\nabla f(\mathbf{x}) = \mathbf{0}$ and that $\mathbf{x}^*$ is indeed a strong local minimum.

(iii) Is $f(\mathbf{x})$ a convex function? Justify your answer.

Solution

(i)

$$\nabla f(\mathbf{x}) = \begin{bmatrix} 200(x_2 - x_1^2)(-2x_1) - 2(1 - x_1) \\ 200(x_2 - x_1^2) \end{bmatrix}$$

$$= \begin{bmatrix} -400x_2x_1 + 400x_1^3 - 2 + 2x_1 \\ 200x_2 - 200x_1^2 \end{bmatrix}$$

$$\mathbf{H}(\mathbf{x}) = \begin{bmatrix} -400x_2 + 1200x_1^2 + 2 & \vdots & -400x_1 \\ -400x_1 & \vdots & 200 \end{bmatrix}.$$

(ii) $\nabla f(\mathbf{x}) = \mathbf{0}$ implies that

$$-400x_2x_1 + 400x_1^3 - 2 + 2x_1 = 0 \qquad (4.1)$$
$$200x_2 - 200x_1^2 = 0. \qquad (4.2)$$

From (4.2) $x_2 = x_1^2$, and substituting into (4.1) yields

$$-400x_1^3 + 400x_1^3 - 2 - 2x_1 = 0$$

giving $x_1 = 1$ and $x_2 = 1$ which is the unique solution.
Therefore at $\mathbf{x}^* = [1, 1]^T$:

$$\mathbf{H}(\mathbf{x}^*) = \begin{bmatrix} 802 & -400 \\ -400 & 200 \end{bmatrix},$$

and since the leading principal minors $\alpha_1 = 802 > 0$, and $\alpha_2 = \det \mathbf{H}(\mathbf{x}^*) = 400 > 0$, it follows by Sylvester's theorem that $\mathbf{H}(\mathbf{x})$ is positive-definite at $\mathbf{x}^*$ (see different equivalent definitions for positive definiteness that can be checked for, for example Fletcher (1987)). Thus $\mathbf{x}^* = [1, 1]^T$ is a strong local minimum since the necessary and sufficient conditions in (1.24) are both satisfied at $\mathbf{x}^*$.

(iii) By inspection, if $x_1 = 0$ and $x_2 = 1$ then $\mathbf{H} = \begin{bmatrix} -398 & 0 \\ 0 & 200 \end{bmatrix}$ and the determinant of the Hessian matrix is less than zero and $\mathbf{H}$ is not positive-definite at this point. Since by Theorem 5.1.2, $f(\mathbf{x})$ is convex over a set X *if and only if* $\mathbf{H}(\mathbf{x})$ is positive semi-definite for all $\mathbf{x}$ in X, it follows that $f(\mathbf{x})$ is not convex in this case.

Problem 4.1.3

Determine whether or not the following function is convex:

$$f(\mathbf{x}) = 4x_1^2 + 3x_2^2 + 5x_3^2 + 6x_1x_2 + x_1x_3 - 3x_1 - 2x_2 + 15.$$

Solution

The function $f(\mathbf{x})$ is convex if and only if the Hessian matrix $\mathbf{H}(\mathbf{x})$ is positive semi-definite at every point $\mathbf{x}$.

Here $\nabla f(\mathbf{x}) = \begin{bmatrix} 8x_1 + 6x_2 - 3 + x_3 \\ 6x_2 + 6x_1 - 2 \\ 10x_3 + x_1 \end{bmatrix}$ and $\mathbf{H}(\mathbf{x}) = \begin{bmatrix} 8 & 6 & 1 \\ 6 & 6 & 0 \\ 1 & 0 & 10 \end{bmatrix}.$

The principal minors are:

$$\alpha_1 = 8 > 0, \quad \alpha_2 = \begin{vmatrix} 8 & 6 \\ 6 & 6 \end{vmatrix} = 12 > 0 \text{ and } \alpha_3 = |\mathbf{H}| = 114 > 0.$$

Thus by Sylvester's theorem $\mathbf{H}(\mathbf{x})$ is positive-definite and thus $f(\mathbf{x})$ is convex.

Problem 4.1.4

Determine all the stationary points of

$$f(\mathbf{x}) = x_1^3 + 3x_1x_2^2 - 3x_1^2 - 3x_2^2 + 4.$$

Classify each point according to whether it corresponds to maximum, minimum or saddle point.

Solution

The first order necessary condition for a stationary point is that

$$\nabla f = \begin{bmatrix} 3x_1^2 + 3x_2^2 - 6x_1 \\ 6x_1x_2 - 6x_2 \end{bmatrix} = \mathbf{0}$$

from which it follows that $6x_2(x_1 - 1) = 0$.

Therefore either $x_2 = 0$ or $x_1 = 1$ which respectively give:

$$\begin{array}{ccc} 3x_1^2 - 6x_1 = 0 & & 3 + 3x_2^2 - 6 = 0 \\ 3x_1(x_1 - 2) = 0 & \text{or} & x_2^2 = 1 \\ x_1 = 0; \; x_1 = 2 & & x_2 = \pm 1. \end{array}$$

Therefore the stationary points are: $(0; 0)$, $(2; 0)$; $(1; 1)$, $(1; -1)$.

The nature of these stationary points may be determined by substituting their coordinates into the Hessian matrix $\mathbf{H}(\mathbf{x}) = \begin{bmatrix} 6x_1 - 6 & 6x_2 \\ 6x_2 & 6x_1 - 6 \end{bmatrix}$ and applying Sylvester's theorem.

The results are listed below.

Point	Hessian	Minors	Nature of $\mathbf{H}$	Type
$(0; 0)$	$\begin{bmatrix} -6 & 0 \\ 0 & -6 \end{bmatrix}$	$\alpha_1 < 0$ $\alpha_2 > 0$	negative-definite	maximum
$(2; 0)$	$\begin{bmatrix} 6 & 0 \\ 0 & 6 \end{bmatrix}$	$\alpha_1 > 0$ $\alpha_2 > 0$	positive-definite	minimum
$(1; 1)$	$\begin{bmatrix} 0 & 6 \\ 6 & 0 \end{bmatrix}$	$\alpha_1 = 0$ $\alpha_2 < 0$	indefinite	saddle
$(1; -1)$	$\begin{bmatrix} 0 & -6 \\ -6 & 0 \end{bmatrix}$	$\alpha_1 = 0$ $\alpha_2 < 0$	indefinite	saddle

Problem 4.1.5

Characterize the stationary points of $f(\mathbf{x}) = x_1^3 + x_2^3 + 2x_1^2 + 4x_2^2 + 6$.

Solution

First determine gradient vector $\boldsymbol{\nabla} f(\mathbf{x})$ and consider $\boldsymbol{\nabla} f(\mathbf{x}) = \mathbf{0}$:

$$3x_1^2 + 4x_1 = x_1(3x_1 + 4) = 0$$
$$3x_2^2 + 8x_2 = x_2(3x_2 + 8) = 0.$$

The solutions are: $(0, 0)$; $\left(0, -\frac{8}{3}\right)$; $\left(-\frac{4}{3}, 0\right)$; $\left(-\frac{4}{3}, -\frac{8}{3}\right)$.

To determine the nature of the stationary points substitute their coordinates in the Hessian matrix:

$$\mathbf{H}(\mathbf{x}) = \begin{bmatrix} 6x_1 + 4 & 0 \\ 0 & 6x_2 + 8 \end{bmatrix}$$

and study the principal minors α_1 and α_2 for each point:

Point $\mathbf{x}$	α_1	α_2	Nature of $\mathbf{H}(\mathbf{x})$	Type	$f(\mathbf{x})$
$(0,0)$	4	32	positive-definite	minimum	6
$\left(0,-\frac{8}{3}\right)$	4	-32	indefinite	saddle	15.48
$\left(-\frac{4}{3},0\right)$	-4	-32	indefinite	saddle	7.18
$\left(-\frac{4}{3},-\frac{8}{3}\right)$	-4	32	negative-definite	maximum	16.66

Problem 4.1.6

Minimize $f(\mathbf{x}) = x_1 - x_2 + 2x_1^2 + 2x_1x_2 + x_2^2$ by means of the basic Newton method. Use initial estimate $\mathbf{x}^0 = [0,0]^T$ for the minimizer.

Solution

$$\nabla f(\mathbf{x}) = \begin{bmatrix} 1 + 4x_1 + 2x_2 \\ -1 + 2x_1 + 2x_2 \end{bmatrix}$$

$$\mathbf{H}(\mathbf{x}) = \left\{ \frac{\partial^2 f}{\partial x_i \partial x_j} \right\} = \begin{bmatrix} 4 & 2 \\ 2 & 2 \end{bmatrix} ; \quad \mathbf{H}^{-1} = \begin{bmatrix} \frac{1}{2} & -\frac{1}{2} \\ -\frac{1}{2} & 1 \end{bmatrix}.$$

First Newton iteration:

$$\begin{aligned}
\mathbf{x}^1 &= \mathbf{x}^0 - \mathbf{H}^{-1} \nabla f(\mathbf{x}^0) \\
&= \begin{bmatrix} 0 \\ 0 \end{bmatrix} - \begin{bmatrix} \frac{1}{2} & -\frac{1}{2} \\ -\frac{1}{2} & 1 \end{bmatrix} \begin{bmatrix} 1 \\ -1 \end{bmatrix} = \begin{bmatrix} -1 \\ \frac{3}{2} \end{bmatrix}
\end{aligned}$$

and

$$\nabla f(\mathbf{x}^1) = \begin{bmatrix} 1 + 4(-1) + 2\left(\frac{3}{2}\right) \\ -1 + 2(-1) + 2\left(\frac{3}{2}\right) \end{bmatrix} = \begin{bmatrix} 0 \\ 0 \end{bmatrix}.$$

Therefore since $\mathbf{H}$ is positive-definite for all $\mathbf{x}$, the sufficient conditions (1.24) for a strong local minimum at $\mathbf{x}^1$ are satisfied and the global minimum is $\mathbf{x}^* = \mathbf{x}^1$.

Problem 4.1.7

Minimize $f(\mathbf{x}) = 3x_1^2 - 2x_1x_2 + x_2^2 + x_1$ by means of the basic Newton method using $\mathbf{x}^0 = [1,1]^T$.

Solution

$$\nabla f(\mathbf{x}) = \begin{bmatrix} \frac{\partial f}{\partial x_1} \\ \frac{\partial f}{\partial x_2} \end{bmatrix} = \begin{bmatrix} 6x_1 - 2x_2 + 1 \\ -2x_1 + 2x_2 \end{bmatrix}$$

$$\mathbf{H}(\mathbf{x}) = \begin{bmatrix} 6 & -2 \\ -2 & 2 \end{bmatrix}; \text{ and } \mathbf{H}^{-1} = \begin{bmatrix} \frac{1}{4} & \frac{1}{4} \\ \frac{1}{4} & \frac{3}{4} \end{bmatrix}.$$

First Newton iteration:

$$\begin{aligned} \mathbf{x}^1 &= \begin{bmatrix} 1 \\ 1 \end{bmatrix} - \mathbf{H}^{-1} \begin{bmatrix} 6 - 2 + 1 \\ -2 + 2 \end{bmatrix} \\ &= \begin{bmatrix} 1 \\ 1 \end{bmatrix} - \begin{bmatrix} \frac{1}{4} & \frac{1}{4} \\ \frac{1}{4} & \frac{3}{4} \end{bmatrix} \begin{bmatrix} 5 \\ 0 \end{bmatrix} = \begin{bmatrix} 1 - \frac{5}{4} \\ 1 - \frac{5}{4} \end{bmatrix} = \begin{bmatrix} -\frac{1}{4} \\ -\frac{1}{4} \end{bmatrix}. \end{aligned}$$

With $\nabla f(\mathbf{x}^1) = \mathbf{0}$ and $\mathbf{H}$ positive-definite the necessary and sufficient conditions (1.24) are satisfied at $\mathbf{x}^1$ and therefore the global minimum is given by

$$\mathbf{x}^* = \begin{bmatrix} -\frac{1}{4}, & -\frac{1}{4} \end{bmatrix}^T \text{ and } f(\mathbf{x}^*) = -0.125.$$

4.2 Line search descent methods

Problem 4.2.1

Minimize $F(\lambda) = -\lambda \cos \lambda$ over the interval $\left[0, \frac{\pi}{2}\right]$ by means of the golden section method.

Solution

The golden ratio is $r = 0.618$ and $F(0) = F(\frac{\pi}{2}) = 0$.

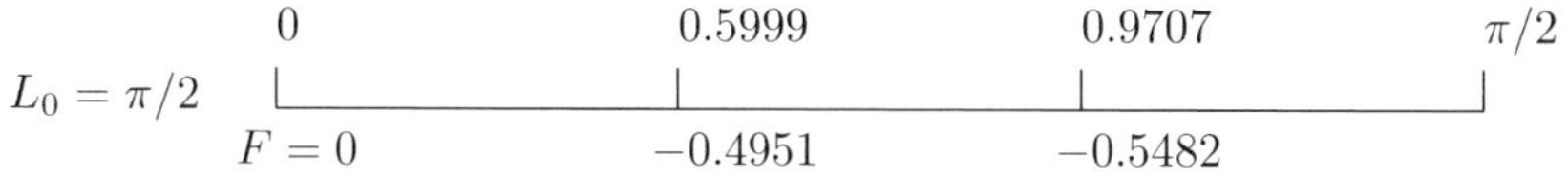

Figure 4.2

The first two interior points are: $rL_0 = 0.9707$ and $r^2 L_0 = 0.5999$ with function values: $F(0.9707) = -0.5482$ and $F(0.5999) = -0.4951$.

Next new point $= 0.5999 + rL_1 = 1.200$ with $F(1.200) = -0.4350$.

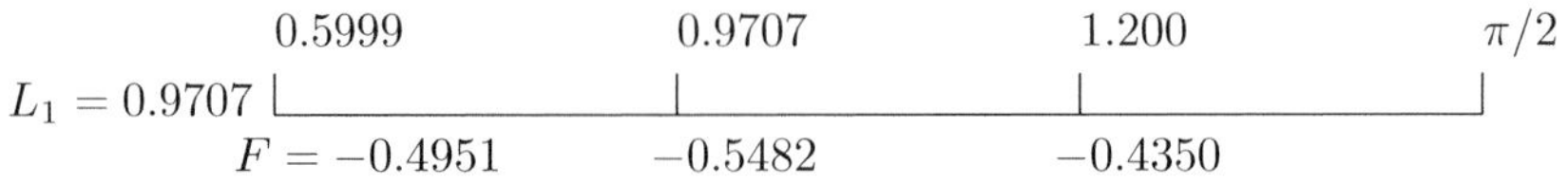

$$
\begin{array}{cccc}
& 0.5999 & 0.9707 & 1.200 & \pi/2 \\
L_1 = 0.9707 & \rule{0pt}{0pt} & & & \\
& F = -0.4951 & -0.5482 & -0.4350 &
\end{array}
$$

Figure 4.3

Next new point $= 0.5999 + r^2 L_2 = 0.8292$ with $F(0.8292) = -0.5601$.

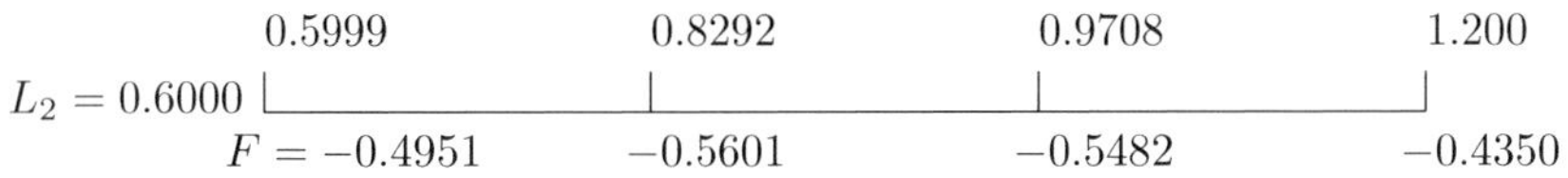

$$
\begin{array}{cccc}
& 0.5999 & 0.8292 & 0.9708 & 1.200 \\
L_2 = 0.6000 & & & & \\
& F = -0.4951 & -0.5601 & -0.5482 & -0.4350
\end{array}
$$

Figure 4.4

Next new point $= 0.5999 + r^2 L_3 = 0.7416$ with $F(0.7416) = -0.5468$.

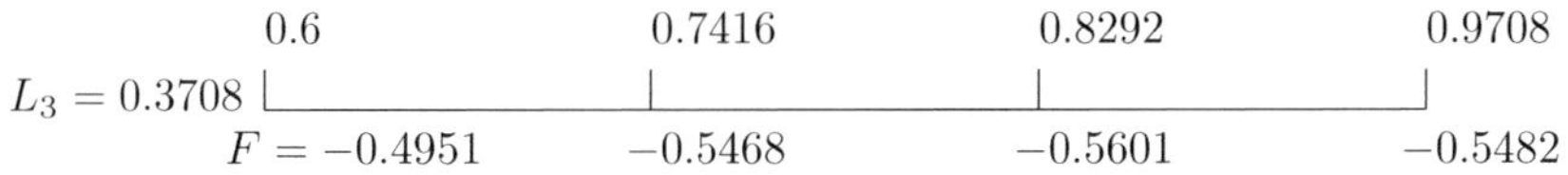

$$
\begin{array}{cccc}
& 0.6 & 0.7416 & 0.8292 & 0.9708 \\
L_3 = 0.3708 & & & & \\
& F = -0.4951 & -0.5468 & -0.5601 & -0.5482
\end{array}
$$

Figure 4.5

Next new point $= 0.7416 + rL_4 = 0.8832$ with $F(0.8832) = -0.5606$.

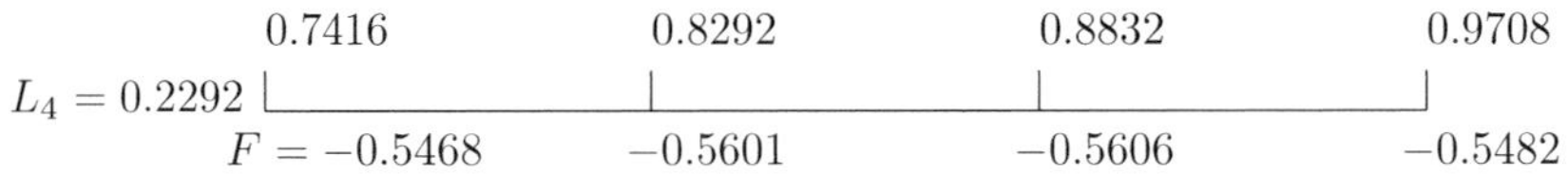

$$
\begin{array}{cccc}
& 0.7416 & 0.8292 & 0.8832 & 0.9708 \\
L_4 = 0.2292 & & & & \\
& F = -0.5468 & -0.5601 & -0.5606 & -0.5482
\end{array}
$$

Figure 4.6

The uncertainty interval is now $[0.8282; 0.9708]$. Stopping here gives $\lambda^* \cong 0.9$ (midpoint of interval) with $F(\lambda^*) \cong -0.5594$ which is taken as the approximate minimum after only 7 function evaluations (indeed the actual $\lambda^* = 0.8603$ gives $F(\lambda^*) = -0.5611$).

Problem 4.2.2

Minimize the function
$F(\lambda) = (\lambda - 1)(\lambda + 1)^2$, where λ is a single real variable, by means of Powell's quadratic interpolation method. Choose $\lambda_0 = 0$ and use $h = 0.1$. Perform *only two iterations*.

Solution

Set up difference table:

λ	$F(\lambda)$	$F[\ ,\ ]$	$F[\ ,\ ,\ ]$
0	-1		
		-0.890	
0.1	-1.089		1.300
		-0.630	
0.2	-1.152		1.694
		-0.135	
$\lambda_m^{(1)} = 0.392$	-1.178		
$\lambda_m^{(2)} = 0.336$			

Turning point λ_m given by $\lambda_m = \dfrac{F[\ ,\ ,\ ](\lambda_0 + \lambda_1) - F[\ ,\ ]}{2F[\ ,\ ,\ ]}$:

First iteration:

$$\lambda_m^{(1)} = \frac{1.3(0.1) + 0.89}{2(1.3)} = 0.392$$

Second iteration:

$$\lambda_m^{(2)} = \frac{1.694(0.3) + 0.63}{2(1.694)} = 0.336.$$

Problem 4.2.3

Apply two steps of the *steepest descent method* to the minimization of
$f(\mathbf{x}) = x_1 - x_2 + 2x_1^2 + 2x_1x_2 + x_2^2$.
Use as starting point $\mathbf{x}^0 = [0, 0]^T$.

Solution

$$\nabla f(\mathbf{x}) = \begin{bmatrix} 1 + 4x_1 + 2x_2 \\ -1 + 2x_1 + 2x_2 \end{bmatrix}.$$

Step 1

The first steepest descent direction $\mathbf{u}^1 = -\nabla f(\mathbf{x}^0) = \begin{bmatrix} -1 \\ 1 \end{bmatrix}$.

Here it is convenient *not* to normalize.

The minimizer along the direction $\mathbf{u}^1$ is given by

$$\mathbf{x}^1 = \mathbf{x}^0 + \lambda_1 \mathbf{u}^1 = \begin{bmatrix} -\lambda_1 \\ \lambda_1 \end{bmatrix}.$$

To find λ_1, minimize with respect to λ the one variable function:

$$F(\lambda) = f(\mathbf{x}^0 + \lambda \mathbf{u}^1) = -\lambda - \lambda + 2\lambda^2 - 2\lambda^2 + \lambda^2.$$

The necessary condition for a minimum is $\frac{dF}{d\lambda} = 2\lambda - 2 = 0$ giving $\lambda_1 = 1$ and with $\frac{d^2 F(\lambda_1)}{d\lambda^2} = 2 > 0$, λ_1 indeed corresponds to the minimum of $F(\lambda)$, and $\mathbf{x}^1 = \begin{bmatrix} -1 \\ 1 \end{bmatrix}$.

Step 2

The next steepest descent direction is

$$\mathbf{u}^2 = -\boldsymbol{\nabla} f(\mathbf{x}^1) = \begin{bmatrix} 1 \\ 1 \end{bmatrix}$$

and

$$\mathbf{x}^2 = \mathbf{x}^1 + \lambda_2 \mathbf{u}^2 = \begin{bmatrix} -1 + \lambda_2 \\ 1 + \lambda_2 \end{bmatrix}.$$

To minimize in the direction of $\mathbf{u}^2$ consider

$$F(\lambda) = f(\mathbf{x}^1 + \lambda \mathbf{u}^2) = (-1+\lambda) - (1+\lambda) + 2(\lambda-1)^2 + 2(\lambda-1)(\lambda+1) + (\lambda+1)^2$$

and apply the necessary condition

$$\frac{dF}{d\lambda} = 10\lambda - 2 = 0.$$

This gives $\lambda_2 = \frac{1}{5}$ and with $\frac{d^2 F(\lambda_2)}{d^2\lambda} = 10 > 0$ (minimum) it follows that

$$\mathbf{x}^2 = \begin{bmatrix} -1 + \frac{1}{5} \\ 1 + \frac{1}{5} \end{bmatrix} = \begin{bmatrix} -0.8 \\ 1.2 \end{bmatrix}.$$

Problem 4.2.4

Apply two steps of the *steepest descent method* to the minimization of $f(\mathbf{x}) = (2x_1 - x_2)^2 + (x_2 + 1)^2$ with $\mathbf{x}^0 = \left[\frac{5}{2}, 2\right]^T$.

Solution
Step 1
$$\nabla f(\mathbf{x}) = \begin{bmatrix} 4(2x_1 - x_2) \\ 4x_2 - 4x_1 + 2 \end{bmatrix},$$

giving $\nabla f(\mathbf{x}^0) = \begin{bmatrix} 4(5-2) \\ 8 - 4(\frac{5}{2}) + 2 \end{bmatrix} = \begin{bmatrix} 12 \\ 0 \end{bmatrix}.$

After normalizing, the first search direction is $\mathbf{u}^1 = \begin{bmatrix} -1 \\ 0 \end{bmatrix}$ and $\mathbf{x}^1 = \begin{bmatrix} \frac{5}{2} - \lambda_1 \\ 2 \end{bmatrix}$. Now minimize with respect to λ:

$$F(\lambda) = f(\mathbf{x}^0 + \lambda \mathbf{u}^1) = \left(2\left(\tfrac{5}{2} - \lambda\right) - 2\right)^2 + (2+1)^2 = (3 - 2\lambda)^2 + 9.$$

The necessary condition $\frac{dF}{d\lambda} = 0$ gives $\lambda_1 = \frac{3}{2}$, and with $\frac{d^2F}{d\lambda^2}(\lambda_1) > 0$ ensuring a minimum, it follows that

$$\mathbf{x}^1 = \begin{bmatrix} \frac{5}{2} - \frac{3}{2} \\ 2 \end{bmatrix} \text{ and } \nabla f(\mathbf{x}^1) = \begin{bmatrix} 4(2-2) \\ 8 - 4 + 2 \end{bmatrix} = \begin{bmatrix} 0 \\ 6 \end{bmatrix}.$$

Step 2
New normalized steepest descent direction: $\mathbf{u}^2 = \begin{bmatrix} 0 \\ -1 \end{bmatrix}$, thus $\mathbf{x}^2 = \begin{bmatrix} 1 \\ 2 - \lambda_2 \end{bmatrix}.$

In the direction of $\mathbf{u}^2$:

$$F(\lambda) = f(\mathbf{x}^1 + \lambda \mathbf{u}^2) = \lambda^2 + (3 - \lambda)^2.$$

Setting $\frac{dF}{d\lambda} = 2\lambda - 2(3 - \lambda) = 0$ gives $\lambda_2 = \frac{3}{2}$ and since $\frac{d^2F}{d\lambda^2}(\lambda_2) > 0$ a minimum is ensured, and therefore

$$\mathbf{x}^2 = \begin{bmatrix} 1 \\ 2 - \frac{3}{2} \end{bmatrix} = \begin{bmatrix} 1 \\ \frac{1}{2} \end{bmatrix}.$$

Problem 4.2.5

Apply the *Fletcher-Reeves method* to the minimization of

$$f(\mathbf{x}) = (2x_1 - x_2)^2 + (x_2 + 1)^2 \text{ with } \mathbf{x}^0 = \left[\tfrac{5}{2}, 2\right]^T.$$

Solution

The first step is identical to that of the steepest descent method given for Problem 4.2.4,

$$\nabla f(\mathbf{x}) = \begin{bmatrix} 4(2x_1 - x_2) \\ 4x_2 - 4x_1 + 2 \end{bmatrix},$$

giving $\mathbf{u}^1 = -\nabla f(\mathbf{x}^0) = [-12, 0]^T$.

For the *second step* the *Fletcher-Reeves* search direction becomes

$$\mathbf{u}^2 = -\nabla f(\mathbf{x}^1) + \frac{\|\nabla f(\mathbf{x}^1)\|^2}{\|\nabla f(\mathbf{x}^0)\|^2} \mathbf{u}^1.$$

Using the data from the first step in Problem 4.2.4, the second search direction becomes

$$\mathbf{u}^2 = \begin{bmatrix} 0 \\ -6 \end{bmatrix} + \frac{36}{144} \begin{bmatrix} -12 \\ 0 \end{bmatrix} = \begin{bmatrix} -3 \\ -6 \end{bmatrix}$$

and

$$\mathbf{x}^2 = \mathbf{x}^1 + \lambda_2 \mathbf{u}^2 = \begin{bmatrix} 1 \\ 2 \end{bmatrix} - \begin{bmatrix} 3\lambda_2 \\ 6\lambda_2 \end{bmatrix} = \begin{bmatrix} 1 - 3\lambda_2 \\ 2 - 6\lambda_2 \end{bmatrix}.$$

In the direction $\mathbf{u}^2$:

$$F(\lambda) = (2(1 - 3\lambda) - 2 + 6\lambda)^2 + (2 - 6\lambda + 1)^2 = (3 - 6\lambda)^2$$

and the necessary condition for a minimum is

$$\frac{dF}{d\lambda} = -12(3 - 6\lambda) = 0 \text{ giving } \lambda_2 = \tfrac{1}{2}$$

and thus with $\frac{d^2 F(\lambda_2)}{d\lambda^2} = 36 > 0$:

$$\mathbf{x}^2 = \begin{bmatrix} 1 - \frac{3}{2} \\ 2 - \frac{6}{2} \end{bmatrix} = \begin{bmatrix} -\frac{1}{2} \\ -1 \end{bmatrix}$$

with

$$\nabla f(\mathbf{x}^2) = \begin{bmatrix} 4\left(2\left(-\frac{1}{2}\right) - 1(-1)\right) \\ 4(-1) - 4\left(-\frac{1}{2}\right) + 2 \end{bmatrix} = \begin{bmatrix} 0 \\ 0 \end{bmatrix}.$$

Since $\nabla f(\mathbf{x}^2) = \mathbf{0}$ and $\mathbf{H} = \begin{bmatrix} 8 & -4 \\ -4 & 4 \end{bmatrix}$ is positive-definite, $\mathbf{x}^2 = \mathbf{x}^*$.

Problem 4.2.6

Minimize $F(\mathbf{x}) = x_1 - x_2 + 2x_1^2 + 2x_1 x_2 + x_2^2$ by using the *Fletcher-Reeves* method.
Use as starting point $\mathbf{x}^0 = [0, 0]^T$.

Solution

Step 1

$$\nabla f(\mathbf{x}) = \begin{bmatrix} 1 + 4x_1 + 2x_2 \\ -1 + 2x_1 + 2x_2 \end{bmatrix}, \quad \mathbf{x}^0 = [0, 0]^T,$$

$$\nabla f(\mathbf{x}^0) = \begin{bmatrix} 1 \\ -1 \end{bmatrix}, \quad \mathbf{u}^1 = -\nabla f(\mathbf{x}^0) = \begin{bmatrix} -1 \\ 1 \end{bmatrix} \text{ and } \mathbf{x}^1 = \mathbf{x}^0 + \lambda_1 \mathbf{u}^1 = \begin{bmatrix} -\lambda_1 \\ \lambda_1 \end{bmatrix}.$$

Therefore

$$F(\lambda) = f(\mathbf{x}^0 + \lambda \mathbf{u}^1) = -\lambda - \lambda + 2\lambda^2 - 2\lambda^2 + \lambda^2 = \lambda^2 - 2\lambda$$

and $\frac{dF}{d\lambda} = 2\lambda - 2 = 0$ giving $\lambda_1 = 1$ and with $\frac{d^2 F(\lambda_1)}{d\lambda^2} = 2 > 0$ ensuring a minimum, it follows that $\mathbf{x}^1 = [-1, 1]^T$.

Step 2

$$\nabla f(\mathbf{x}^1) = \begin{bmatrix} -1 \\ -1 \end{bmatrix}$$

$$\mathbf{u}^2 = -\nabla f(\mathbf{x}^1) + \frac{\|\nabla f(\mathbf{x}^1)\|^2}{\|\nabla f(\mathbf{x}^0)\|^2} \mathbf{u}^1 = \begin{bmatrix} 1 \\ 1 \end{bmatrix} + \frac{2}{2} \begin{bmatrix} -1 \\ 1 \end{bmatrix} = \begin{bmatrix} 0 \\ 2 \end{bmatrix}$$

$$\mathbf{x}^2 = \begin{bmatrix} -1 \\ 1 \end{bmatrix} + \lambda_2 \begin{bmatrix} 0 \\ 2 \end{bmatrix} = [-1, 1 + 2\lambda_2]^T.$$

Thus

$$F(\lambda) = -1 - (1 + 2\lambda) + 2(-1)^2 + 2(-1)(1 + 2\lambda) + (1 + 2\lambda)^2 = 4\lambda^2 - 2\lambda - 1$$

with $\frac{dF}{d\lambda} = 8\lambda - 2 = 0$ giving $\lambda_2 = \frac{1}{4}$ and therefore, with $\frac{d^2 F}{d\lambda^2} = 8 > 0$:

$$\mathbf{x}^2 = \begin{bmatrix} -1 \\ 1 \end{bmatrix} + \frac{1}{4} \begin{bmatrix} 0 \\ 2 \end{bmatrix} = [-1, 1.5]^T.$$

This results in

$$\nabla f(\mathbf{x}^2) = \begin{bmatrix} 1 - 4 + 2(1.5) \\ -1 - 2 + 2(1.5) \end{bmatrix} = [0, 0]^T$$

and since $\mathbf{H}$ is positive-definite, the optimum solution is $\mathbf{x}^* = \mathbf{x}^2 = [-1, 1.5]^T$.

Problem 4.2.7

Minimize $f(\mathbf{x}) = x_1^2 - x_1 x_2 + 3x_2^2$ with $\mathbf{x}^0 = (1, 2)^T$ by means of the *Fletcher-Reeves* method.

Solution

$$\nabla f(\mathbf{x}) = \begin{bmatrix} 2x_1 - x_2 \\ 6x_2 - x_1 \end{bmatrix}.$$

Step 1

Since $\nabla f(\mathbf{x}^0) = \begin{bmatrix} 2(1) - 2 \\ 6(2) - 1 \end{bmatrix} = \begin{bmatrix} 0 \\ 11 \end{bmatrix}$, $\mathbf{u}^1 = -\nabla f(\mathbf{x}^0) = [0, -11]^T$
and

$$\mathbf{x}^1 = [1, 2]^T + \lambda_1[0, -11]^T.$$

This results in

$$F(\lambda) = 1 - 1(2 - 11\lambda) + 3(2 - 11\lambda)^2$$

with

$$\frac{dF}{d\lambda} = 11 + 6(2 - 11\lambda)(-11) = 0 \text{ giving } \lambda_1 = \tfrac{1}{6}.$$

Thus, with $\frac{d^2 F(\lambda_1)}{d\lambda^2} > 0$: $\mathbf{x}^1 = [1, 2]^T + \tfrac{1}{6}[0, -11]^T = \left[1, \tfrac{1}{6}\right]^T$.

Step 2

$\nabla f(\mathbf{x}^1) = \begin{bmatrix} \tfrac{11}{6} \\ 0 \end{bmatrix}$ and thus

$$\mathbf{u}^2 = -\nabla f(\mathbf{x}^1) + \frac{\|\nabla f(\mathbf{x}^1)\|^2}{\|\nabla f(\mathbf{x}^0)\|^2} \mathbf{u}^1 = \begin{bmatrix} -\tfrac{11}{6} \\ 0 \end{bmatrix} + \tfrac{1}{36} \begin{bmatrix} 0 \\ -11 \end{bmatrix} = \begin{bmatrix} -\tfrac{11}{6} \\ -\tfrac{11}{36} \end{bmatrix}$$

giving

$$\mathbf{x}^2 = \mathbf{x}^1 + \lambda_2 \mathbf{u}^2 = \begin{bmatrix} 1 \\ \tfrac{1}{6} \end{bmatrix} - \lambda_2 \begin{bmatrix} \tfrac{11}{6} \\ \tfrac{11}{36} \end{bmatrix} = \begin{bmatrix} \left(1 - \lambda_2 \tfrac{11}{6}\right) \\ \tfrac{1}{6}\left(1 - \lambda_2 \tfrac{11}{6}\right) \end{bmatrix}$$

Thus

$$F(\lambda) = \left(1 - \tfrac{11}{6}\lambda\right)^2 - \tfrac{1}{6}\left(1 - \tfrac{11}{6}\lambda\right)^2 + \tfrac{3}{36}\left(1 - \tfrac{11}{6}\lambda\right)^2$$

$$= \left(1 - \tfrac{1}{6} + \tfrac{1}{12}\right)\left(1 - \tfrac{11}{6}\lambda\right)^2$$

and $\frac{dF}{d\lambda} = \tfrac{11}{6}\left(1 - \tfrac{11}{6}\lambda\right)\left(-\tfrac{11}{6}\right) = 0$ gives $\lambda_2 = \tfrac{6}{11}$, and with $\frac{d^2 F(\lambda_2)}{d\lambda^2} > 0$ gives $\mathbf{x}^2 = [0, 0]^T$. With $\nabla f(\mathbf{x}^2) = \mathbf{0}$, and since $\mathbf{H}$ is positive-definite for all $\mathbf{x}$, this is the optimal solution.

Problem 4.2.8

Obtain the first updated matrix $\mathbf{G}_1$ when applying the DFP method to the minimization of $f(\mathbf{x}) = 4x_1^2 - 40x_1 + x_2^2 - 12x_2 + 136$ with starting point $\mathbf{x}^0 = [8, 9]^T$.

Solution

Factorizing $f(\mathbf{x})$ gives $f(\mathbf{x}) = 4(x_1 - 5)^2 + (x_2 - 6)^2$ and $\nabla f(\mathbf{x}) = \begin{bmatrix} 8x_1 - 40 \\ 2x_2 - 12 \end{bmatrix}$.

Step 1

Choose $\mathbf{G}_0 = \mathbf{I}$, then for $\mathbf{x}^0 = [8, 9]^T$, $\nabla f(\mathbf{x}^0) = [24, 6]^T$.

$$\mathbf{u}^1 = -\mathbf{I}\nabla f(\mathbf{x}^0) = -\begin{bmatrix} 24 \\ 6 \end{bmatrix}$$

and $\mathbf{x}^1 = \mathbf{x}^0 + \lambda_2\mathbf{u}^1 = \begin{bmatrix} 8 - 24\lambda_2 \\ 9 - 6\lambda_2 \end{bmatrix}$.

The function to be minimized with respect to λ is

$$F(\lambda) = f(\mathbf{x}^0 + \lambda\mathbf{u}^1) = 4(8 - 24\lambda - 5)^2 + (9 - 6\lambda - 6)^2.$$

The necessary condition for a minimum, $\frac{dF}{d\lambda} = -8(24)(3-24\lambda)+2(-6)(3-6\lambda) = 0$ yields $\lambda_1 = 0.1308$ and thus with $\frac{d^2 F(\lambda_1)}{d\lambda^2} > 0$:

$$\mathbf{x}^1 = \begin{bmatrix} 8 - 0.1308(24) \\ 9 - 0.1308(6) \end{bmatrix} = \begin{bmatrix} 4.862 \\ 8.215 \end{bmatrix} \text{ with } \nabla f(\mathbf{x}^1) = \begin{bmatrix} -1.10 \\ 4.43 \end{bmatrix}.$$

The DFP update now requires the following quantities:

$$\mathbf{v}^1 = \lambda_1 \mathbf{u}^1 = \mathbf{x}^1 - \mathbf{x}^0 = \begin{bmatrix} 4.862 - 8 \\ 8.215 - 9 \end{bmatrix} = \begin{bmatrix} -3.14 \\ -0.785 \end{bmatrix}$$

and $\mathbf{y}^1 = \nabla f(\mathbf{x}^1) - \nabla f(\mathbf{x}^0) = \begin{bmatrix} -25.10 \\ -1.57 \end{bmatrix}$

giving $\mathbf{v}^{1T}\mathbf{y}^1 = [-3.14, -0.785] \begin{bmatrix} -25.10 \\ -1.57 \end{bmatrix} = 80.05$

and $\mathbf{y}^{1T}\mathbf{y}^1 = [(25.10)^2 + (-1.57)^2] = 632.47$

to be substituted in the update formula (2.18):

$$
\begin{aligned}
\mathbf{G}_1 &= \mathbf{G}_0 + \frac{\mathbf{v}^1\mathbf{v}^{1T}}{\mathbf{v}^{1T}\mathbf{y}^1} - \frac{\mathbf{y}^1\mathbf{y}^{1T}}{\mathbf{y}^{1T}\mathbf{y}^1} \\
&= \begin{bmatrix} 1 & 0 \\ 0 & 1 \end{bmatrix} + \frac{1}{80.05}\begin{bmatrix} -3.14 \\ -0.785 \end{bmatrix}[-3.14; -0.785] \\
&\quad - \frac{1}{632.47}\begin{bmatrix} -25.10 \\ -1.57 \end{bmatrix}[-25.10; -1.57] \\
&= \begin{bmatrix} 1 & 0 \\ 0 & 1 \end{bmatrix} + \frac{1}{80.05}\begin{bmatrix} 9.860 & 2.465 \\ 2.465 & 0.6161 \end{bmatrix} - \frac{1}{632.47}\begin{bmatrix} 630.01 & 3941 \\ 39.41 & 2.465 \end{bmatrix} \\
&= \begin{bmatrix} 0.127 & -0.032 \\ -0.032 & 1.004 \end{bmatrix}.
\end{aligned}
$$

Problem 4.2.9

Determine the first updated matrix $\mathbf{G}_1$ when applying the DFP method to the minimization of $f(\mathbf{x}) = 3x_1^2 - 2x_1x_2 + x_2^2 + x_1$ with $\mathbf{x}^0 = [1, 1]^T$.

Solution

$$\nabla f(\mathbf{x}) = \begin{bmatrix} 6x_1 - 2x_2 + 1 \\ -2x_1 + 2x_2 \end{bmatrix}.$$

Step 1

$\nabla f(\mathbf{x}^0) = [5, 0]^T$ and $\mathbf{G}_0 = \mathbf{I}$ which results in

$$\mathbf{x}^1 = \mathbf{x}^0 + \lambda_1 \mathbf{u}^1 = \mathbf{x}^0 - \lambda_1 \mathbf{G}_0 \nabla f(\mathbf{x}^0) = \begin{bmatrix} 1 \\ 1 \end{bmatrix} - \lambda_1 \begin{bmatrix} 1 & 0 \\ 0 & 1 \end{bmatrix} \begin{bmatrix} 5 \\ 0 \end{bmatrix} = \begin{bmatrix} 1 - 5\lambda_1 \\ 1 \end{bmatrix}.$$

For the function $F(\lambda) = 3(1-5\lambda)^2 - 2(1-5\lambda) + 1 + (1-5\lambda)$ the necessary condition for a minimum is $\frac{\partial F}{\partial \lambda} = 6(1-5\lambda)(-5) + 10 - 5 = 0$, which gives $\lambda_1 = \frac{25}{150} = \frac{1}{6}$, and since $\frac{d^2 F(\lambda_1)}{d\lambda^2} > 0$:

$$\mathbf{x}^1 = \begin{bmatrix} 1 - \frac{5}{6} \\ 1 \end{bmatrix} = \begin{bmatrix} \frac{1}{6} \\ 1 \end{bmatrix}$$

and $\mathbf{v}^1 = \mathbf{x}^1 - \mathbf{x}^0 = \begin{bmatrix} \frac{1}{6} \\ 1 \end{bmatrix} - \begin{bmatrix} 1 \\ 1 \end{bmatrix} = \begin{bmatrix} -\frac{5}{6} \\ 0 \end{bmatrix}$,

with $\mathbf{y}^1 = \nabla f(\mathbf{x}^1) - \nabla f(\mathbf{x}^0) = \begin{bmatrix} 6\left(\frac{1}{6}\right) - 2 + 1 \\ -2\left(\frac{1}{6}\right) + 2 \end{bmatrix} - \begin{bmatrix} 5 \\ 0 \end{bmatrix} = \begin{bmatrix} -5 \\ \frac{5}{3} \end{bmatrix}$.

It follows that

$$\mathbf{v}^1 \mathbf{v}^{1T} = \begin{bmatrix} -\frac{5}{6} \\ 0 \end{bmatrix} \begin{bmatrix} -\frac{5}{6}, 0 \end{bmatrix} = \begin{bmatrix} \frac{25}{36} & 0 \\ 0 & 0 \end{bmatrix}, \quad \mathbf{v}^{1T}\mathbf{y}^1 = \begin{bmatrix} -\frac{5}{6}, 0 \end{bmatrix} \begin{bmatrix} -5 \\ \frac{5}{3} \end{bmatrix} = \frac{25}{6},$$

$$\mathbf{y}^1 \mathbf{y}^{1T} = \begin{bmatrix} -5 \\ \frac{5}{3} \end{bmatrix} \begin{bmatrix} -5, \frac{5}{3} \end{bmatrix} = \begin{bmatrix} 25 & -\frac{25}{3} \\ -\frac{25}{3} & \frac{25}{9} \end{bmatrix}$$

and $\mathbf{y}^{1T}\mathbf{y}^1 = \begin{bmatrix} -5, \frac{5}{3} \end{bmatrix} \begin{bmatrix} -5 \\ \frac{5}{3} \end{bmatrix} = 25 + \frac{25}{9} = \frac{250}{9}$.

In the above computations $\mathbf{G}_0$ has been taken as $\mathbf{G}_0 = \mathbf{I}$.

Substituting the above results in the update formula (2.18) yields $\mathbf{G}_1$ as follows:

$$\begin{aligned}
\mathbf{G}_1 &= \mathbf{G}_0 + \mathbf{A}_1 + \mathbf{B}_1 \\
&= \begin{bmatrix} 1 & 0 \\ 0 & 1 \end{bmatrix} + \frac{6}{25}\begin{bmatrix} \frac{25}{36} & 0 \\ 0 & 1 \end{bmatrix} - \frac{9}{250}\begin{bmatrix} 25 & -\frac{25}{3} \\ -\frac{25}{3} & \frac{25}{9} \end{bmatrix} \\
&= \begin{bmatrix} 1 & 0 \\ 0 & 1 \end{bmatrix} + \begin{bmatrix} \frac{1}{6} & 0 \\ 0 & 1 \end{bmatrix} - \frac{9}{10}\begin{bmatrix} 1 & -\frac{1}{3} \\ -\frac{1}{3} & \frac{1}{9} \end{bmatrix} \\
&= \begin{bmatrix} \frac{7}{9} - \frac{9}{10} & \frac{3}{10} \\ \frac{3}{10} & \frac{9}{10} \end{bmatrix} = \begin{bmatrix} \frac{4}{15} & \frac{3}{10} \\ \frac{3}{10} & \frac{9}{10} \end{bmatrix}.
\end{aligned}$$

Comparing $\mathbf{G}_1$ with $\mathbf{H}^{-1} = \begin{bmatrix} \frac{1}{4} & \frac{1}{4} \\ \frac{1}{4} & \frac{1}{4} \end{bmatrix}$ shows that after only one iteration a reasonable approximation to the inverse has already been obtained.

Problem 4.2.10

Apply the DFP-method to the minimization $f(\mathbf{x}) = x_1 - x_2 + 2x_1^2 + 2x_1 x_2 + x_2^2$ with starting point $\mathbf{x}^0 = [0, 0]^T$.

Solution

Step 1

$\mathbf{G}_0 = \mathbf{I} = \begin{bmatrix} 1 & 0 \\ 0 & 1 \end{bmatrix}$ and $\nabla f(\mathbf{x}^0) = \begin{bmatrix} 1 + 4x_1 + 2x_2 \\ -1 + 2x_1 + 2x_2 \end{bmatrix} = \begin{bmatrix} 1 \\ -1 \end{bmatrix}$ gives

$\mathbf{u}^1 = -\mathbf{G}_0 \nabla f(\mathbf{x}^0) = -\mathbf{I}\begin{bmatrix} 1 \\ -1 \end{bmatrix} = \begin{bmatrix} -1 \\ 1 \end{bmatrix}$ and thus

$\mathbf{x}^1 = \mathbf{x}^0 + \lambda_1 \mathbf{u}^1 = \begin{bmatrix} -\lambda_1 \\ \lambda_1 \end{bmatrix}.$

Thus $F(\lambda) = \lambda^2 - 2\lambda$, and $\frac{dF}{d\lambda} = 0$ yields $\lambda_1 = 1$, and with $\frac{d^2 F(\lambda_1)}{d\lambda^2} > 0$:

$\mathbf{x}^1 = \begin{bmatrix} -1 \\ 1 \end{bmatrix}$ and $\nabla f(\mathbf{x}^1) = \begin{bmatrix} -1 \\ -1 \end{bmatrix}.$

Now, using $\mathbf{v}^1 = \begin{bmatrix} -1 \\ 1 \end{bmatrix}$ and $\mathbf{y}^1 = \nabla f(\mathbf{x}^1) - \nabla f(\mathbf{x}^0) = \begin{bmatrix} -2 \\ 0 \end{bmatrix}$

in the update formula (2.18), gives

$$\mathbf{A}_1 = \frac{\mathbf{v}^1 \mathbf{v}^{1T}}{\mathbf{v}^{1T} \mathbf{y}^1} = \frac{\begin{bmatrix} -1 \\ 1 \end{bmatrix} [-1\ 1]}{2} = \frac{1}{2} \begin{bmatrix} 1 & -1 \\ -1 & 1 \end{bmatrix}, \text{ and since } \mathbf{G}_0 \mathbf{y}^1 = \mathbf{y}^1:$$

$$\mathbf{B}_1 = -\frac{\mathbf{y}^1 \mathbf{y}^{1T}}{\mathbf{y}^{1T} \mathbf{y}^1} = -\frac{1}{4} \begin{bmatrix} 4 & 0 \\ 0 & 0 \end{bmatrix}.$$

Substitute the above in (2.18):

$$\mathbf{G}_1 = \mathbf{G}_0 + \mathbf{A}_1 + \mathbf{B}_1 = \begin{bmatrix} 1 & 0 \\ 0 & 1 \end{bmatrix} + \frac{1}{2} \begin{bmatrix} 1 & -1 \\ -1 & 1 \end{bmatrix} - \frac{1}{4} \begin{bmatrix} 4 & 0 \\ 0 & 0 \end{bmatrix}$$

$$= \begin{bmatrix} \frac{1}{2} & -\frac{1}{2} \\ -\frac{1}{2} & \frac{3}{2} \end{bmatrix}.$$

Step 2

New search direction $\mathbf{u}^2 = -\mathbf{G}_1 \nabla f(\mathbf{x}^1) = -\begin{bmatrix} \frac{1}{2} & -\frac{1}{2} \\ -\frac{1}{2} & \frac{3}{2} \end{bmatrix} \begin{bmatrix} -1 \\ -1 \end{bmatrix} =$

$\begin{bmatrix} 0 \\ 1 \end{bmatrix}$ and therefore

$$\mathbf{x}^2 = \begin{bmatrix} -1 \\ 1 \end{bmatrix} + \lambda_2 \begin{bmatrix} 0 \\ 1 \end{bmatrix} = [-1, 1 + \lambda_2]^T.$$

Minimizing $F(\lambda) = -1 - (1 + \lambda) + 2 - 2(1 + \lambda) + (1 + \lambda)^2$ implies

$\frac{dF}{d\lambda} = -1 - 2 + 2(1 + \lambda) = 0$ which gives $\lambda_2 = \frac{1}{2}$.

Thus, since $\frac{d^2 F(\lambda_2)}{d\lambda^2} > 0$, the minimum is given by $\mathbf{x}^2 = \begin{bmatrix} -1 \\ 1 \end{bmatrix} +$

$\frac{1}{2} \begin{bmatrix} 0 \\ 1 \end{bmatrix} = \begin{bmatrix} -1 \\ \frac{3}{2} \end{bmatrix}$ with $\nabla f(\mathbf{x}^2) = \begin{bmatrix} 0 \\ 0 \end{bmatrix}$ and therefore $\mathbf{x}^2$ is optimal.

4.3   Standard methods for constrained optimization

4.3.1   Penalty function problems

Problem 4.3.1.1

An alternative interpretation to Problem 3.5.1.3 is to determine the shortest distance from the origin to the plane $x_1 + x_2 + x_3 = 1$ by means of the penalty function method.

Solution

Notice that the problem is equivalent to the problem:

minimize $f(\mathbf{x}) = x_1^2 + x_2^2 + x_3^2$ such that $x_1 + x_2 + x_3 = 1$.

The appropriate penalty function is $P = x_1^2 + x_2^2 + x_3^2 + \rho(x_1 + x_2 + x_3 - 1)^2$.

The necessary conditions at an unconstrained minimum of P are

$$\frac{\partial P}{\partial x_1} = 2x_1 + 2\rho(x_1 + x_2 + x_3 - 1) = 0$$

$$\frac{\partial P}{\partial x_2} = 2x_2 + 2\rho(x_1 + x_2 + x_3 - 1) = 0$$

$$\frac{\partial P}{\partial x_3} = 2x_3 + 2\rho(x_1 + x_2 + x_3 - 1) = 0.$$

Clearly $x_1 = x_2 = x_3$, and it follows that $x_1 = -\rho(3x_1 - 1)$, i.e.

$x_1(\rho) = \frac{\rho}{1+3\rho} = \frac{1}{3+\frac{1}{\rho}}$ and $\lim_{\rho \to \infty} x_1(\rho) = \frac{1}{3}$.

The shortest distance is therefore$= \sqrt{x_1^2 + x_2^2 + x_3^2} = \sqrt{3(\frac{1}{3})^2} = \frac{1}{\sqrt{3}}$.

Problem 4.3.1.2

Apply the *penalty function method* to the problem:

minimize $f(\mathbf{x}) = (x_1 - 1)^2 + (x_2 - 2)^2$

such that

$h(\mathbf{x}) = x_2 - x_1 - 1 = 0, \ g(\mathbf{x}) = x_1 + x_2 - 2 \leq 0, \ -x_1 \leq 0, \ -x_2 \leq 0.$

Solution

The appropriate penalty function is

$$P = (x_1 - 1)^2 + (x_2 - 2)^2 + \rho(x_2 - x_1 - 1)^2 + \beta_1(x_1 + x_2 - 2)^2 + \beta_2 x_1^2 + \beta_3 x_2^2$$

where $\rho \gg 0$ and $\beta_j = \rho$ if the corresponding inequality constraint is violated, otherwise $\beta_j = 0$.

Clearly the unconstrained minimum $[1, 2]^T$ violates the constraint $g(\mathbf{x}) \leq 0$, therefore assume that $\mathbf{x}$ is in the first quadrant but outside the feasible region, i.e. $\beta_1 = \rho$. The penalty function then becomes

$$P = (x_1 - 1)^2 + (x_2 - 2)^2 + \rho(x_2 - x_1 - 1)^2 + \rho(x_1 + x_2 - 2)^2.$$

The necessary conditions at the unconstrained minimum of P are:

$$\frac{\partial P}{\partial x_1} = 2(x_1 - 1) - 2\rho(x_2 - x_1 - 1) + 2\rho(x_2 + x_1 - 2) = 0$$

$$\frac{\partial P}{\partial x_2} = 2(x_2 - 1) + 2\rho(x_2 - x_1 - 1) + 2\rho(x_2 + x_1 - 2) = 0.$$

The first condition is $x_1(2 + 4\rho) - 2 - 2\rho = 0$, from which it follows that

$$x_1(\rho) = \frac{2\rho + 2}{4\rho + 2} = \frac{2 + \frac{2}{\rho}}{4 + \frac{2}{\rho}} \quad \text{and} \quad \lim_{\rho \to \infty} x_1(\rho) = \tfrac{1}{2}.$$

The second condition is $x_2(2 + 4\rho) - 4 - 6\rho = 0$, which gives

$$x_2(\rho) = \frac{6\rho + 4}{4\rho + 2} = \frac{6 + \frac{4}{\rho}}{4 + \frac{2}{\rho}} \quad \text{and} \quad \lim_{\rho \to \infty} x_2(\rho) = \tfrac{3}{2}.$$

The optimum is therefore $\mathbf{x}^* = [\tfrac{1}{2}, \tfrac{3}{2}]^T$.

Problem 4.3.1.3

Apply the *penalty function method* to the problem:

minimize $f(\mathbf{x}) = x_1^2 + 2x_2^2$ such that $g(\mathbf{x}) = 1 - x_1 - x_2 \leq 0$.

Solution

The penalty function is $P = x_1^2 + 2x_2^2 + \beta(1 - x_1 - x_2)^2$. Again the unconstrained minimum clearly violates the constraint, therefore assume the constraint is violated in the penalty function, i.e. $\beta = \rho$. The necessary conditions at an unconstrained minimum of P are

$$\frac{\partial P}{\partial x_1} = 2x_1 - 2\rho(1 - x_1 - x_2) = 0$$

$$\frac{\partial P}{\partial x_2} = 4x_2 - 2\rho(1 - x_1 - x_2) = 0.$$

These conditions give $x_1 = 2x_2$, and solving further yields

$$x_2(\rho) = \frac{2\rho}{4 + 6\rho} = \frac{2\rho}{\rho\left(\frac{4}{\rho} + 6\right)} \text{ and thus } \lim_{\rho\to\infty} x_2(\rho) = x_2^* = \tfrac{1}{3}, \text{ and also }$$

$x_1^* = \tfrac{2}{3}$.

Problem 4.3.1.4

Minimize $f(\mathbf{x}) = 2x_1^2 + x_2^2$ such that $g(\mathbf{x}) = 5 - x_1 + 3x_2 \leq 0$ by means of the *penalty function* approach.

Solution

The unconstrained solution, $x_1^* = x_2^* = 0$, violates the constraint, therefore it is active and P becomes $P = 2x_1^2 + x_2^2 + \rho(5 - x_1 + 3x_2)^2$ with the necessary conditions for an unconstrained minimum:

$$\frac{\partial P}{\partial x_1} = 4x_1 - 2\rho(5 - x_1 + 3x_2) = 0$$

$$\frac{\partial P}{\partial x_2} = 2x_2 + 6\rho(5 - x_1 + 3x_2) = 0.$$

It follows that $x_2 = -6x_1$ and substituting into the first condition yields

$$x_1(\rho) = \frac{10\rho}{4 + 38\rho} = \frac{10\rho}{\rho\left(\frac{4}{\rho} + 38\right)} \text{ and } \lim_{\rho\to\infty} x_1(\rho) = x_1^* = \tfrac{10}{38} = 0.2632. \text{ This }$$

gives $x_2^* = -1.5789$ with $f(\mathbf{x}^*) = 2.6316$.

4.3.2   The Lagrangian method applied to equality constrained problems

Problem 4.3.2.1

Determine the minima and maxima of $f(\mathbf{x}) = x_1 x_2$, such that $x_1^2 + x_2^2 = 1$, by means of the Lagrangian method.

Solution

Here $L = x_1 x_2 + \lambda(x_1^2 + x_2^2 - 1)$ and therefore the stationary conditions are

$$\frac{\partial L}{\partial x_1} = x_2 + 2x_1\lambda = 0, \qquad x_2 = -2x_1\lambda$$

$$\frac{\partial L}{\partial x_2} = x_1 + 2x_2\lambda = 0, \qquad x_1 = -2x_2\lambda.$$

From the equality it follows that $1 = x_1^2 + x_2^2 = 4x_1^2\lambda^2 + 4x_2^2\lambda^2 = 4\lambda^2$ giving $\lambda = \pm\frac{1}{2}$.

Choosing $\lambda = \frac{1}{2}$ gives $x_2 = -x_1$, $2x_1^2 = 1$, and $x_1 = \pm\frac{1}{\sqrt{2}}$.

This results in the possibilities:

$$x_1 = \frac{1}{\sqrt{2}}, \; x_2 = -\frac{1}{\sqrt{2}} \Rightarrow f^* = -\frac{1}{2}$$

or

$$x_1 = -\frac{1}{\sqrt{2}}, \; x_2 = \frac{1}{\sqrt{2}} \Rightarrow f^* = -\frac{1}{2}.$$

Alternatively choosing $\lambda = -\frac{1}{2}$ gives $x_2 = x_1$, $2x_1^2 = 1$, and $x_1 = \pm\frac{1}{\sqrt{2}}$ and the possibilities:

$$x_1 = \frac{1}{\sqrt{2}}, \; x_2 = \frac{1}{\sqrt{2}} \Rightarrow f^* = \frac{1}{2}$$

or

$$x_1 = -\frac{1}{\sqrt{2}}, \; x_2 = -\frac{1}{\sqrt{2}} \Rightarrow f^* = \frac{1}{2}.$$

These possibilities are sketched in Figure 4.7.

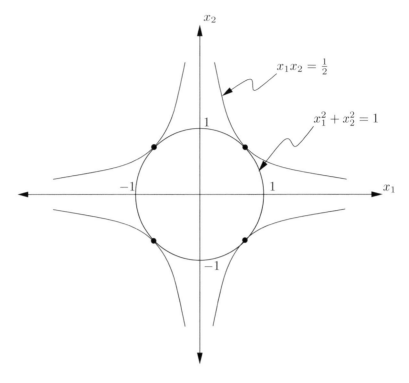

Figure 4.7

Problem 4.3.2.2

Determine the dimensions, radius r and height h, of the solid cylinder of minimum total surface area which can be cast from a solid metallic sphere of radius r_0.

Solution

This problem is equivalent to: minimize $f(r, h) = 2\pi rh + 2\pi r^2$ such that $\pi r^2 h = \frac{4}{3}\pi r_0^3$.

The Lagrangian is $L(r, h, \lambda) = -2\pi rh - 2\pi r^2 + \lambda(\pi r^2 h - \frac{4}{3}\pi r_0^3)$. The necessary conditions for stationary points are

$$\frac{\partial L}{\partial r} = -2\pi h - 4\pi r + \lambda 2\pi r h = 0$$

$$\frac{\partial L}{\partial h} = -2\pi r + \lambda \pi r^2 = 0$$

$$\frac{\partial L}{\partial \lambda} = \pi r^2 h - \frac{4}{3}\pi r_0^3 = 0.$$

The second condition gives $\lambda \pi r^2 = 2\pi r$, i.e. $\lambda = \frac{2}{r}$, and substituting in the first condition yields $r = \frac{h}{2}$. Substituting this value in the equality gives $\pi r^2 2r - \frac{4}{3}\pi r_0^3 = 0$, i.e.

$$r = r_0\left(\tfrac{2}{3}\right)^{\frac{1}{3}}, \ h = 2r_0\left(\tfrac{2}{3}\right)^{\frac{1}{3}} \text{ and } \lambda = \tfrac{2}{r}.$$

Problem 4.3.2.3

Minimize $f(\mathbf{x}) = -2x_1 - x_2 - 10$ such that $h(\mathbf{x}) = x_1 - 2x_2^2 - 3 = 0$. Show whether or not the candidate point, obtained via the Lagrangian method, is indeed a constrained *minimum*.

Solution

$$L(\mathbf{x}, \lambda) = -2x_1 - x_2 - 10 + \lambda(x_1 - 2x_2^2 - 3)$$

and the necessary stationary conditions are

$$\frac{\partial L}{\partial x_1} = -2 + \lambda = 0,$$

$$\frac{\partial L}{\partial x_2} = -1 - 4\lambda x_2 = 0$$

$$\frac{\partial L}{\partial \lambda} = x_1 - 2x_2^2 - 3 = 0.$$

Solving gives the candidate point $\lambda^* = 2$, $x_2^* = -\frac{1}{8}$, $x_1^* = 3.03$ with $f(\mathbf{x}^*) = -16.185$.

To prove that this point, $\mathbf{x}^* = [3.03, -\frac{1}{8}]^T$, indeed corresponds to a minimum requires the following argument. By Taylor, for any step $\boldsymbol{\Delta}\mathbf{x}$ compatible with the constraint, i.e. $h(\mathbf{x}) = 0$, it follows that

$$f(\mathbf{x}^* + \boldsymbol{\Delta}\mathbf{x}) = f(\mathbf{x}^*) + \boldsymbol{\nabla}^T f(\mathbf{x}^*)\boldsymbol{\Delta}\mathbf{x} \tag{4.3}$$

and $h(\mathbf{x}^* + \boldsymbol{\Delta}\mathbf{x}) = h(\mathbf{x}^*) + \boldsymbol{\nabla}^T h(\mathbf{x}^*)\boldsymbol{\Delta}\mathbf{x} + \frac{1}{2}\boldsymbol{\Delta}\mathbf{x}^T \boldsymbol{\nabla}^2 h(\mathbf{x}^*)\boldsymbol{\Delta}\mathbf{x} = 0.$

The latter equation gives

$$0 = [1, -4x_2]\begin{bmatrix} \Delta x_1 \\ \Delta x_2 \end{bmatrix} + \frac{1}{2}[\Delta x_1 \Delta x_2]\begin{bmatrix} 0 & 0 \\ 0 & -4 \end{bmatrix}\begin{bmatrix} \Delta x_1 \\ \Delta x_2 \end{bmatrix}, \text{ i.e.}$$

$$\Delta x_1 = 4x_2 \Delta x_2 + 2\Delta x_2^2 \tag{4.4}$$

and (4.3) gives

$$\Delta f = [-2 \ -1]\begin{bmatrix} \Delta x_1 \\ \Delta x_2 \end{bmatrix} = -2\Delta x_1 - \Delta x_2. \tag{4.5}$$

Substituting (4.4) into (4.5) results in $\Delta f = -2(4x_2\Delta x_2 + 2\Delta x_2^2) - \Delta x_2$ and setting $x_2 = x_2^* = \frac{1}{8}$ gives $\Delta f(\mathbf{x}^*) = \Delta x_2 - 4\Delta x_2^2 - \Delta x_2 = -4\Delta x_2^2 < 0$ for all Δx_2. Thus $\mathbf{x}^*$ is *not* a minimum, but in fact a constrained *maximum*. In fact this problem does not have a minimum as it is unbounded.

Problem 4.3.2.4

Minimize $f(\mathbf{x}) = 3x_1^2 + x_2^2 + 2x_1 x_2 + 6x_1 + 2x_2$ such that $h(\mathbf{x}) = 2x_1 - x_2 - 4 = 0$. Show that the candidate point you obtain is indeed a local constrained minimum.

Solution

The Lagrangian is given by

$$L(\mathbf{x}, \lambda) = 3x_1^2 + x_2^2 + 2x_1 x_2 + 6x_1 + 2x_2 + \lambda(2x_1 - x_2 - 4)$$

and the associated necessary stationary conditions are

$$\frac{\partial L}{\partial x_1} = 6x_1 + 2x_2 + 6 + 2\lambda = 0$$

$$\frac{\partial L}{\partial x_2} = 2x_2 + 2x_1 + 2 - \lambda = 0$$

$$\frac{\partial L}{\partial \lambda} = 2x_1 - x_2 - 4 = 0.$$

Solving gives $x_1^* = \frac{7}{11}$, $x_2^* = -\frac{30}{11}$, $\lambda^* = -\frac{24}{11}$.

To prove that the above point indeed corresponds to a local minimum, the following further argument is required. Here

$$\nabla f = \begin{bmatrix} 6x_1 + 2x_2 + 6 \\ 2x_2 + 2x_1 + 2 \end{bmatrix}, \quad \nabla h = \begin{bmatrix} 2 \\ -1 \end{bmatrix}, \quad \nabla^2 h = \mathbf{0} \text{ and } \mathbf{H} = \begin{bmatrix} 6 & 2 \\ 2 & 2 \end{bmatrix}$$

is positive-definite. Now for any step $\Delta \mathbf{x}$ compatible with the constraint $h(\mathbf{x}) = 0$ it follows that the changes in the constraint function and objective function are respectively given by

$$\Delta h = \nabla^T h \Delta \mathbf{x} = [2 \ -1] \begin{bmatrix} \Delta x_1 \\ \Delta x_2 \end{bmatrix} = 2\Delta x_1 - \Delta x_2 = 0, \text{ i.e. } 2\Delta x_1 = \Delta x_2 \tag{4.6}$$

and

$$\Delta f = \nabla^T f(\mathbf{x}^*)\Delta \mathbf{x} + \tfrac{1}{2}\Delta \mathbf{x}^T \mathbf{H}(\mathbf{x}^*)\Delta \mathbf{x}. \tag{4.7}$$

Also since $\nabla f(\mathbf{x}^*) = \tfrac{24}{11}\begin{bmatrix} 2 \\ -1 \end{bmatrix}$, the first term in (4.7) may be written as

$$\nabla^T f(\mathbf{x}^*)\Delta \mathbf{x} = \tfrac{24}{11}[2, -1]\begin{bmatrix} \Delta x_1 \\ \Delta x_2 \end{bmatrix} = \tfrac{24}{11}(2\Delta x_1 - \Delta x_2) = 0 \text{ (from (4.6))}.$$

Finally, substituting the latter expression into (4.7) gives, for any step $\Delta \mathbf{x}$ at $\mathbf{x}^*$ compatible with the constraint, the change in function value as $\Delta f = 0 + \tfrac{1}{2}\Delta \mathbf{x}^T \mathbf{H}(\mathbf{x}^*)\Delta \mathbf{x} > 0$ since $\mathbf{H}$ is positive-definite. The point $\mathbf{x}^*$ is therefore a strong local constrained minimum.

Problem 4.3.2.5

Maximize $f = xyz$ such that $\left(\tfrac{x}{a}\right)^2 + \left(\tfrac{y}{b}\right)^2 + \left(\tfrac{z}{c}\right)^2 = 1$.

Solution

$$L = xyz + \lambda \left(\left(\tfrac{x}{a}\right)^2 + \left(\tfrac{y}{b}\right)^2 + \left(\tfrac{z}{c}\right)^2 - 1 \right)$$

with necessary conditions:

$$\frac{\partial L}{\partial x} = yz + 2\lambda \frac{x}{a^2} = 0$$

$$\frac{\partial L}{\partial y} = xz + 2\lambda \frac{y}{b^2} = 0$$

$$\frac{\partial L}{\partial z} = xy + 2\lambda \frac{z}{c^2} = 0.$$

Solving the above together with the given equality, yields $\lambda = -\tfrac{3}{2}xyz$, $x = \tfrac{1}{\sqrt{3}}a$, $y = \tfrac{1}{\sqrt{3}}b$ and $z = \tfrac{1}{\sqrt{3}}c$, and thus $f^* = \tfrac{1}{3\sqrt{3}}abc$.

Problem 4.3.2.6

Minimize $(x_1 - 1)^3 + (x_2 - 1)^2 + 2x_1x_2$ such that $x_1 + x_2 = 2$.

Solution

With $h(\mathbf{x}) = x_1 + x_2 - 2 = 0$ the Lagrangian is given by

$$L(\mathbf{x}, \lambda) = (x_1 - 1)^3 + (x_2 - 1)^2 + 2x_1x_2 + \lambda(x_1 + x_2 - 2)$$

with necessary conditions:

$$\frac{\partial L}{\partial x_1} = 3(x_1 - 1)^2 + 2x_2 + \lambda = 0$$

$$\frac{\partial L}{\partial x_2} = 2(x_2 - 1) + 2x_1 + \lambda = 0$$

$$x_1 + x_2 = 2.$$

Solving gives $\lambda^* = -2$ and the possible solutions

$$x_1^* = 1, \ x_2^* = 1 \text{ or } x_1^* = \tfrac{5}{3}, \ x_2^* = \tfrac{1}{3}.$$

Analysis of the solutions:

For any $\boldsymbol{\Delta}\mathbf{x}$ consistent with the constraint:

$$\Delta h = 0 = \boldsymbol{\nabla}^T h \boldsymbol{\Delta}\mathbf{x} = [1, 1] \begin{bmatrix} \Delta x_1 \\ \Delta x_2 \end{bmatrix}, \ \text{i.e. } \Delta x_1 + \Delta x_2 = 0$$

and

$$\Delta f = \boldsymbol{\nabla}^T f \boldsymbol{\Delta}\mathbf{x} + \tfrac{1}{2} \boldsymbol{\Delta}\mathbf{x} \mathbf{H}(\overline{\mathbf{x}}) \boldsymbol{\Delta}\mathbf{x} \text{ where } \overline{\mathbf{x}} = \mathbf{x}^* + \theta \boldsymbol{\Delta}\mathbf{x}, \ 0 \le \theta \le 1.$$

For both candidate points $\mathbf{x}^*$ above, $\boldsymbol{\nabla}^T f = [2, 2]$ and thus: $\boldsymbol{\nabla}^T f \boldsymbol{\Delta}\mathbf{x} = 2(\Delta x_1 + \Delta x_2) = 0$. Considering only $\boldsymbol{\Delta}\mathbf{x} \mathbf{H}(\overline{\mathbf{x}}) \boldsymbol{\Delta}\mathbf{x}$, it is clear that as $\boldsymbol{\Delta}\mathbf{x} \to \mathbf{0}$, $\overline{\mathbf{x}} \to \mathbf{x}^*$, and $\Delta f > 0$ if $\mathbf{H}(\mathbf{x}^*)$ is positive-definite.

$\mathbf{H} = \begin{bmatrix} 6(x_1 - 1) & 2 \\ 2 & 2 \end{bmatrix}$ and thus at $\mathbf{x}^* = [1, 1]^T$, $\mathbf{H} = \begin{bmatrix} 0 & 2 \\ 2 & 2 \end{bmatrix}$ is not

positive-definite, and at $\mathbf{x}^* = \left[\tfrac{5}{3}, \tfrac{1}{3}\right]^T$, $\mathbf{H} = \begin{bmatrix} 4 & 2 \\ 2 & 2 \end{bmatrix}$ is positive-definite.

Therefore the point $\mathbf{x}^* = \left[\tfrac{5}{3}, \tfrac{1}{3}\right]^T$ is a strong local constrained minimum.

Problem 4.3.2.7

Determine the dimensions of a cylindrical can of maximum volume subject to the condition that the total surface area be equal to 24π. Show that the answer indeed corresponds to a maximum. (x_1 = radius, x_2 = height)

Solution

The problem is equivalent to:

minimize $f(x_1, x_2) = -\pi x_1^2 x_2$ such that $2\pi x_1^2 + 2\pi x_1 x_2 = A_0 = 24\pi$.

Thus with $h(\mathbf{x}) = 2\pi x_1^2 + 2\pi x_1 x_2 - 24\pi = 0$ the appropriate Lagrangian is

$$L(\mathbf{x}, \lambda) = -\pi x_1^2 x_2 + \lambda(2\pi x_1^2 + 2\pi x_1 x_2 - 24\pi)$$

with necessary conditions for a local minimum:

$$\frac{\partial L}{\partial x_1} = -2\pi x_1 x_2 + 4\pi \lambda x_1 + 2\pi \lambda x_2 = 0$$

$$\frac{\partial L}{\partial x_2} = -\pi x_1^2 + 2\pi \lambda x_1 = 0$$

$$\text{and } h(\mathbf{x}) = 2\pi x_1^2 + 2\pi x_1 x_2 - 24\pi = 0.$$

Solving gives $x_1^* = 2$, $x_2^* = 4$ with $\lambda^* = 1$ and $f^* = -16\pi$.

Analysis of the solution:

At $\mathbf{x}^*$ for change $\Delta\mathbf{x}$ compatible with the constraint it is required to show that

$$\Delta f = \nabla^T f \Delta\mathbf{x} + \tfrac{1}{2}\Delta\mathbf{x}^T \mathbf{H} \Delta\mathbf{x} > 0 \tag{4.8}$$

in the limit as $\Delta\mathbf{x} \to \mathbf{0}$ and $h(\mathbf{x}^* + \Delta\mathbf{x}) = 0$.

For $f(\mathbf{x})$ at $\mathbf{x}^*$ the gradient vector and Hessian are given by

$$\nabla f = \begin{bmatrix} -2\pi x_1 x_2 \\ -\pi x_1^2 \end{bmatrix} = -4\pi \begin{bmatrix} 4 \\ 1 \end{bmatrix} \text{ and}$$

$$\mathbf{H} = \begin{bmatrix} -2\pi x_2 & -2\pi x_1 \\ -2\pi x_1 & 0 \end{bmatrix} = \begin{bmatrix} -8\pi & -4\pi \\ -4\pi & 0 \end{bmatrix}.$$

For satisfaction of the constraint: $\Delta h = \nabla^T h \Delta \mathbf{x} + \frac{1}{2}\Delta \mathbf{x}^T \nabla^2 h \Delta \mathbf{x} = 0$, i.e.

$$\nabla^T h \Delta \mathbf{x} = -\frac{1}{2}\Delta \mathbf{x}^T \nabla^2 h \Delta \mathbf{x} \qquad (4.9)$$

with $\nabla h = \begin{bmatrix} 4\pi x_1 + 2\pi x_2 \\ 2\pi x_1 \end{bmatrix} = 4\pi \begin{bmatrix} 4 \\ 1 \end{bmatrix}$

and where $\nabla^2 h = \begin{bmatrix} 4\pi & 2\pi \\ 2\pi & 0 \end{bmatrix}$. Now clearly $\nabla h = -\nabla f$ at the candidate point.

It therefore follows that $\nabla f^T \Delta \mathbf{x} = -\nabla^T h \Delta \mathbf{x} = \frac{1}{2}\Delta \mathbf{x}^T \nabla^2 h \Delta \mathbf{x}$.

Substituting in (4.8):

$$\begin{aligned}
\Delta f &= \tfrac{1}{2}\Delta \mathbf{x}^T (\nabla^2 h + \mathbf{H})\Delta \mathbf{x} \\
&= \tfrac{1}{2}\Delta \mathbf{x}^T \begin{bmatrix} -4\pi & -2\pi \\ -2\pi & 0 \end{bmatrix} \Delta \mathbf{x} = -\pi [\Delta x_1 \ \Delta x_2] \begin{bmatrix} 2 & 1 \\ 1 & 0 \end{bmatrix} \begin{bmatrix} \Delta x_1 \\ \Delta x_2 \end{bmatrix} \\
&= -\pi [2\Delta x_1^2 + 2\Delta x_1 \Delta x_2].
\end{aligned}$$

From (4.9) in the limit as $\Delta \mathbf{x} \to \mathbf{0}$, $\Delta x_2 = -4\Delta x_1$, and thus

$\Delta f = -\pi[2\Delta x_1^2 + 2\Delta x_1(-4\Delta x_1)] = 6\pi \Delta x_1^2 > 0$, as expected for a constrained local minimum.

Problem 4.3.2.8

Minimize $f(\mathbf{x}) = x_1^2 + x_2^2 + \cdots + x_n^2$ such that $h(\mathbf{x}) = x_1 + x_2 + \cdots + x_n - 1 = 0$.

Solution

The Lagrangian is

$$L = (x_1^2 + x_2^2 + \cdots + x_n^2) + \lambda(x_1 + x_2 + \cdots + x_n - 1)$$

with necessary conditions:

$$\frac{\partial L}{\partial x_i} = 2x_i + \lambda = 0, \ i = 1, \ldots, n.$$

Thus $\sum_{i=1}^{n} 2x_i + n\lambda = 0 \Rightarrow \lambda = -\frac{2}{n}$ and

$$2x_i - \frac{2}{n} = 0 \Rightarrow x_i = \frac{1}{n}, \ i = 1, 2, \ldots, n.$$

Therefore the distance $= \sqrt{f(\mathbf{x}^*)} = \sqrt{\frac{n}{n^2}} = \frac{1}{\sqrt{n}}$.

Test for a minimum at $\mathbf{x}^*$:

$$\Delta f = \boldsymbol{\nabla}^T f(\mathbf{x}^*) \boldsymbol{\Delta} \mathbf{x} + \tfrac{1}{2} \boldsymbol{\Delta} \mathbf{x}^T \mathbf{H}(\bar{\mathbf{x}}) \boldsymbol{\Delta} \mathbf{x} \text{ and } \boldsymbol{\nabla}^T f = [2x_1, \dots, 2x_n]^T$$

and $\mathbf{H}(\bar{\mathbf{x}}) = 2 \begin{bmatrix} 1 & & & \\ & 1 & & \\ & & 1 & \\ & & & 1 \end{bmatrix}$ which is positive-definite, $\Delta f > 0$ if

$\boldsymbol{\nabla} f(\mathbf{x}^*) \boldsymbol{\Delta} \mathbf{x} \geq 0$.

For $\boldsymbol{\Delta} \mathbf{x}$ such that $\Delta h = \boldsymbol{\nabla}^T h \boldsymbol{\Delta} \mathbf{x} = 0$, gives $\Delta x_1 + \Delta x_2 + \cdots + \Delta x_n = 0$.

Thus $\boldsymbol{\nabla}^T f(\mathbf{x}^*) \boldsymbol{\Delta} \mathbf{x} = 2[x_1^*, \dots, x_n^*]^T \boldsymbol{\Delta} \mathbf{x} = \frac{2}{n}(\Delta x_1 + \Delta x_2 + \cdots + \Delta x_n) = 0$ and therefore $\mathbf{x}^*$ is indeed a constrained minimum.

Problem 4.3.2.9

Minimize $x_1^2 + x_2^2 + x_3^2$ such that $x_1 + 3x_2 + 2x_3 - 12 = 0$.

Solution

$$L = x_1^2 + x_2^2 + x_3^2 + \lambda(x_1 + 3x_2 + 2x_3 - 12)$$

$$\frac{\partial L}{\partial x_1} = 2x_1 + \lambda = 0$$

$$\frac{\partial L}{\partial x_2} = 2x_2 + 3\lambda = 0$$

$$\frac{\partial L}{\partial x_3} = 2x_3 + 2\lambda = 0$$

$$x_1 + 3x_2 + 2x_3 = 12$$

with solution: $\lambda^* = -\frac{12}{7}$, $x_1^* = \frac{6}{7}$, $x_2^* = \frac{18}{7}$, $x_3^* = \frac{12}{7}$ with $f^* = 10.286$.

Test Δf for all $\boldsymbol{\Delta} \mathbf{x}$ compatible with the constraint: $\Delta f = \boldsymbol{\nabla} f^T \boldsymbol{\Delta} \mathbf{x} + \tfrac{1}{2} \boldsymbol{\Delta} \mathbf{x}^T \mathbf{H} \boldsymbol{\Delta} \mathbf{x}$, $\boldsymbol{\nabla} f = [2x_1, 2x_2, 2x_3]^T$, $\mathbf{H} = \begin{bmatrix} 2 & & 0 \\ & 2 & \\ 0 & & 2 \end{bmatrix}$ positive-definite

and

$$\Delta h = \boldsymbol{\nabla}^T h \boldsymbol{\Delta} \mathbf{x} + \tfrac{1}{2} \boldsymbol{\Delta} \mathbf{x}^T \boldsymbol{\nabla}^2 h \boldsymbol{\Delta} \mathbf{x} = 0, \Rightarrow \Delta x_1 + 3\Delta x_2 + 2\Delta x_3 = 0.$$

Therefore

$$\begin{aligned} \Delta f(\mathbf{x}^*) &= 2x_1^* \Delta x_1 + 2x_2^* \Delta x_2 + 2x_3^* \Delta x_3 + \tfrac{1}{2} \boldsymbol{\Delta} \mathbf{x}^T \mathbf{H} \boldsymbol{\Delta} \mathbf{x} \\ &= 2\tfrac{6}{7}(\Delta x_1 + 3\Delta x_2 + 2\Delta x_3) + \tfrac{1}{2} \boldsymbol{\Delta} \mathbf{x}^T \mathbf{H} \boldsymbol{\Delta} \mathbf{x} = 0 + \tfrac{1}{2} \boldsymbol{\Delta} \mathbf{x}^T \mathbf{H} \boldsymbol{\Delta} \mathbf{x} > 0. \end{aligned}$$

Thus $\mathbf{x}^*$ is indeed a local minimum.

4.3.3 Solution of inequality constrained problems via auxiliary variables

Problem 4.3.3.1

Consider the problem:

Minimize $f(\mathbf{x}) = x_1^2 + x_2^2 - 3x_1x_2$ such that $x_1^2 + x_2^2 - 6 \le 0$.

Solution

Introducing an auxiliary variable θ, the problem is transformed to an equality constrained problem with $x_1^2 + x_2^2 - 6 + \theta^2 = 0$. Since $\theta^2 \ge 0$ the original equality constraint is satisfied for all values of θ for which the new equality constraint is satisfied. Solve the new equality constrained problem with additional variable θ by the Lagrange method:

$$L(\mathbf{x}, \theta, \lambda) = x_1^2 + x_2^2 - 3x_1x_2 + \lambda(x_1^2 + x_2^2 - 6 + \theta^2)$$

with necessary conditions:

$$\frac{\partial L}{\partial x_1} = 2x_1 - 3x_2 + 2\lambda x_1 = 0$$

$$\frac{\partial L}{\partial x_2} = 2x_2 - 3x_1 + 2\lambda x_2 = 0$$

$$\frac{\partial L}{\partial \theta} = 2\theta\lambda = 0$$

$$\frac{\partial L}{\partial \lambda} = x_1^2 + x_2^2 - 6 + \theta^2 = 0.$$

The third equation implies three possibilities:

(i) $\lambda = 0$, $\theta \ne 0$, $\Rightarrow x_1^* = x_2^* = 0$, $\theta^2 = 6$ and $f(\mathbf{x}) = 0$.

(ii) $\theta = 0$, $\lambda \ne 0$, then from the first two conditions:

$$2\lambda = -2 + \frac{3x_2}{x_1} = -2 + \frac{3x_1}{x_2} \Rightarrow x_1^2 = x_2^2.$$

Thus $x_1 = \pm x_2$ and from the last condition it follows that $x_1 = \pm\sqrt{3}$.

Choosing $x_1^* = -x_2^*$ gives $f(\mathbf{x}^*) = 15$, and choosing $x_1^* = x_2^* = \pm\sqrt{3}$ gives $f(\mathbf{x}^*) = -3$.

(iii) $\theta = 0$ and $\lambda = 0$ leads to a contradiction.

The constrained minimum therefore corresponds to case (ii) above.

Problem 4.3.3.2

Minimize $f(\mathbf{x}) = 2x_1^2 - 3x_2^2 - 2x_1$ such that $x_1^2 + x_2^2 \leq 1$.

Solution

Introduce auxiliary variable θ such that $x_1^2 + x_2^2 + \theta^2 = 1$. The Lagrangian is then

$$L(\mathbf{x}, \theta) = 2x_1^2 - 3x_2^2 - 2x_1 + \lambda(x_1^2 + x_2^2 + \theta^2 - 1)$$

and the necessary conditions for a minimum:

$$\frac{\partial L}{\partial x_1} = 4x_1 - 2 + 2\lambda x_1 = 0, \quad \frac{\partial L}{\partial x_2} = -6x_2 + 2\lambda x_2 = 0, \quad \frac{\partial L}{\partial \theta} = 2\lambda\theta = 0$$

$$\text{and } x_1^2 + x_2^2 + \theta^2 = 1.$$

The possibilities are:

either (i) $\lambda = 0$, $\theta \neq 0$ or (ii) $\lambda \neq 0$, $\theta = 0$ or (iii) $\lambda = 0$ and $\theta = 0$. Considering each possibility in turn gives the following:

(i) $\lambda = 0 \Rightarrow x_1^* = \frac{1}{2}$, $x_2^* = 0$, $\theta^{*2} = \frac{3}{4}$ and $f(x_1^*, x_2^*) = -\frac{1}{2}$.

(ii) $\theta = 0 \Rightarrow x_1^2 + x_2^2 = 1$; $4x_1 - 2 + 2\lambda x_1 = 0$; $x_2(-6 + 2\lambda) = 0$ giving firstly for $x_2^* = 0 \Rightarrow x_1^* = \pm 1$; $\lambda^* = -1$; -3 and $f^*(1,0) = 0$; $f^*(-1,0) = 4$, or secondly for $\lambda^* = 3$ it follows that $x_1^* = \frac{1}{5}$ and $x_2^* = \pm\sqrt{\frac{24}{25}}$, for which in both cases $f^* = -3.2$.

(iii) leads to a contradiction.

Inspection of the alternatives above gives the global minima at $x_1^* = \frac{1}{5}$ and $x_2^* = \pm\sqrt{\frac{24}{25}}$ with $f^* = -3.2$ and maximum at $x_1^* = -1$, $x_2^* = 0$, with $f^* = 4$.

Problem 4.3.3.3

Maximise $f(\mathbf{x}) = -x_1^2 + x_1 x_2 - 2x_2^2 + x_1 + x_2$ such that $2x_1 + x_2 \leq 1$, by using auxiliary variables.

Solution

Here solve the minimization problem: $\min_{\mathbf{x}} f = x_1^2 - x_1 x_2 + 2x_2^2 - x_1 - x_2$ such that $1 - 2x_1 - x_2 \geq 0$.

Introducing an auxiliary variable it follows that

$$L(\mathbf{x}, \theta) = x_1^2 - x_1 x_2 + 2x_2^2 - x_1 - x_2 + \lambda(1 - 2x_1 - x_2 - \theta^2)$$

with necessary conditions:

$$\frac{\partial L}{\partial x_1} = 2x_1 - x_2 - 1 - 2\lambda = 0$$

$$\frac{\partial L}{\partial x_2} = -x_1 + 4x_2 - 1 - \lambda = 0$$

$$\frac{\partial L}{\partial \theta} = -2\lambda\theta = 0$$

$$\frac{\partial L}{\partial \lambda} = 1 - 2x_1 - x_2 - \theta^2 = 0.$$

The possibilities are:

  (i) $\lambda = 0$ and $\theta \neq 0 \Rightarrow 2x_1 - x_2 - 1 = 0$ and $-x_1 + 4x_2 - 1 = 0$, giving $x_1 = \frac{5}{7}$ and $x_2 = \frac{3}{7}$. Substituting in the equality $1 - 2x_1 - x_2 - \theta^2 = 0 \Rightarrow \theta^2 = -\frac{6}{7} < 0$, which is not possible for θ real.

  (ii) $\theta = 0$ and $\lambda \neq 0$ gives $1 - 2x_1 - x_2 = 0$ and solving together with the first two necessary conditions gives $x_2^* = \frac{3}{11}$, $x_1^* = \frac{4}{11}$ and $\lambda^* = -\frac{3}{11}$.

4.3.4 Solution of inequality constrained problems via the Karush-Kuhn-Tucker conditions

Problem 4.3.4.1

Minimize $f(\mathbf{x}) = 3x_1 + x_2$
subject to $g(\mathbf{x}) = x_1^2 + x_2^2 - 5 \leq 0$.

Solution

For $L(\mathbf{x}, \lambda) = 3x_1 + x_2 + \lambda(x_1^2 + x_2^2 - 5)$ the KKT conditions are:

$$\frac{\partial L}{\partial x_1} = 3 + 2\lambda x_1 = 0$$

$$\frac{\partial L}{\partial x_2} = 1 + 2\lambda x_2 = 0$$

$$x_1^2 + x_2^2 - 5 \leq 0$$

$$\lambda(x_1^2 + x_2^2 - 5) = 0$$

$$\lambda \geq 0.$$

From $\lambda(x_1^2 + x_2^2 - 5) = 0$ it follows that either $\lambda = 0$ or $x_1^2 + x_2^2 - 5 = 0$.

If $\lambda = 0$ then the first two conditions are not satisfied and therefore $\lambda \neq 0$ and we have the equality $x_1^2 + x_2^2 - 5 = 0$. It now follows that $x_1 = -\frac{3}{2\lambda}$ and $x_2 = -\frac{1}{2\lambda}$.

Substituting these values into the equality yields

$$\left(-\frac{3}{2\lambda}\right)^2 + \left(-\frac{1}{2\lambda}\right)^2 = 5,$$

which implies that $\lambda^* = +\sqrt{\frac{1}{2}} > 0$. The optimal solution is thus $x_1^* = -\frac{3}{\sqrt{2}}$, $x_2^* = -\frac{1}{\sqrt{2}}$ with $f(\mathbf{x}^*) = -\frac{10}{\sqrt{2}}$. By inspection, one could argue that no interior optimum can exist since the objective function is linear.

Problem 4.3.4.2

Minimize $x_1^2 + x_2^2 - 14x_1 - 6x_2 - 7$ such that $x_1 + x_2 \leq 2$ and $x_1 + 2x_2 \leq 3$.

Solution

$$L(\mathbf{x}, \boldsymbol{\lambda}) = x_1^2 + x_2^2 - 14x_1 - 6x_2 - 7 + \lambda_1(x_1 + x_2 - 2) + \lambda_2(x_1 + 2x_2 - 3)$$

with KKT conditions:

$$\frac{\partial L}{\partial x_1} = 2x_1 - 14 + \lambda_1 + \lambda_2 = 0$$

$$\frac{\partial L}{\partial x_2} = 2x_2 - 6 + \lambda_1 + 2\lambda_2 = 0$$

$$\lambda_1(x_1 + x_2 - 2) = 0$$

$$\lambda_2(x_1 + 2x_2 - 3) = 0$$

$$\lambda \geq 0.$$

The possibilities are:

(i) The choice $\lambda_1 \neq 0$ and $\lambda_2 \neq 0$ gives

$$x_1 + x_2 - 2 = 0, \ x_1 + 2x_2 - 3 = 0 \ \Rightarrow \ x_2 = 1, \ x_1 = 1, \text{ with } f = -25.$$

Both constraints are satisfied but $\lambda_2 = -8 < 0$ with $\lambda_2 = 20$.

(ii) $\lambda_1 \neq 0$ and $\lambda_2 = 0$ gives $x_1 + x_2 - 2 = 0$ and

$$2x_1 - 14 + \lambda_1 = 0$$
$$2x_2 - 6 + \lambda_1 = 0$$

which yield

$$\lambda_1 = 8 \geq 0 \text{ and } x_1 = 3, \ x_2 = -1, \text{ with } f = -33$$

and both constraints are satisfied.

(iii) $\lambda_1 = 0$ and $\lambda_2 \neq 0$ gives $x_1 + 2x_2 - 3 = 0$ and it follows that

$\lambda_2 = 4 \geq 0$ and $x_1 = 5, \ x_2 = -1$, with $f = -45$. However, the first constraint is violated.

(iv) The final possibility $\lambda_1 = \lambda_2 = 0$ gives $x_1 = 7$ and $x_2 = 3$ with $f = -65$ but both the first and second constraints are violated.

The unique optimum solution is therefore given by possibility (ii).

Problem 4.3.4.3

Minimize $f(\mathbf{x}) = x_1^2 + x_2^2 - 4x_1 - 6x_2 + 13$ such that $x_1 + x_2 \geq 7$ and $x_1 - x_2 \leq 2$.

Solution

We note that in this case the KKT conditions are also sufficient since the problem is convex.

$$L(\mathbf{x}, \boldsymbol{\lambda}) = (x_1^2 + x_2^2 - 4x_1 - 6x_2 + 13) + \lambda_1(-x_1 - x_2 + 7) + \lambda_2(x_1 - x_2 - 2)$$

with KKT conditions:

$$\frac{\partial L}{\partial x_1} = 2x_1 - 4 - \lambda_1 + \lambda_2 = 0$$

$$\frac{\partial L}{\partial x_2} = 2x_2 - 6 - \lambda_1 - \lambda_2 = 0$$

$$\lambda_1(-x_1 - x_2 + 7) = 0$$

$$\lambda_2(x_1 - x_2 - 2) = 0$$

$$\lambda \geq 0.$$

The possibilities are:

(i) $\lambda_1 = \lambda_2 = 0 \Rightarrow x_1 = 2, x_2 = 3$ with $x_1 + x_2 = 2 + 3 < 7$ and thus the first constraint is not satisfied.

(ii) $\lambda_1 \neq 0, \lambda_2 = 0 \Rightarrow 2x_1 - 4 - \lambda_1 = 0, 2x_2 - 6 - \lambda_1 = 0$ and $x_1 - x_2 + 1 = 0$.

The above implies $-x_1 - x_2 + 7 = 0, \lambda_2 = 0, \lambda_1 = 2, x_2 = 4,$ $x_1 = 3$ with $f = 2$. This point satisfies all the KKT conditions and is therefore a local minimum.

(iii) $\lambda_1 = 0$ and $\lambda_2 \neq 0 \Rightarrow 2x_1 - 4 + \lambda_2 = 0, 2x_2 - 6 - \lambda_2 = 0,$ $2x_1 + 2x_2 - 10 = 0, x_1 - x_2 - 2 = 0$ and solving yields

$$x_1 = \tfrac{7}{2}, \lambda_1 = 0, x_2 = \tfrac{3}{2} \text{ and } \lambda_2 = -3.$$

This point violates the condition that $\lambda_2 \geq 0$, and is therefore not a local minimum.

(iv) $\lambda_1 \neq 0$ and $\lambda_2 \neq 0 \Rightarrow 2x_1 - 4 - \lambda_1 + \lambda_2 = 0, 2x_2 - 6 - \lambda_1 - \lambda_2 = 0,$ and $-x_1 - x_2 + 7 = 0$, and $x_1 - x_2 - 2 = 0 \Rightarrow x_2 = \tfrac{5}{2}, x_1 = \tfrac{9}{2}$ with $\lambda_1 = 2$ and $\lambda_2 = -3 < 0$ which violates the condition $\lambda \geq 0$.

The unique optimum solution therefore corresponds to possibility (ii), i.e. $x_1^* = 3, \lambda_1^* = 2, x_2^* = 4, \lambda_2^* = 0$ with $f^* = 2$.

Problem 4.3.4.4

Minimize $x_1^2 + 2(x_2 + 1)^2$ such that $-x_1 + x_2 = 2, -x_1 - x_2 - 1 \leq 0$.

Solution

Here the Lagrangian is given by

$$L(\mathbf{x}, \lambda, \mu) = x_1^2 + 2(x_2 + 1)^2 + \lambda(-x_1 - x_2 - 1) + \mu(-x_1 + x_2 - 2)$$

with KKT conditions:

$$
\begin{aligned}
\frac{\partial L}{\partial x_1} = 2x_1 - \lambda - \mu &= 0 \\
\frac{\partial L}{\partial x_2} = 4(x_2 + 1) - \lambda + \mu &= 0 \\
-x_1 + x_2 - 2 &= 0 \\
-x_1 - x_2 - 1 &\leq 0 \\
\lambda(-x_1 - x_2 - 1) &= 0 \\
\lambda &\geq 0.
\end{aligned}
$$

Possibilities:

(i) $\lambda \neq 0 \Rightarrow -x_1 - x_2 - 1 = 0$ and with $-x_1 + x_2 - 2 = 0 \Rightarrow x_1 = -\frac{3}{2}$, $x_2 = \frac{1}{2}$. Substituting into the first two conditions give $\lambda = \frac{3}{2}$ and $\mu = -\frac{9}{2}$, and thus all conditions are satisfied.

(ii) $\lambda = 0 \Rightarrow 2x_1 - \mu = 0$ and $4x_2 + 4 + \mu = 0$ giving $2x_1 + 4x_2 + 4 = 0$ and with $-x_1 + x_2 - 2 = 0$ it follows that $x_2 = 0$ and $x_1 = -2$. However, $-x_1 - x_2 - 1 = 1$ which does not satisfy the inequality.

Case (i) therefore represents the optimum solution with $f^* = \frac{27}{4}$.

Problem 4.3.4.5

Determine the shortest distance from the origin to the set defined by:

$$4 - x_1 - x_2 \leq 0, \quad 5 - 2x_1 - x_2 \leq 0.$$

Solution

$$L(\mathbf{x}, \boldsymbol{\lambda}) = x_1^2 + x_2^2 + \lambda_1(4 - x_1 - x_2) + \lambda_2(5 - 2x_1 - x_2)$$

with KKT conditions:

$$\frac{\partial L}{\partial x_1} = 2x_1 - \lambda_1 - 2\lambda_2 = 0$$

$$\frac{\partial L}{\partial x_2} = 2x_2 - \lambda_1 - \lambda_2 = 0$$

$$\lambda_1 g_1 + \lambda_2 g_2 = 0$$

$$\lambda_1, \lambda_2 \geq 0.$$

The possibilities are:

(i) $\lambda_1 = \lambda_2 = 0 \Rightarrow x_1 = x_2 = 0$ and both constraints are violated.

(ii) $\lambda_1 = 0, \lambda_2 \neq 0 \Rightarrow 5 - 2x_1 - x_2 = 0$ which gives $x_2 = 1, x_1 = 2$ and $\lambda_2 = 2 > 0$, but this violates constraint g_1.

(iii) $\lambda_1 \neq 0, \lambda_2 = 0 \Rightarrow x_1 = x_2 = 2$ and $\lambda_1 = 4 > 0$.

(iv) $\lambda_1 \neq 0, \lambda_2 \neq 0 \Rightarrow g_1 = 0$ and $g_2 = 0$ which implies that $\lambda_2 = -4 < 0, x_1 = 1, x_2 = 3$ which violates the non-negativity condition on λ_2.

The solution $\mathbf{x}^*$ therefore corresponds to case (iii) with shortest distance $= \sqrt{8}$.

Problem 4.3.4.6

Minimize $x_1^2 + x_2^2 - 2x_1 - 2x_2 + 2$

such that $-2x_1 - x_2 + 4 \leq 0$ and $-x_1 - 2x_2 + 4 \leq 0$.

Solution

$$L(\mathbf{x}, \boldsymbol{\lambda}) = x_1^2 + x_2^2 - 2x_1 - 2x_2 + 2 + \lambda_1(-2x_1 - x_2 + 4) + \lambda_2(-x_1 - 2x_2 + 4)$$

with KKT conditions:

$$\frac{\partial L}{\partial x_1} = 2x_1 - 2 - 2\lambda_1 - \lambda_2 = 0$$

$$\frac{\partial L}{\partial x_2} = 2x_2 - 2 - \lambda_1 - 2\lambda_2 = 0$$

$$g_1 = -2x_1 - x_2 + 4 \leq 0$$

$$g_2 = -x_1 - 2x_2 + 4 \leq 0$$

$$\lambda_1 g_1 = 0; \quad \lambda_2 g_2 = 0$$

$$\lambda_1 \geq 0; \quad \lambda_2 \geq 0.$$

The fifth conditions give the following possibilities:

(i) $\lambda_1 = \lambda_2 = 0 \Rightarrow 2x_1 - 2 = 0, 2x_2 - 2 = 0, x_1 = x_2 = 1$, not valid since both g_1 and $g_2 > 0$.

(ii) $\lambda_1 = 0$ and $g_2 = 0 \Rightarrow 2x_1 - 2 - \lambda_2 = 0, 2x_2 - 2 - 2\lambda_2 = 0$ and $-x_1 - 2x_2 + 4 = 0$ which yield

$$x_1 = \tfrac{6}{5}, \ x_2 = \tfrac{7}{5}, \ \lambda_2 = \tfrac{2}{5} > 0$$

but, not valid since $g_1 = 0.2 > 0$.

(iii) $g_1 = 0, \lambda_2 = 0 \Rightarrow 2x_1 - 2 - 2\lambda_1 = 0, 2x_2 - 2 - \lambda_1 = 0$ which together with $g_1 = 0$ give $x_1 = \tfrac{7}{5}, x_2 = \tfrac{6}{5}$ and $\lambda_1 = \tfrac{2}{5} > 0$, but not valid since $g_2 = 0.2 > 0$.

(iv) $g_1 = g_2 = 0 \Rightarrow x_1 = x_2 = \tfrac{4}{3}, \lambda_1 = \lambda_2 = \tfrac{2}{9} > 0$.

Since (iv) satisfies all the conditions it corresponds to the optimum solution with $f(\mathbf{x}^*) = \tfrac{2}{9}$.

Problem 4.3.4.7

Minimize $\left(x_1 - \tfrac{9}{4}\right)^2 + (x_2 - 2)^2$

such that $g_1(\mathbf{x}) = x_1^2 - x_2 \leq 0$, $g_2(\mathbf{x}) = x_1 + x_2 - 6 \leq 0$, and $x_1, x_2 \geq 0$.

Is the point $\overline{\mathbf{x}} = \left(\tfrac{3}{2}, \tfrac{9}{4}\right)$ a local minimum?

Solution

$$L(\mathbf{x}, \boldsymbol{\lambda}) = \left(x_1 - \tfrac{9}{4}\right)^2 + (x_2 - 2)^2 + \lambda_1(x_1^2 - x_2) + \lambda_2(x_1 + x_2 - 6) - \lambda_3 x_1 - \lambda_4 x_2$$

with KKT conditions:

$$\frac{\partial L}{\partial x_1} = 2\left(x_1 - \frac{9}{4}\right) + 2\lambda_1 x_1 + \lambda_2 - \lambda_3 \; = \; 0$$

$$\frac{\partial L}{\partial x_2} = 2(x_2 - 2) - \lambda_1 + \lambda_2 - \lambda_4 \; = \; 0$$

$$g_1 \; \leq \; 0 \text{ and } g_2 \; \leq \; 0$$

$$-x_1 \; \leq \; 0 \text{ and } -x_2 \; \leq \; 0$$

$$\lambda_1(x_1^2 - x_2) = 0; \quad \lambda_2(x_1 + x_2 - 6) \; = \; 0$$

$$\lambda_3 x_1 = 0; \quad \lambda_4 x_2 \; = \; 0$$

$$\lambda_1 \geq \lambda_2 \geq \lambda_3 \geq \lambda_4 \; \geq \; 0.$$

At $\bar{\mathbf{x}} = \left(\frac{3}{2}, \frac{9}{4}\right)$, x_1 and $x_2 \neq 0$ $\Rightarrow$ $\lambda_3 = \lambda_4 = 0$ and $g_1 = \left(\frac{3}{2}\right)^2 - \frac{9}{4} = 0$ and $g_2 = \frac{3}{2} + \frac{9}{4} - 6 = \frac{15}{4} - 6 < 0$ and thus $\lambda_2 = 0$.

Also from the first condition it follows that since $2\left(\frac{3}{2} - \frac{9}{4}\right) + 2\lambda_1 \frac{3}{2} = 0$ that $\lambda_1 = \frac{1}{2} > 0$. This value also satisfies the second condition.

Since $\bar{\mathbf{x}}$ satisfies all the KKT conditions, and all the constraints are convex, it is indeed the global constrained optimum.

Problem 4.3.4.8

Minimize $x_1^2 + x_2^2 - 8x_1 - 10x_2$ such that $3x_1 + 2x_2 - 6 \leq 0$.

Solution

$$L(\mathbf{x}, \lambda) = x_1^2 + x_2^2 - 8x_1 - 10x_2 + \lambda(3x_1 + 2x_2 - 6)$$

with KKT conditions:

$$\frac{\partial L}{\partial x_1} = 2x_1 - 8 + 3\lambda \; = \; 0$$

$$\frac{\partial L}{\partial x_2} = 2x_2 - 10 + 2\lambda \; = \; 0$$

$$\lambda(3x_1 + 2x_2 - 6) \; = \; 0.$$

The possibilities are:

(i) $\lambda = 0$ then $x_1 = 4$ and $x_2 = 5$ giving $3x_1 + 2x_2 - 6 = 16 > 0$ and thus this point is not valid.

(ii) $3x_1 + 2x_2 - 6 = 0$ then $\lambda = \frac{32}{13}$ and $x_1 = \frac{4}{13}$, $x_2 = \frac{33}{13}$.

Thus the solution corresponds to case (ii).

Problem 4.3.4.9

Minimize $f(\mathbf{x}) = x_1^2 - x_2$ such that $x_1 \geq 1$, $x_1^2 + x_2^2 \leq 26$.

Solution

$$L(\mathbf{x}, \boldsymbol{\lambda}) = x_1^2 - x_2 + \lambda_1(1 - x_1) + \lambda_2(x_1^2 + x_2^2 - 26)$$

with KKT conditions:

$$\frac{\partial L}{\partial x_1} = 2x_1 - \lambda_1 + 2\lambda_2 x_2 = 0$$

$$\frac{\partial L}{\partial x_2} = -1 + 2\lambda_2 x_2 = 0$$

$$\lambda_1(1 - x_1) = 0$$

$$\lambda_2(x_1^2 + x_2^2 - 26) = 0$$

$$x_1^2 + x_2^2 - 26 \leq 0$$

$$1 - x_1 \leq 0$$

$$\lambda_1 \geq 0; \quad \lambda_2 \geq 0.$$

Investigate the possibilities implied by the third and fourth conditions.

By inspection the possibility $1 - x_1 = 0$ and $x_1^2 + x_2^2 - 26 = 0$ yields $x_1 = 1$; $x_2 = 5$. This gives $\lambda_2 = \frac{1}{10} > 0$ and $\lambda_1 = 2 > 0$ which satisfies the last condition. Thus all the conditions are now satisfied. Since $f(\mathbf{x})$ and all the constraints functions are convex the conditions are also sufficient and $\mathbf{x}^* = [1, 5]^T$ with $f(\mathbf{x}^*) = -4$ is a constrained minimum.

4.3.5   Solution of constrained problems via the dual problem formulation

Problem 4.3.5.1

Minimize $x_1^2 + 2x_2^2$ such that $x_2 \geq -x_1 + 2$.

Solution

The constraint in standard form is $2 - x_1 - x_2 \le 0$ and the Lagrangian therefore:

$$L(\mathbf{x}, \lambda) = x_1^2 + 2x_2^2 + \lambda(2 - x_1 - x_2).$$

For a given value of λ the stationary conditions are

$$\frac{\partial L}{\partial x_1} = 2x_1 - \lambda = 0$$

$$\frac{\partial L}{\partial x_2} = 4x_2 - \lambda = 0$$

giving $x_1 = \frac{\lambda}{2}$ and $x_2 = \frac{\lambda}{4}$. Since the Hessian of L, $\mathbf{H}_L$ is positive-definite the solution corresponds to a minimum.

Substituting the solution into L gives the dual function:

$$h(\lambda) = \tfrac{\lambda^2}{4} + \tfrac{\lambda^2}{8} + \lambda\left(2 - \tfrac{\lambda}{2} - \tfrac{\lambda}{4}\right) = -\tfrac{3}{8}\lambda^2 + 2\lambda.$$

Since $\frac{d^2h}{d\lambda^2} < 0$ the maximum occurs where $\frac{dh}{d\lambda} = -\tfrac{6}{8}\lambda + 2 = 0$, i.e. $\lambda^* = \tfrac{8}{3}$ with

$$h(\lambda^*) = -\tfrac{3}{8}\left(\tfrac{8}{3}\right)^2 + 2\left(\tfrac{8}{3}\right) = \tfrac{8}{3}.$$

Thus $\mathbf{x}^* = \left(\tfrac{4}{3}, \tfrac{2}{3}\right)^T$ with $f(\mathbf{x}^*) = \tfrac{8}{3}$.

Problem 4.3.5.2

Minimize $x_1^2 + x_2^2$ such that $2x_1 + x_2 \le -4$.

Solution

$$L(\mathbf{x}, \lambda) = x_1^2 + x_2^2 + \lambda(2x_1 + x_2 + 4)$$

and the necessary conditions for a minimum with respect to $\mathbf{x}$ imply

$$\frac{\partial L}{\partial x_1} = 2x_1 + 2\lambda = 0$$

$$\text{and } \frac{\partial L}{\partial x_2} = 2x_2 + \lambda = 0$$

giving $x_1 = -\lambda$ and $x_2 = -\tfrac{\lambda}{2}$. Since $\mathbf{H}_L$ is positive-definite the solution is indeed a minimum with respect to $\mathbf{x}$. Substituting in L gives

$$h(\lambda) = -\tfrac{5}{4}\lambda^2 + 4\lambda.$$

Since $\frac{d^2h}{d\lambda^2} = -\frac{5}{2} < 0$ the maximum occurs where $\frac{dh}{d\lambda} = -\frac{5}{2}\lambda + 4 = 0$, i.e. where $\lambda^* = \frac{8}{5} > 0$ and $f(\mathbf{x}^*) = h(\lambda^*) = \frac{16}{5}$.

The solution is thus $x_1^* = -\frac{8}{5}$; $x_2^* = -\frac{4}{5}$ which satisfies the KKT conditions.

Problem 4.3.5.3

Minimize $2x_1^2 + x_2^2$ such that $x_1 + 2x_2 \geq 1$.

Solution

$$L(\mathbf{x}, \lambda) = 2x_1^2 + x_2^2 + \lambda(1 - x_1 - 2x_2)$$

with stationary conditions

$$\frac{\partial L}{\partial x_1} = 4x_1 - \lambda = 0$$

$$\frac{\partial L}{\partial x_2} = 2x_2 - 2\lambda = 0$$

giving $x_1 = \frac{\lambda}{4}$ and $x_2 = \lambda$. Since $\mathbf{H}_L$ is positive-definite the solution is a minimum and

$$h(\lambda) = \frac{1}{8}\lambda^2 + \lambda^2 + \lambda - \frac{1}{4}\lambda^2 - 2\lambda^2 = -\frac{9}{8}\lambda^2 + \lambda.$$

Since $\frac{d^2h}{d\lambda^2} < 0$ the maximum occurs where $\frac{dh}{d\lambda} = -\frac{9}{4}\lambda + 1 = 0$, i.e. $\lambda = \frac{4}{9}$ and since $\lambda > 0$, it follows that λ is an element of the duality function set D defined by equation (3.23).

Thus

$$h(\lambda^*) = -\frac{9}{8}\left(\frac{4}{9}\right)^2 + \frac{4}{9} = \frac{2}{9}$$
$$\text{and } x_1^* = \frac{1}{4}\left(\frac{4}{9}\right) = \frac{1}{9}, \quad x_2^* = \frac{4}{9}.$$

Test: $f(\mathbf{x}^*) = 2\left(\frac{1}{9}\right)^2 + \left(\frac{4}{9}\right)^2 = \frac{2}{81} + \frac{16}{81} = \frac{18}{81} = \frac{2}{9} = h(\lambda^*)$.

Problem 4.3.5.4

Minimize $(x_1 - 1)^2 + (x_2 - 2)^2$ such that $x_1 - x_2 \leq 1$ and $x_1 + x_2 \leq 2$.

Solution

$$L(\mathbf{x}, \boldsymbol{\lambda}) = (x_1 - 1)^2 + (x_2 - 2)^2 + \lambda_1(x_1 - x_2 - 1) + \lambda_2(x_1 + x_2 - 2)$$

and it follows that for a fixed choice of $\boldsymbol{\lambda} = [\lambda_1, \lambda_2]^T$ the stationary conditions are:

$$\frac{\partial L}{\partial x_1} = 2(x_1 - 1) + \lambda_1 + \lambda_2 = 0$$

$$\frac{\partial L}{\partial x_2} = 2(x_2 - 2) - \lambda_1 + \lambda_2 = 0$$

which give $x_1 = 1 - \frac{1}{2}(\lambda_1 + \lambda_2)$ and $x_2 = 2 + \frac{1}{2}(\lambda_1 - \lambda_2)$. $\mathbf{H}_L = \begin{bmatrix} 2 & 0 \\ 0 & 2 \end{bmatrix}$ is positive-definite and therefore the solution is a minimum with respect to $\mathbf{x}$. Substituting the solution into L gives

$$h(\boldsymbol{\lambda}) = -\frac{\lambda_1^2}{2} - \frac{\lambda_2^2}{2} + \lambda_2.$$

The necessary conditions for a maximum are

$$\frac{\partial h}{\partial \lambda_1} = -\frac{2\lambda_1}{2} = 0, \quad \text{i.e. } \lambda_1^* = 0$$

$$\frac{\partial h}{\partial \lambda_2} = -\frac{2\lambda_2}{2} + 1 = 0, \quad \text{i.e. } \lambda_2^* = 1$$

and since the Hessian of h with respect to $\boldsymbol{\lambda}$ is given by $\mathbf{H}_h = \begin{bmatrix} -1 & 0 \\ 0 & -1 \end{bmatrix}$ which is negative-definite the solution indeed corresponds to a maximum, with $h(\boldsymbol{\lambda}^*) = \frac{1}{2} = f(\mathbf{x}^*)$ and thus $x_1^* = \frac{1}{2}$ and $x_2^* = \frac{3}{2}$.

4.3.6 Quadratic programming problems

Problem 4.3.6.1

Minimize $f(\mathbf{x}) = x_1^2 + x_2^2 + x_3^2$ such that $h_1(\mathbf{x}) = x_1 + x_2 + x_3 = 0$ and $h_2(\mathbf{x}) = x_1 + 2x_2 + 3x_3 - 1 = 0$.

Solution

Here, for the equality constrained problem the solution is obtained via the Lagrangian method with

$$L(\mathbf{x}, \boldsymbol{\lambda}) = x_1^2 + x_2^2 + x_3^2 + \lambda_1(x_1 + x_2 + x_3) + \lambda_2(x_1 + 2x_2 + 3x_3 - 1).$$

The necessary conditions for a minimum give:

$$\frac{\partial L}{\partial x_1} = 2x_1 + \lambda_1 + \lambda_2 = 0, \qquad x_1 = -\tfrac{1}{2}(\lambda_1 + \lambda_2)$$

$$\frac{\partial L}{\partial x_2} = 2x_2 + \lambda_1 + 2\lambda_2 = 0, \qquad x_2 = -\tfrac{1}{2}(\lambda_1 + 2\lambda_2)$$

$$\frac{\partial L}{\partial x_3} = 2x_3 + \lambda_1 + 3\lambda_2 = 0, \qquad x_3 = -\tfrac{1}{2}(\lambda_1 + 3\lambda_2).$$

Substituting into the equality constraints gives:

$$-\left(\tfrac{\lambda_1+\lambda_2}{2} + \tfrac{\lambda_1+2\lambda_2}{2} + \tfrac{\lambda_1+3\lambda_2}{2}\right) = 0, \text{ i.e. } \lambda_1 + 2\lambda_2 = 0$$

and

$$\tfrac{1}{2}(\lambda_1 + \lambda_2) + (\lambda_1 + 2\lambda_2) + \tfrac{3}{2}(\lambda_1 + 3\lambda_2) = -1, \text{ i.e. } 3\lambda_1 + 7\lambda_2 = -1.$$

Solving for the λ's: $\lambda_2 = -1$ and $\lambda_1 = 2$.

The candidate solution is therefore: $x_1^* = -\tfrac{1}{2}$, $x_2^* = 0$, $x_3^* = \tfrac{1}{2}$.

For the further analysis:

$$f(\mathbf{x} + \mathbf{\Delta x}) = f(\mathbf{x}) + \mathbf{\nabla}^T f \mathbf{\Delta x} + \tfrac{1}{2}\mathbf{\Delta x H \Delta x}$$

where $\mathbf{\nabla} f = (2x_1, 2x_2, 2x_3)^T$ and thus $\mathbf{\nabla} f(\mathbf{x}^*) = [-1, 0, 1]^T$ and $\mathbf{H} = \begin{bmatrix} 2 & 0 & 0 \\ 0 & 2 & 0 \\ 0 & 0 & 2 \end{bmatrix}$ is positive-definite.

For changes consistent with the constraints:

$$0 = \Delta h_1 = \mathbf{\nabla}^T h_1 \mathbf{\Delta x} + \tfrac{1}{2}\mathbf{\Delta x}^T \mathbf{\nabla}^2 h_1 \mathbf{\Delta x}, \quad \text{with} \quad \mathbf{\nabla} h_1 = \begin{bmatrix} 1 \\ 1 \\ 1 \end{bmatrix}$$

$$0 = \Delta h_2 = \mathbf{\nabla}^T h_2 \mathbf{\Delta x} + \tfrac{1}{2}\mathbf{\Delta x}^T \mathbf{\nabla}^2 h_2 \mathbf{\Delta x}, \quad \text{with} \quad \mathbf{\nabla} h_2 = \begin{bmatrix} 1 \\ 2 \\ 3 \end{bmatrix}.$$

It follows that $\Delta x_1 + \Delta x_2 + \Delta x_3 = 0$ and $\Delta x_1 + 2\Delta x_2 + 3\Delta x_3 = 0$ giving $-\Delta x_1 + \Delta x_3 = 0$ and thus

$$\Delta f(\mathbf{x}^*) = -\Delta x_1 + \Delta x_3 + \tfrac{1}{2}\mathbf{\Delta x}^T \mathbf{H \Delta x} = \tfrac{1}{2}\mathbf{\Delta x}^T \mathbf{H \Delta x} \geq 0.$$

The candidate point $\mathbf{x}^*$ above is therefore a constrained minimum.

Problem 4.3.6.2

Minimize $f(\mathbf{x}) = -2x_1 - 6x_2 + x_1^2 - 2x_1x_2 + 2x_2^2$ such that $x_1 \geq 0$, $x_2 \geq 0$ and $g_1(\mathbf{x}) = x_1 + x_2 \leq 2$ and $g_2(\mathbf{x}) = -x_1 + 2x_2 \leq 2$.

Solution

In matrix form the problem is:

minimize $f(\mathbf{x}) = \frac{1}{2}\mathbf{x}^T\mathbf{A}\mathbf{x} + \mathbf{b}^T\mathbf{x}$ where $\mathbf{A} = \begin{bmatrix} 2 & -2 \\ -2 & 4 \end{bmatrix}$, $\mathbf{b} = \begin{bmatrix} -2 \\ -6 \end{bmatrix}$

subject to the specified linear constraints.

First determine the unconstrained minimum:

$$\mathbf{x}^* = -\mathbf{A}^{-1}\mathbf{b} = -\frac{1}{4}\begin{bmatrix} 4 & 2 \\ 2 & 2 \end{bmatrix}\begin{bmatrix} -2 \\ -6 \end{bmatrix} = \begin{bmatrix} 5 \\ 4 \end{bmatrix}.$$

Test for violation of constraints:

$$x_1, x_2 > 0, \quad x_1 + x_2 = 9 > 2; \quad -x_1 + 2x_2 = -5 + 8 = 3 > 2.$$

Thus two constraints are violated. Considering each separately active assume firstly that $x_1 + x_2 = 2$. For this case $L = f(\mathbf{x}) + \lambda g_1(\mathbf{x}) = \frac{1}{2}\mathbf{x}^T\mathbf{A}\mathbf{x} + \mathbf{b}^T\mathbf{x} + \lambda(x_1 + x_2 - 2)$. The necessary conditions are:

$$\begin{bmatrix} \mathbf{A} & \begin{matrix} 1 \\ 1 \end{matrix} \\ 1 \quad 1 & 0 \end{bmatrix}\begin{bmatrix} \mathbf{x} \\ \lambda \end{bmatrix} = \begin{bmatrix} -\mathbf{b} \\ 2 \end{bmatrix},$$

i.e. solve

$$\begin{bmatrix} 2 & -2 & 1 \\ -2 & 4 & 1 \\ 1 & 1 & 0 \end{bmatrix}\begin{bmatrix} x_1 \\ x_2 \\ \lambda \end{bmatrix} = \begin{bmatrix} 2 \\ 6 \\ 2 \end{bmatrix}.$$

Let D be the determinant of the linear system matrix and D_i be the determinant of the linear system matrix with the ith column replaced by the right-hand side column vector. This, by Cramer's rule, gives $x_1 = \frac{D_1}{D} = \frac{4}{5}$; $x_2 = \frac{D_2}{D} = \frac{6}{5}$ and $\lambda = \frac{D_3}{D} = \frac{14}{5} > 0$, and therefore

$$x_1, x_2 > 0 \text{ and } -x_1 + 2x_2 = -\frac{4}{5} + \frac{2.6}{5} = \frac{8}{5} < 2.$$

Thus with all the constraints satisfied and $\lambda > 0$ the KKT sufficient conditions apply and $\mathbf{x}^* = \left[\frac{4}{5}; \frac{6}{5}\right]$ is a local constrained minimum. Indeed, since the problem is convex it is in fact the global minimum and no further investigation is required.

Problem 4.3.6.3

Minimize $f(\mathbf{x}) = x_1^2 + 4x_2^2 - 2x_1 + 8x_2$ such that $5x_1 + 2x_2 \leq 4$ and $x_1, \ x_2 \geq 0$.

Solution

Try various possibilities and test for satisfaction of the KKT conditions.

(i) For the unconstrained solution: solve $\nabla f(\mathbf{x}) = \mathbf{0}$, i.e. $2x_1 - 2 = 0$, $8x_2 + 8 = 0$ giving $x_1 = 1$; $x_2 = -1$ which violates the second non-negativity constraint.

(ii) Setting $x_2 = 0$ results in $x_1 = 1$ which violates the constraint $5x_1 + 2x_2 \leq 4$.

(iii) Similarly, setting $x_1 = 0$ results in $x_2 = -1$, which violates the second non-negativity constraint.

(iv) Setting $5x_1 + 2x_2 = 4$ active gives

$$L(\mathbf{x}, \boldsymbol{\lambda}) = x_1^2 + 4x_2^2 - 2x_1 + 8x_2 + \lambda(5x_1 + 2x_2 - 4)$$

with necessary conditions:

$$\frac{\partial L}{\partial x_1} = 2x_1 - 2 + 5\lambda = 0$$

$$\frac{\partial L}{\partial x_2} = 8x_2 + 8 + 2\lambda = 0$$

$$\frac{\partial L}{\partial \lambda} = 5x_1 + 2x_2 - 4 = 0.$$

Solving gives $\lambda = -\frac{1}{13}$, $x_1 = 1.92$ and $x_2 = -0.98 < 0$, which also violates a non-negativity constraint.

(v) Now setting $x_2 = 0$ in addition to $5x_1 + 2x_2 - 4 = 0$ results in an additional term $-\lambda_2 x_2$ in L giving

$$\frac{\partial L}{\partial x_1} = 2x_1 - 2 + 5\lambda_1 = 0 \text{ and } \frac{\partial L}{\partial \lambda} = 8x_2 + 8 + 2\lambda_1 - \lambda_2 = 0.$$

Solving together with the equalities gives $x_1 = \frac{4}{5}$; $x_2 = 0$ which satisfies all the constraints and with $\lambda_1 = \frac{2}{25} > 0$ and $\lambda_2 = \frac{4}{25} > 0$.

(vi) Finally setting $x_1 = 0$ and $5x_1 + 2x_2 - 4 = 0$ leads to the solution $x_1 = 0$, $x_2 = 2$ but with both associated λ's negative.

Thus the solution corresponds to case (v) i.e.:

$$x_1 = \tfrac{4}{5}, \; x_2 = 0 \text{ with } f^* = -\tfrac{24}{25}.$$

Problem 4.3.6.4

Minimize $f(\mathbf{x}) = \frac{1}{2}\mathbf{x}^T \mathbf{A}\mathbf{x} + \mathbf{b}^T\mathbf{x}$, $\mathbf{A} = \begin{bmatrix} 1 & -1 \\ -1 & 2 \end{bmatrix}$, $\mathbf{b} = \begin{bmatrix} -2 \\ 1 \end{bmatrix}$ such that

$$\begin{aligned} x_1, \; x_2 &\geq 0 \\ x_1 + x_2 &\leq 3 \\ 2x_1 - x_2 &\leq 4. \end{aligned}$$

Solution

First determine the unconstrained minimum:

$$\mathbf{x}^0 = -\mathbf{A}^{-1}\mathbf{b} = - \begin{bmatrix} 2 & 1 \\ 1 & 1 \end{bmatrix} \begin{bmatrix} -2 \\ 1 \end{bmatrix} = \begin{bmatrix} 3 \\ 1 \end{bmatrix}$$

$x_1, \; x_2 > 0$, $x_1 + x_2 = 3 + 1 = 4 > 3$ (constraint violation), and $2x_1 - x_2 = 5 > 4$ (constraint violation). Now consider the violated constraints separately active.

Firstly for $x_1 + x_2 - 3 = 0$ the Lagrangian is

$$L(\mathbf{x}, \lambda) = \tfrac{1}{2}\mathbf{x}^T \mathbf{A}\mathbf{x} + \mathbf{b}^T\mathbf{x} + \lambda(x_1 + x_2 - 3)$$

with necessary conditions:

$$\begin{bmatrix} \mathbf{A} & \begin{matrix} 1 \\ 1 \end{matrix} \\ 1 \quad 1 & 0 \end{bmatrix} \begin{bmatrix} \mathbf{x} \\ \lambda \end{bmatrix} = \begin{bmatrix} 2 \\ -1 \\ 3 \end{bmatrix}, \text{ i.e. } \begin{bmatrix} 1 & -1 & 1 \\ -1 & 2 & 1 \\ 1 & 1 & 0 \end{bmatrix} \begin{bmatrix} x_1 \\ x_2 \\ \lambda \end{bmatrix} = \begin{bmatrix} 2 \\ -1 \\ 3 \end{bmatrix}.$$

The solution of which is given by Cramer's rule:

$$x_1 = \frac{D_1}{D} = \frac{-12}{-5} = 2.4, \quad x_2 = \frac{D_2}{D} = \frac{-3}{-5} = 0.6 \text{ and } \lambda = \frac{D_3}{D} = \frac{-1}{-5} = 0.2.$$

This solution, however violates the last inequality constraint.

Similarly, setting $2x_1 - x_2 - 4 = 0$ gives the solution $\begin{bmatrix} x_1 \\ x_2 \\ \lambda \end{bmatrix} = \begin{bmatrix} 2.4 \\ 0.8 \\ 0.2 \end{bmatrix}$,

which violates the first constraint.

Finally try both constraints simultaneously active. This results in a system of four linear equations in four unknowns, the solution of which (do self) is $x_1 = 2\frac{1}{3}$, $x_2 = \frac{2}{3}$ and $\lambda_1 = \lambda_2 = 0.11 > 0$, which satisfies the KKT conditions. Thus $\mathbf{x}^* = \left[2\frac{1}{3}, \frac{2}{3} \right]^T$ with $f(\mathbf{x}^*) = 1.8$.

4.3.7   Application of the gradient projection method

Problem 4.3.7.1

Apply the gradient projection method to the following problem:

minimize $f(\mathbf{x}) = x_1^2 + x_2^2 + x_3^2 - 2x_1$

such that $2x_1 + x_2 + x_3 = 7$; $x_1 + x_2 + 2x_3 = 6$ and given initial starting point $\mathbf{x}^0 = [2, 2, 1]^T$. Perform the minimization for the first projected search direction only.

Solution

$$\mathbf{P} = \left(\mathbf{I} - \mathbf{A}^T (\mathbf{A}\mathbf{A}^T)^{-1} \mathbf{A} \right), \quad \mathbf{A} = \begin{bmatrix} 2 & 1 & 1 \\ 1 & 1 & 2 \end{bmatrix},$$

$$\mathbf{A}\mathbf{A}^T = \begin{bmatrix} 6 & 5 \\ 5 & 6 \end{bmatrix}, \quad (\mathbf{A}\mathbf{A}^T)^{-1} = \frac{1}{11} \begin{bmatrix} 6 & -5 \\ -5 & 6 \end{bmatrix},$$

$$(\mathbf{A}\mathbf{A}^T)^{-1} \mathbf{A} = \frac{1}{11} \begin{bmatrix} 6 & -5 \\ -5 & 6 \end{bmatrix} \begin{bmatrix} 2 & 1 & 1 \\ 1 & 1 & 2 \end{bmatrix} = \frac{1}{11} \begin{bmatrix} 7 & 1 & -4 \\ -4 & 1 & 7 \end{bmatrix},$$

$$\mathbf{A}^T (\mathbf{A}\mathbf{A}^T)^{-1} \mathbf{A} = \frac{1}{11} \begin{bmatrix} 2 & 1 \\ 1 & 1 \\ 1 & 2 \end{bmatrix} \begin{bmatrix} 7 & 1 & -4 \\ -4 & 1 & 7 \end{bmatrix} = \frac{1}{11} \begin{bmatrix} 10 & 3 & -1 \\ 3 & 2 & 3 \\ -1 & 3 & 10 \end{bmatrix}.$$

It now follows that

$$\mathbf{P} = \frac{1}{11} \begin{bmatrix} 1 & -3 & 1 \\ -3 & 9 & -3 \\ 1 & -3 & 1 \end{bmatrix} \quad \text{and} \quad \boldsymbol{\nabla} f(\mathbf{x}^0) = \begin{bmatrix} 2 \\ 4 \\ 2 \end{bmatrix}.$$

This gives the projected steepest descent direction as $-\mathbf{P}\nabla f = -\frac{1}{11}\begin{bmatrix} -8 \\ 24 \\ -8 \end{bmatrix}$

and therefore select the descent search direction as $\mathbf{u}^1 = [1, -3, 1]^T$.

Along the line through $\mathbf{x}^0$ in the direction $\mathbf{u}^1$, $\mathbf{x}$ is given by

$$\mathbf{x} = \mathbf{x}^0 + \lambda \begin{bmatrix} 1 \\ -3 \\ 1 \end{bmatrix} = \begin{bmatrix} 2+\lambda \\ 2-3\lambda \\ 1+\lambda \end{bmatrix} \text{ and}$$

$$f(\mathbf{x}^0 + \lambda\mathbf{u}) = F(\lambda) = (2+\lambda)^2 + (2-3\lambda)^2 + (1+\lambda)^2 - 2(2+\lambda).$$

For a minimum $\frac{dF}{d\lambda} = 2(2+\lambda) - 6(2-3\lambda) + 2(1+\lambda) - 2 = 0$, which gives $\lambda_1 = \frac{4}{11}$. Thus the next iterate is

$$\mathbf{x}^1 = \left(2 + \tfrac{4}{11}; 2 - \tfrac{12}{11}; 1 + \tfrac{4}{11}\right)^T = \left(2\tfrac{4}{11}; \tfrac{10}{11}; 1\tfrac{4}{11}\right)^T.$$

Problem 4.3.7.2

Apply the gradient projection method to the minimization of

$f(\mathbf{x}) = x_1^2 + x_2^2 - 2x_1 - 4x_2$ such that $h(\mathbf{x}) = x_1 + 4x_2 - 5 = 0$

with starting point $\mathbf{x}^0 = [1; 1]^T$.

Solution

The projection matrix is $\mathbf{P} = \mathbf{I} - \mathbf{A}^T(\mathbf{A}\mathbf{A}^T)^{-1}\mathbf{A}$.

Here $\mathbf{A} = [1 \ 4]$ and therefore

$$\mathbf{P} = \begin{bmatrix} 1 & 0 \\ 0 & 1 \end{bmatrix} - \begin{bmatrix} 1 \\ 4 \end{bmatrix}\left([1 \ 4]\begin{bmatrix} 1 \\ 4 \end{bmatrix}\right)^{-1}[1 \ 4] = \tfrac{1}{17}\begin{bmatrix} 16 & -4 \\ -4 & 1 \end{bmatrix}.$$

With $\nabla f(\mathbf{x}^0) = \begin{bmatrix} 2x_1^0 - 2 \\ 2x_2^0 - 4 \end{bmatrix} = \begin{bmatrix} 0 \\ -2 \end{bmatrix}$ the search direction is

$$\mathbf{u}^1 = -\mathbf{P}\nabla f(\mathbf{x}^0) = -\tfrac{1}{17}\begin{bmatrix} 16 & -4 \\ -4 & 1 \end{bmatrix}\begin{bmatrix} 0 \\ -2 \end{bmatrix} = -\tfrac{1}{17}\begin{bmatrix} 8 \\ -2 \end{bmatrix}$$

or more conveniently $\mathbf{u}^1 = \begin{bmatrix} -4 \\ 1 \end{bmatrix}$.

Thus $\mathbf{x} = \mathbf{x}^0 + \lambda \mathbf{u}^1 = \begin{bmatrix} 1 - 4\lambda \\ 1 + \lambda \end{bmatrix}$ for the line search. Along the search line

$$F(\lambda) = (1 - 4\lambda)^2 + (1 + \lambda)^2 - 2(1 - 4\lambda) + 4(1 + \lambda) = 17\lambda^2 - 2\lambda - 4$$

with minimum occurring where $\frac{dF}{d\lambda} = 34\lambda - 2 = 0$, giving $\lambda_1 = \frac{1}{17}$.

Thus $\mathbf{x}^1 = \mathbf{x}^0 + \frac{1}{17}\mathbf{u}^1 = \begin{bmatrix} 1 \\ 1 \end{bmatrix} + \frac{1}{17}\begin{bmatrix} -4 \\ 1 \end{bmatrix} = \begin{bmatrix} 0.7647 \\ 1.0588 \end{bmatrix}$.

Next, compute $\mathbf{u}^2 = -\mathbf{P}\boldsymbol{\nabla}f(\mathbf{x}^1) = -\frac{1}{17}\begin{bmatrix} 16 & -4 \\ -4 & 1 \end{bmatrix}\begin{bmatrix} 0.4706 \\ 1.8824 \end{bmatrix} = \begin{bmatrix} 0.0 \\ 0.0 \end{bmatrix}$.

Since the projected gradient equals $\mathbf{0}$, the point $\mathbf{x}^1$ is the optimum $\mathbf{x}^*$, with $f(\mathbf{x}^*) = -4.059$.

4.3.8   Application of the augmented Lagrangian method

Problem 4.3.8.1

Minimize $f(\mathbf{x}) = 6x_1^2 + 4x_1x_2 + 3x_2^2$ such that $h(\mathbf{x}) = x_1 + x_2 - 5 = 0$, by means of the augmented Lagrangian multiplier method.

Solution

Here $\mathcal{L}(\mathbf{x}, \lambda, \rho) = 6x_1^2 + 4x_1x_2 + 3x_2^2 + \lambda(x_1 + x_2 - 5) + \rho(x_1 + x_2 - 5)^2$ with necessary conditions for a stationary point:

$$\frac{\partial \mathcal{L}}{\partial x_1} = 0 \quad \Rightarrow \quad x_1(12 + 2\rho) + x_2(4 + 2\rho) = 10\rho - \lambda$$

$$\frac{\partial \mathcal{L}}{\partial x_2} = 0 \quad \Rightarrow \quad x_1(4 + 2\rho) + x_2(6 + 2\rho) = 10\rho - \lambda.$$

With $\rho = 1$ the above become $14x_1 + 6x_2 = 10 - \lambda = 6x_1 + 8x_2$ and it follows that

$$x_2 = 4x_1 \text{ and } x_1 = \frac{10 - \lambda}{38}.$$

The iterations now proceed as follows.

Iteration 1: $\lambda^1 = 0$

$$x_1 = \frac{10}{38} = 0.2632, \quad x_2 = 1.0526$$

$$h = x_1 + x_2 - 5 = 0.2632 + 1.0526 - 5 = -3.6842$$

and for the next iteration

$$\lambda^2 = \lambda^1 + 2\rho h = 0 + (2)(1)(-3.6842) = -7.3684.$$

Iteration 2:

$$x_1 = \frac{10-\lambda}{38} = \frac{10+7.3684}{38} = 0.4571$$

and it follows that $x_2 = 1.8283$ and thus $h = 0.4571 + 1.8283 - 5 = -2.7146$ with the next multiplier value given by

$$\lambda^3 = \lambda^2 + (2)(1)h = -7.3684 + 2(-2.7146) = -12.7978.$$

Iteration 3:

Now increase ρ to 10; then $x_1 = \frac{100-\lambda}{128}$ and $x_2 = 4x_1$.

Thus $x_1 = \frac{100+12.7978}{128} = 0.8812$ and $x_2 = 3.5249$ with

$$\lambda^4 = -12.7978 + 2(10)(-0.5939) = -24.675.$$

Iteration 4:

$$x_1 = \frac{100+24.675}{128} = 0.9740, \quad x_2 = 3.8961 \text{ with } \lambda^5 = -27.27.$$

Iteration 5: Iteration 5 gives $x_1 = 0.9943$, $x_2 = 3.9772$ with $\lambda^6 = -27.84$ and rapid convergence is obtained to the solution $\mathbf{x}^* = [1, 4]^T$.

4.3.9 Application of the sequential quadratic programming method

Problem 4.3.9.1

Minimize $f(\mathbf{x}) = 2x_1^2 + 2x_2^2 - 2x_1x_2 - 4x_1 - 6x_2$, such that $h(\mathbf{x}) = 2x_1^2 - x_2 = 0$ by means of the SQP method and starting point $\mathbf{x}^0 = [0, 1]^T$.

Solution
Here $L(\mathbf{x}, \lambda) = f(\mathbf{x}) + \lambda h(\mathbf{x})$ and the necessary conditions for a stationary point are

$$\frac{\partial L}{\partial x_1} = 4x_1 - 2x_2 - 4 + \lambda 4x_1 = 0$$

$$\frac{\partial L}{\partial x_2} = 4x_2 - 2x_1 - 6 - \lambda = 0$$

$$h(\mathbf{x}) = 2x_1^2 - x_2 = 0.$$

This system of non-linear equations may be written in vector form as $\mathbf{c}(\mathbf{X}) = \mathbf{0}$, where $\mathbf{X} = [x_1, x_2, \lambda]^T$. If an approximate solution $(\mathbf{x}^k, \lambda^k)$ to this system is available a possible improvement may be obtained by solving the linearized system via Newton's method:

$$\left[\frac{\partial \mathbf{c}}{\partial \mathbf{x}}\right]^T \left[\begin{array}{c} \mathbf{x} - \mathbf{x}^k \\ \lambda - \lambda^k \end{array}\right] = -\mathbf{c}(\mathbf{x}^k, \lambda^k),$$

where $\frac{\partial \mathbf{c}}{\partial \mathbf{x}}$ is the Jacobian of the system $\mathbf{c}$.

In detail this linear system becomes

$$\left[\begin{array}{ccc} 4 & -2 & 4x_1 \\ -2 & 4 & -1 \\ 4x_1 & -1 & 0 \end{array}\right] \left[\begin{array}{c} \mathbf{x} - \mathbf{x}^k \\ \lambda - \lambda^k \end{array}\right] = -\left[\begin{array}{c} c_1 \\ c_2 \\ c_3 \end{array}\right]$$

and for the approximation $\mathbf{x}^0 = [0, 1]^T$ and $\lambda^0 = 0$, the system may be written as

$$\left[\begin{array}{ccc} 4 & -2 & 0 \\ -2 & 4 & -1 \\ 0 & -1 & 0 \end{array}\right] \left[\begin{array}{c} s_1 \\ s_2 \\ \Delta\lambda \end{array}\right] = -\left[\begin{array}{c} -6 \\ -2 \\ -1 \end{array}\right]$$

which has the solution $s_1 = 1$, $s_2 = -1$ and $\Delta\lambda = -8$. Thus after the first iteration $x_1^1 = 0 + 1 = 1$, $x_2^1 = 1 - 1 = 0$ and $\lambda^1 = 0 - 8 = -8$.

The above Newton step is equivalent to the solution of the following quadratic programming (QP) problem set up at $\mathbf{x}^0 = [0, 1]^T$ with $\lambda^0 = 0$ (see Section 3.5.3):

minimize $F(\mathbf{s}) = \frac{1}{2}\mathbf{s}^T \left[\begin{array}{cc} 4 & -2 \\ -2 & 4 \end{array}\right] \mathbf{s} + [-6 \ -2]\mathbf{s} - 4$ such that $s_2 + 1 = 0$.

Setting $s_2 = -1$ gives $F(\mathbf{s}) = F(s_1) = 2s_1^2 - 4s_1 - 2$ with minimum at $s_1 = 1$. Thus after the first iteration $x_1^1 = s_1 = 1$ and $x_2^1 = s_2 + 1 = 0$, giving $\mathbf{x}^1 = [1, 0]^T$, which corresponds to the Newton step solution.

Here, in order to simplify the computation, λ is not updated. Continue by setting up the next QP: with $f(\mathbf{x}^1) = -2$; $\boldsymbol{\nabla} f(\mathbf{x}^1) = [0, -8]^T$; $h(\mathbf{x}^1) = 2$; $\boldsymbol{\nabla} h(\mathbf{x}^1) = [4, -1]^T$ the QP becomes:

$$\begin{aligned} \text{minimize } F(\mathbf{s}) &= \frac{1}{2}\mathbf{s}^T \left[\begin{array}{cc} 4 & -2 \\ -2 & 4 \end{array}\right] \mathbf{s} + [0, -8]\mathbf{s} - 2 \\ &= 2s_1^2 - 2s_1 s_2 + 2s_2^2 - 8s_2 - 2 \end{aligned}$$

with constraint

$$h(\mathbf{s}) = [4, -1] \begin{bmatrix} s_1 \\ s_2 \end{bmatrix} + 2 = 0$$
$$= 4s_1 - s_2 + 2 = 0.$$

Substituting $s_2 = 4s_1 + 2$ in $F(\mathbf{s})$ results in the unconstrained minimization of the single variable function:

$$F(s_1) = 26s_1^2 - 4s_1 - 10.$$

Setting $\frac{dF}{ds_1} = 52s_1 - 4 = 0$, gives $s_1 = 0.07692$ and $s_2 = 2.30769$. Thus $\mathbf{x}^2 = [1 + s_1, 0 + s_2] = [1.07692, 2.30769]^T$. Continuing in this manner yields $\mathbf{x}^3 = [1.06854, 2.2834]^T$ and $\mathbf{x}^4 = [1.06914, 2.28575]^T$, which represents rapid convergence to the exact optimum $\mathbf{x}^* = [1.06904, 2.28569]^T$, with $f(\mathbf{x}^*) = -10.1428$. Substituting $\mathbf{x}^*$ into the necessary conditions for a stationary point of L, gives the value of $\lambda^* = 1.00468$.

Chapter 5

SOME BASIC OPTIMIZATION THEOREMS

5.1 Characterization of functions and minima

Theorem 5.1.1

If $f(\mathbf{x})$ is a differentiable function over the convex set $X \subseteq \mathbb{R}^n$ then $f(\mathbf{x})$ is convex over X *if and only if*

$$f(\mathbf{x}^2) \geq f(\mathbf{x}^1) + \boldsymbol{\nabla}^T f(\mathbf{x}^1)(\mathbf{x}^2 - \mathbf{x}^1) \tag{5.1}$$

for all $\mathbf{x}^1, \mathbf{x}^2 \in X$.

Proof

If $f(\mathbf{x})$ is convex over X then by the definition (1.9), for all $\mathbf{x}^1, \mathbf{x}^2 \in X$ and for all $\lambda \in (0, 1]$

$$f(\lambda \mathbf{x}^2 + (1 - \lambda)\mathbf{x}^1) \leq \lambda f(\mathbf{x}^2) + (1 - \lambda)f(\mathbf{x}^1) \tag{5.2}$$

i.e.

$$\frac{f(\mathbf{x}^1 + \lambda(\mathbf{x}^2 - \mathbf{x}^1)) - f(\mathbf{x}^1)}{\lambda} \leq f(\mathbf{x}^2) - f(\mathbf{x}^1). \tag{5.3}$$

© Springer International Publishing AG, part of Springer Nature 2018
J.A. Snyman and D.N. Wilke, *Practical Mathematical Optimization,*
Springer Optimization and Its Applications 133,
https://doi.org/10.1007/978-3-319-77586-9_5

Taking the limit as $\lambda \to 0$ it follows that

$$\left.\frac{df(\mathbf{x}^1)}{d\lambda}\right|_{\mathbf{x}^2-\mathbf{x}^1} \leq f(\mathbf{x}^2) - f(\mathbf{x}^1). \tag{5.4}$$

The directional derivative on the left hand side may also, by (1.16) be written as

$$\left.\frac{df(\mathbf{x}^1)}{d\lambda}\right|_{\mathbf{x}^2-\mathbf{x}^1} = \boldsymbol{\nabla}^T f(\mathbf{x})(\mathbf{x}^2 - \mathbf{x}^1). \tag{5.5}$$

Substituting (5.5) into (5.4) gives $f(\mathbf{x}^2) \geq f(\mathbf{x}^1) + \boldsymbol{\nabla}^T f(\mathbf{x}^1)(\mathbf{x}^2 - \mathbf{x}^1)$, i.e. (5.1) is true.

Conversely, if (5.1) holds, then for $\mathbf{x} = \lambda\mathbf{x}^2 + (1-\lambda)\mathbf{x}^1 \in X,\ \lambda \in (0,1]$:

$$f(\mathbf{x}^2) \geq f(\mathbf{x}) + \boldsymbol{\nabla}^T f(\mathbf{x})(\mathbf{x}^2 - \mathbf{x}) \tag{5.6}$$
$$f(\mathbf{x}^1) \geq f(\mathbf{x}) + \boldsymbol{\nabla}^T f(\mathbf{x})(\mathbf{x}^1 - \mathbf{x}). \tag{5.7}$$

Multiplying (5.6) by λ and (5.7) by $(1-\lambda)$ and adding gives

$$\lambda f(\mathbf{x}^2) + (1-\lambda)f(\mathbf{x}^1) - f(\mathbf{x}) \geq \boldsymbol{\nabla}^T f(\mathbf{x})(\lambda(\mathbf{x}^2 - \mathbf{x}) + (1-\lambda)(\mathbf{x}^1 - \mathbf{x})) = 0$$

since $\lambda(\mathbf{x}^2 - \mathbf{x}) + (1-\lambda)(\mathbf{x}^1 - \mathbf{x}) = \mathbf{0}$ and it follows that

$$f(\mathbf{x}) = f(\lambda\mathbf{x}^2 + (1-\lambda)\mathbf{x}^1) \leq \lambda f(\mathbf{x}^2) + (1-\lambda)f(\mathbf{x}^1),$$

i.e. $f(\mathbf{x})$ is convex and the theorem is proved. $\qquad\square$

(Clearly if $f(\mathbf{x})$ is to be strictly convex, the strict inequality $>$ applies in (5.1).)

Theorem 5.1.2

If $f(\mathbf{x}) \in C^2$ over an open convex set $X \subseteq \mathbb{R}^n$, then $f(\mathbf{x})$ is convex *if and only if* $\mathbf{H}(\mathbf{x})$ is positive semi-definite for all $\mathbf{x} \in X$.

Proof

If $\mathbf{H}(\mathbf{x})$ is positive semi-definite for all $\mathbf{x} \in X$, then for all $\mathbf{x}^1,\ \mathbf{x}^2$ in X it follows by the Taylor expansion (1.21) that

$$\begin{aligned} f(\mathbf{x}^2) = f(\mathbf{x}^1 + (\mathbf{x}^2 - \mathbf{x}^1)) = f(\mathbf{x}^1) + \boldsymbol{\nabla}^T f(\mathbf{x}^1)(\mathbf{x}^2 - \mathbf{x}^1) \\ + \tfrac{1}{2}(\mathbf{x}^2 - \mathbf{x}^1)^T \mathbf{H}(\bar{\mathbf{x}})(\mathbf{x}^2 - \mathbf{x}^1) \end{aligned} \tag{5.8}$$

where $\bar{\mathbf{x}} = \mathbf{x}^1 + \theta(\mathbf{x}^2 - \mathbf{x}^1))$, $\theta \in [0, 1]$. Since $\mathbf{H}(\mathbf{x})$ is positive semi-definite it follows directly from (5.8) that $f(\mathbf{x}^2) \geq f(\mathbf{x}^1) + \nabla^T f(\mathbf{x}^1)(\mathbf{x}^2 - \mathbf{x}^1)$ and therefore by Theorem 5.1.1 $f(\mathbf{x})$ is convex.

Conversely if $f(\mathbf{x})$ is convex, then from (5.1) for all $\mathbf{x}^1$, $\mathbf{x}^2$ in X

$$f(\mathbf{x}^2) \geq f(\mathbf{x}^1) + \nabla^T f(\mathbf{x}^1)(\mathbf{x}^2 - \mathbf{x}^1). \tag{5.9}$$

Also the Taylor expansion (5.8) above applies, and comparison of (5.8) and (5.9) implies that

$$\frac{1}{2}(\mathbf{x}^2 - \mathbf{x}^1)^T \mathbf{H}(\bar{\mathbf{x}})(\mathbf{x}^2 - \mathbf{x}^1) \geq 0. \tag{5.10}$$

Clearly since (5.10) must apply for all $\mathbf{x}^1$, $\mathbf{x}^2$ in X and since $\bar{\mathbf{x}}$ is assumed to vary continuously with $\mathbf{x}^1$, $\mathbf{x}^2$, (5.10) must be true for any $\bar{\mathbf{x}}$ in X, i.e. $\mathbf{H}(\mathbf{x})$ is positive semi-definite for all $\mathbf{x} \in X$, which concludes the proof. $\square$

Theorem 5.1.3

If $f(\mathbf{x}) \in C^2$ over an open convex set $X \subseteq \mathbb{R}^n$, then if the Hessian matrix $\mathbf{H}(\mathbf{x})$ is positive-definite for all $\mathbf{x} \in X$, then $f(\mathbf{x})$ is *strictly* convex over X.

Proof

For any $\mathbf{x}^1$, $\mathbf{x}^2$ in X it follows by the Taylor expansion (1.21) that

$$f(\mathbf{x}^2) = f(\mathbf{x}^1 + (\mathbf{x}^2 - \mathbf{x}^1))$$
$$= f(\mathbf{x}^1) + \nabla^T f(\mathbf{x}^1)(\mathbf{x}^2 - \mathbf{x}^1) + \frac{1}{2}(\mathbf{x}^2 - \mathbf{x}^1)^T \mathbf{H}(\bar{\mathbf{x}})(\mathbf{x}^2 - \mathbf{x}^1)$$

where $\bar{\mathbf{x}} = \mathbf{x}^1 + \theta(\mathbf{x}^2 - \mathbf{x}^1)$, $\theta \in [0, 1]$. Since $\mathbf{H}(\mathbf{x})$ is positive-definite it follows directly that $f(\mathbf{x}^2) > f(\mathbf{x}^1) + \nabla^T f(\mathbf{x}^1)(\mathbf{x}^2 - \mathbf{x}^1)$ and therefore by Theorem 5.1.1 $f(\mathbf{x})$ is strictly convex and the theorem is proved. $\square$

Theorem 5.1.4

If $f(\mathbf{x}) \in C^2$ over the convex set $X \subseteq \mathbb{R}^n$ and $f(\mathbf{x})$ is convex over X, then any interior local minimum of $f(\mathbf{x})$ is a global minimum.

Proof
If $f(\mathbf{x})$ is convex then by (5.1):

$$f(\mathbf{x}^2) \geq f(\mathbf{x}^1) + \nabla^T f(\mathbf{x}^1)(\mathbf{x}^2 - \mathbf{x}^1)$$

for all $\mathbf{x}^1$, $\mathbf{x}^2 \in X$. In particular for any point $\mathbf{x}^2 = \mathbf{x}$ and in particular $\mathbf{x}^1 = \mathbf{x}^*$ an interior local minimum, it follows that

$$f(\mathbf{x}) \geq f(\mathbf{x}^*) + \boldsymbol{\nabla}^T f(\mathbf{x}^*)(\mathbf{x} - \mathbf{x}^*).$$

Since the necessary condition $\boldsymbol{\nabla}^T f(\mathbf{x}^*) = \mathbf{0}$ applies at $\mathbf{x}^*$, the above reduces to $f(\mathbf{x}) \geq f(\mathbf{x}^*)$ and therefore $\mathbf{x}^*$ is a global minimum. □

Theorem 5.1.5

More generally, let $f(\mathbf{x})$ be strictly convex on the convex set X, but $f(\mathbf{x})$ not necessarily $\in C^2$, then a strict local minimum is the global minimum.

Proof

Let $\mathbf{x}^0$ be the strict local minimum in a δ-neighbourhood and $\mathbf{x}^*$ the global minimum, $\mathbf{x}^0$, $\mathbf{x}^* \in X$ and assume that $\mathbf{x}^0 \neq \mathbf{x}^*$. Then there exists an ε, $0 < \varepsilon < \delta$, with $f(\mathbf{x}) > f(\mathbf{x}^0)$ for all $\mathbf{x}$ such that $\|\mathbf{x} - \mathbf{x}^0\| < \varepsilon$.

A convex combination of $\mathbf{x}^0$ and $\mathbf{x}^*$ is given by

$$\widehat{\mathbf{x}} = \lambda \mathbf{x}^* + (1 - \lambda)\mathbf{x}^0.$$

Note that if $\lambda \to 0$ then $\widehat{\mathbf{x}} \to \mathbf{x}^0$.

As $f(\mathbf{x})$ is strictly convex it follows that

$$f(\widehat{\mathbf{x}}) < \lambda f(\mathbf{x}^*) + (1 - \lambda)f(\mathbf{x}^0) \leq f(\mathbf{x}^0) \text{ for all } \lambda \in (0, 1).$$

In particular this holds for λ arbitrarily small and hence for λ such that $\|\widehat{\mathbf{x}}(\lambda) - \mathbf{x}^0\| < \varepsilon$. But as $\mathbf{x}^0$ is the strict local minimum

$$f(\widehat{\mathbf{x}}) > f(\mathbf{x}^0) \text{ for all } \widehat{\mathbf{x}} \text{ with } \|\widehat{\mathbf{x}} - \mathbf{x}^0\| < \varepsilon.$$

This contradicts the fact that

$$f(\widehat{\mathbf{x}}) < f(\mathbf{x}^0)$$

that followed from the convexity and the assumption that $\mathbf{x}^0 \neq \mathbf{x}^*$. It follows therefore that $\mathbf{x}^0 \equiv \mathbf{x}^*$ which completes the proof. □

5.2 Equality constrained problem

Theorem 5.2.1

In problem (3.5), let f and $h_j \in C^1$ and assume that the Jacobian matrix $\left[\dfrac{\partial \mathbf{h}(\mathbf{x}^*)}{\partial \mathbf{x}}\right]$ is of rank r. Then the *necessary conditions* for a interior local minimum $\mathbf{x}^*$ of the equality constrained problem (3.5) is that $\mathbf{x}^*$ must coincide with the stationary point $(\mathbf{x}^*, \boldsymbol{\lambda}^*)$ of the Lagrange function L, i.e. that there exits a $\boldsymbol{\lambda}^*$ such that

$$\frac{\partial L}{\partial x_i}(\mathbf{x}^*, \boldsymbol{\lambda}^*) = 0, \ i = 1, 2, \ldots, n; \ \ \frac{\partial L}{\partial \lambda_j}(\mathbf{x}^*, \boldsymbol{\lambda}^*) = 0, \ j = 1, 2, \ldots, r.$$

Note: Here the elements of the Jacobian matrix are taken as $\left[\dfrac{\partial \mathbf{h}}{\partial \mathbf{x}}\right]_{ij} = \dfrac{\partial h_j}{\partial x_i}$, i.e. $\left[\dfrac{\partial \mathbf{h}}{\partial \mathbf{x}}\right] = [\boldsymbol{\nabla} h_1, \boldsymbol{\nabla} h_2, \ldots, \boldsymbol{\nabla} h_r]$.

Proof

Since f and $h_j \in C^1$, it follows for an interior local minimum at $\mathbf{x} = \mathbf{x}^*$, that

$$df = \boldsymbol{\nabla}^T f(\mathbf{x}^*) d\mathbf{x} = 0 \tag{5.11}$$

since $df \geq 0$ for $d\mathbf{x}$ and for $-d\mathbf{x}$, for all perturbations $d\mathbf{x}$ which are consistent with the constraints, i.e. for all $d\mathbf{x}$ such that

$$dh_j = \boldsymbol{\nabla}^T h_j(\mathbf{x}^*) d\mathbf{x} = 0, \ j = 1, 2, \ldots, r. \tag{5.12}$$

Consider the Lagrange function

$$L(\mathbf{x}, \boldsymbol{\lambda}) = f(\mathbf{x}) + \sum_{j=1}^{r} \lambda_j h_j(\mathbf{x}).$$

The differential of L is given by

$$dL = df + \sum_{j=1}^{r} \lambda_j dh_j$$

and it follows from (5.11) and (5.12) that at $\mathbf{x}^*$

$$dL = \frac{\partial L}{\partial x_1} dx_1 + \frac{\partial L}{\partial x_2} dx_2 + \cdots + \frac{\partial L}{\partial x_n} dx_n = 0 \tag{5.13}$$

for all $d\mathbf{x}$ such that $\mathbf{h}(\mathbf{x}) = \mathbf{0}$, i.e. $\mathbf{h}(\mathbf{x}^* + d\mathbf{x}) = \mathbf{0}$.

Choose Lagrange multipliers λ_j, $j = 1, \ldots, r$ such that at $\mathbf{x}^*$

$$\frac{\partial L}{\partial x_j}(\mathbf{x}^*, \boldsymbol{\lambda}) = \frac{\partial f}{\partial x_j}(\mathbf{x}^*) + \left[\frac{\partial \mathbf{h}}{\partial x_j}(\mathbf{x}^*)\right]^T \boldsymbol{\lambda} = 0, \quad j = 1, 2, \ldots, r. \quad (5.14)$$

The solution of this system provides the vector $\boldsymbol{\lambda}^*$. Here the r variables, x_j, $j = 1, 2, \ldots, r$, may be any appropriate set of r variables from the set x_i, $i = 1, 2, \ldots, n$. A unique solution for $\boldsymbol{\lambda}^*$ exists as it is assumed that $\left[\dfrac{\partial \mathbf{h}(\mathbf{x}^*)}{\partial \mathbf{x}}\right]$ is of rank r. It follows that (5.13) can now be written as

$$dL = \frac{\partial L}{\partial x_{r+1}}(\mathbf{x}^*, \boldsymbol{\lambda}^*)dx_{r+1} + \cdots + \frac{\partial L}{\partial x_n}(\mathbf{x}^*, \boldsymbol{\lambda}^*)dx_n = 0. \quad (5.15)$$

Again consider the constraints $h_j(\mathbf{x}) = 0$, $j = 1, 2, \ldots, r$. If these equations are considered as a system of r equations in the unknowns x_1, x_2, $\ldots$, x_r, these dependent unknowns can be solved for in terms of x_{r+1}, $\ldots$, x_n. Hence the latter $n - r$ variables are the independent variables. For any choice of these independent variables, the other dependent variables x_1, $\ldots$, x_r, are determined by solving $\mathbf{h}(\mathbf{x}) = [h_1(\mathbf{x}), h_2(\mathbf{x}), \ldots, h_r(\mathbf{x})]^T = \mathbf{0}$. In particular x_{r+1} to x_n may by varied one by one at $\mathbf{x}^*$, and it follows from (5.15) that

$$\frac{\partial L}{\partial x_j}(\mathbf{x}^*, \boldsymbol{\lambda}^*) = 0, \quad j = r + 1, \ldots, n$$

and, together with (5.14) and the constraints $\mathbf{h}(\mathbf{x}) = \mathbf{0}$, it follows that the necessary conditions for an interior local minimum can be written as

$$\begin{aligned} \frac{\partial L}{\partial x_i}(\mathbf{x}^*, \boldsymbol{\lambda}^*) &= 0, \quad i = 1, 2, \ldots, n \\ \frac{\partial L}{\partial \lambda_j}(\mathbf{x}^*, \boldsymbol{\lambda}^*) &= 0, \quad j = 1, 2, \ldots, r \end{aligned} \quad (5.16)$$

or

$$\boldsymbol{\nabla}_{\mathbf{x}} L(\mathbf{x}^*, \boldsymbol{\lambda}^*) = \mathbf{0} \text{ and } \boldsymbol{\nabla}_{\boldsymbol{\lambda}} L(\mathbf{x}^*, \boldsymbol{\lambda}^*) = \mathbf{0}.$$

$\square$

Note that (5.16) provides $n + r$ equations in the $n + r$ unknowns x_1^*, x_2^*, $\ldots$, x_n^*, λ_1^*, $\ldots$, λ_r^*. The solutions of these, in general non-linear, equations will be candidate solutions for problem (3.5).

As a general rule, if possible, it is advantageous to solve explicitly for any of the variables in the equality constraints in terms of the others, so as to reduce the number of variables and constraints in (3.5).

Note on the existence of $\boldsymbol{\lambda}^*$

Up to this point it has been assumed that $\boldsymbol{\lambda}^*$ does indeed exist. In Theorem 5.2.1 it is assumed that the equations

$$\frac{\partial L}{\partial x_j}(\mathbf{x}^*, \boldsymbol{\lambda}) = 0$$

apply for a certain appropriate set of r variables from the set x_i, $i = 1, 2, \ldots, n$. Thus $\boldsymbol{\lambda}$ may be solved for to find $\boldsymbol{\lambda}^*$, via the linear system

$$\frac{\partial f}{\partial x_j}(\mathbf{x}^*) + \left[\frac{\partial \mathbf{h}}{\partial x_j}\right]^T \boldsymbol{\lambda} = 0, \quad j = 1, 2, \ldots, r.$$

This system can be solved if there exists a $r \times r$ submatrix $H_r \subset \left[\frac{\partial \mathbf{h}}{\partial \mathbf{x}}\right]^*$, evaluated at $\mathbf{x}^*$, such that H_r is non-singular. This is the same as requiring that $\left[\frac{\partial \mathbf{h}}{\partial \mathbf{x}}\right]^*$ be of rank r at the optimal point $\mathbf{x}^*$. This result is interesting and illuminating but, for obvious reasons, of little practical value. It does, however, emphasise the fact that it may not be assumed that multipliers will exist for every problem.

Theorem 5.2.2

If it is assumed that:

(i) $f(\mathbf{x})$ has a bounded local minimum at $\mathbf{x}^*$ (associated with this minimum there exists a $\boldsymbol{\lambda}^*$ found by solving (5.16)); and

(ii) if $\boldsymbol{\lambda}$ is chosen arbitrarily in the neighbourhood of $\boldsymbol{\lambda}^*$ then $L(\mathbf{x}, \boldsymbol{\lambda})$ has a local minimum $\mathbf{x}^0$ with respect to $\mathbf{x}$ in the neighbourhood of $\mathbf{x}^*$,

then for the classical equality constrained minimization problem (3.5), the Lagrange function $L(\mathbf{x}, \boldsymbol{\lambda})$ has a saddle point at $(\mathbf{x}^*, \boldsymbol{\lambda}^*)$. Note that assumption (ii) implies that

$$\boldsymbol{\nabla}_\mathbf{x} L(\mathbf{x}^0, \boldsymbol{\lambda}) = \mathbf{0}$$

and that a local minimum may indeed be expected if the Hessian matrix of L is positive-definite at $(\mathbf{x}^*, \boldsymbol{\lambda}^*)$.

Proof

Consider the neighbourhood of $(\mathbf{x}^*, \boldsymbol{\lambda}^*)$. As a consequence of assumption (ii), applied with $\boldsymbol{\lambda} = \boldsymbol{\lambda}^*$ ($\mathbf{x}^0 = \mathbf{x}^*$), it follows that

$$L(\mathbf{x}, \boldsymbol{\lambda}^*) \geq L(\mathbf{x}^*, \boldsymbol{\lambda}^*) = f(\mathbf{x}^*) = L(\mathbf{x}^*, \boldsymbol{\lambda}). \tag{5.17}$$

The equality on the right holds as $\mathbf{h}(\mathbf{x}^*) = \mathbf{0}$. The relationship (5.17) shows that $L(\mathbf{x}, \boldsymbol{\lambda})$ has a degenerate saddle point at $(\mathbf{x}^*, \boldsymbol{\lambda}^*)$; for a regular saddle point it holds that

$$L(\mathbf{x}, \boldsymbol{\lambda}^*) \geq L(\mathbf{x}^*, \boldsymbol{\lambda}^*) \geq L(\mathbf{x}^*, \boldsymbol{\lambda}).$$

□

Theorem 5.2.3

If the above assumptions (i) and (ii) in Theorem 5.2.2 hold, then it follows for the bounded minimum $\mathbf{x}^*$, that

$$f(\mathbf{x}^*) = \max_{\boldsymbol{\lambda}} \left(\min_{\mathbf{x}} L(\mathbf{x}, \boldsymbol{\lambda}) \right).$$

Proof

From (ii) it is assumed that for a given $\boldsymbol{\lambda}$ the

$$\min_{\mathbf{x}} L(\mathbf{x}, \boldsymbol{\lambda}) = h(\boldsymbol{\lambda}) \text{ (the dual function)}$$

exists, and in particular for $\boldsymbol{\lambda} = \boldsymbol{\lambda}^*$

$$\min_{\mathbf{x}} L(\mathbf{x}, \boldsymbol{\lambda}^*) = h(\boldsymbol{\lambda}^*) = f(\mathbf{x}^*) \quad \text{(from (ii) } \mathbf{x}^0 = \mathbf{x}^*).$$

Let X denote the set of points such that $\mathbf{h}(\mathbf{x}) = \mathbf{0}$ for all $\mathbf{x} \in X$. It follows that

$$\min_{\mathbf{x} \in X} L(\mathbf{x}, \boldsymbol{\lambda}) = \min_{\mathbf{x} \in X} f(\mathbf{x}) = f(\mathbf{x}^*) = h(\boldsymbol{\lambda}^*) \tag{5.18}$$

and

$$h(\boldsymbol{\lambda}) = \min_{\mathbf{x}} L(\mathbf{x}, \boldsymbol{\lambda}) \leq \min_{\mathbf{x} \in X} L(\mathbf{x}, \boldsymbol{\lambda}) = \min_{\mathbf{x} \in X} f(\mathbf{x}) = f(\mathbf{x}^*) = h(\boldsymbol{\lambda}^*).$$

$$\tag{5.19}$$

Hence $h(\boldsymbol{\lambda}^*)$ is the maximum of $h(\boldsymbol{\lambda})$. Combining (5.18) and (5.19) yields that

$$f(\mathbf{x}^*) = h(\boldsymbol{\lambda}^*) = \max_{\boldsymbol{\lambda}} h(\boldsymbol{\lambda}) = \max_{\boldsymbol{\lambda}} \left(\min_{\mathbf{x}} L(\mathbf{x}, \boldsymbol{\lambda}) \right)$$

which completes the proof. □

5.3 Karush-Kuhn-Tucker theory

Theorem 5.3.1

In the problem (3.18) let f and $g_i \in C^1$, and given the existence of the Lagrange multipliers $\boldsymbol{\lambda}^*$, then the following conditions have to be satisfied at the point $\mathbf{x}^*$ that corresponds to the solution of the primal problem (3.18):

$$\frac{\partial f}{\partial x_j}(\mathbf{x}^*) + \sum_{i=1}^{m} \lambda_i^* \frac{\partial g_i}{\partial x_j}(\mathbf{x}^*) = 0, \quad j = 1, 2, \ldots, n$$

$$\begin{aligned}
g_i(\mathbf{x}^*) &\leq 0, \quad i = 1, 2, \ldots, m \\
\lambda_i^* g_i(\mathbf{x}^*) &= 0, \quad i = 1, 2, \ldots, m \qquad (5.20) \\
\lambda_i^* &\geq 0, \quad i = 1, 2, \ldots, m
\end{aligned}$$

or in more compact notation:

$$\begin{aligned}
\nabla_{\mathbf{x}} L(\mathbf{x}^*, \boldsymbol{\lambda}^*) &= \mathbf{0} \\
\nabla_{\boldsymbol{\lambda}} L(\mathbf{x}^*, \boldsymbol{\lambda}^*) &\leq \mathbf{0} \\
\boldsymbol{\lambda}^{*T} \mathbf{g}(\mathbf{x}^*) &= 0 \\
\boldsymbol{\lambda}^* &\geq \mathbf{0}.
\end{aligned}$$

These conditions are known as the *Karush-Kuhn-Tucker (KKT) stationary conditions*.

Proof

First convert the inequality constraints in (3.18) into equality constraints by introducing the slack variables s_i:

$$\begin{aligned}
g_i(\mathbf{x}) + s_i &= 0, \quad i = 1, 2, \ldots, m \qquad (5.21) \\
s_i &\geq 0.
\end{aligned}$$

Define the corresponding Lagrange function

$$L(\mathbf{x}, \mathbf{s}, \boldsymbol{\lambda}) = f(\mathbf{x}) + \sum_{i=1}^{m} \lambda_i (g_i(\mathbf{x}) + s_i).$$

Assume that the solution to (3.18) with the constraints (5.21) is given by $\mathbf{x}^*$, $\mathbf{s}^*$.

Now distinguish between the two possibilities:

(i) Let $s_i^* > 0$ for all i. In this case the problem is identical to the interior minimization problem with equality constraints which is solved using Lagrange multipliers. Here there are m additional variables $s_1, s_2, \ldots, s_m$. Hence the necessary conditions for the minimum are

$$\frac{\partial L}{\partial x_j}(\mathbf{x}^*, \mathbf{s}^*, \boldsymbol{\lambda}^*) = \frac{\partial f}{\partial x_j}(\mathbf{x}^*) + \sum_{i=1}^{m} \lambda_i^* \frac{\partial g_i}{\partial x_j}(\mathbf{x}^*) = 0, \quad j = 1, \ldots, n$$

$$\frac{\partial L}{\partial s_i}(\mathbf{x}^*, \mathbf{s}^*, \boldsymbol{\lambda}^*) = \lambda_i^* = 0, \quad i = 1, \ldots, m.$$

As $s_i^* > 0$, it also follows that $g_i(\mathbf{x}^*) < 0$ and with the fact that $\lambda_i^* = 0$ this yields

$$\lambda_i^* g_i(\mathbf{x}^*) = 0.$$

Consequently all the conditions of the theorem hold for the case $s_i^* > 0$ for all i.

(ii) Let $s_i^* = 0$ for $i = 1, 2, \ldots, p$ and $s_i^* > 0$ for $i = p + 1, \ldots, m$.

In this case the solution may be considered to be the solution of an equivalent minimization problem with the following equality constraints:

$$\begin{aligned} g_i(\mathbf{x}) &= 0, \quad i = 1, 2, \ldots, p \\ g_i(\mathbf{x}) + s_i &= 0, \quad i = p + 1, \ldots, m. \end{aligned}$$

Again apply the regular Lagrange theory and it follows that

$$\frac{\partial L}{\partial x_j}(\mathbf{x}^*, \mathbf{s}^*, \boldsymbol{\lambda}^*) = \frac{\partial f}{\partial x_j}(\mathbf{x}^*) + \sum_{i=1}^{m} \lambda_i^* \frac{\partial g_i}{\partial x_j}(\mathbf{x}^*) = 0, \quad j = 1, \ldots, n$$

$$\frac{\partial L}{\partial s_i}(\mathbf{x}^*, \mathbf{s}^*, \boldsymbol{\lambda}^*) = \lambda_i^* = 0, \quad i = p + 1, \ldots, m.$$

As $g_i(\mathbf{x}^*) = 0$ for $i = 1, 2, \ldots, p$ it follows that

$$\lambda_i^* g_i(\mathbf{x}^*) = 0, \quad i = 1, 2, \ldots, m.$$

Obviously

$$g_i(\mathbf{x}^*) < 0, \quad i = p+1, \ldots, m$$

and since $g_i(\mathbf{x}^*) = 0$ for $i = 1, 2, \ldots, p$ it follows that

$$g_i(\mathbf{x}^*) \leq 0, \quad i = 1, 2, \ldots, m.$$

However, no information concerning λ_i^*, $i = 1, \ldots, p$ is available. This information is obtain from the following additional argument.

Consider feasible changes from $\mathbf{x}^*$, $\mathbf{s}^*$ in all the variables $x_1, \ldots,$ $x_n, s_1, \ldots, s_m$. Again consider m of these as dependent variables and the remaining n as independent variables. If $p \leq n$ then $s_1, s_2, \ldots, s_p$ can always be included in the set of independent variables. (Find $\boldsymbol{\lambda}^*$ by putting the partial derivatives of L at $\mathbf{x}^*$, $\mathbf{s}^*$ with respect to the dependent variables, equal to zero and solving for $\boldsymbol{\lambda}^*$.)

As $ds_i > 0$ ($s_i^* = 0$) must apply for feasible changes in the independent variables $s_1, s_2, \ldots, s_p$, it follows that in general for changes which are consistent with the equality constraints, that

$$df \geq 0, \qquad \text{(See Remark 2 below.)}$$

for changes involving $s_1, \ldots, s_p$. Thus if these independent variables are varied one at a time then, since all the partial derivatives of L with respect to the dependent variables must be equal to zero, that

$$df = \frac{\partial L}{\partial s_i}(\mathbf{x}^*, \boldsymbol{\lambda}^*)ds_i = \lambda_i^* ds_i \geq 0, \quad i = 1, 2, \ldots, p.$$

As $ds_i > 0$ it follows that $\lambda_i \geq 0$, $i = 1, 2, \ldots, p$. Thus, since it has already been proved that $\lambda_i^* = 0$, $i = p+1, \ldots, m$, it follows that indeed $\lambda_i^* \geq 0$, $i = 1, 2, \ldots, m$.

This completes the proof of the theorem. $\qquad\qquad\qquad\square$

Remark 1 Obviously, if an equality constraint $h_k(\mathbf{x}) = 0$ is also prescribed explicitly, then s_k does not exist and nothing is known of the sign of λ_k^* as $\dfrac{\partial L}{\partial s_k}$ does not exist.

Exercise   Give a brief outline of how you will obtain the necessary conditions if the equality constraints $h_j(\mathbf{x}) = 0$, $j = 1, 2, \ldots, r$ are added explicitly and L is defined by $L = f + \boldsymbol{\lambda}^T\mathbf{g} + \boldsymbol{\mu}^T\mathbf{h}$.

Remark 2   The constraints (5.21) imply that for a feasible change $ds_i > 0$ there will be a change $d\mathbf{x}$ from $\mathbf{x}^*$ and hence

$$df = \boldsymbol{\nabla}^T f \, d\mathbf{x} \geq 0.$$

If the condition $ds_i > 0$ does not apply, then a negative change $ds_i < 0$, equal in magnitude to the positive change considered above, would result in a corresponding change $-d\mathbf{x}$ and hence $df = \boldsymbol{\nabla}^T f(-d\mathbf{x}) \geq 0$, i.e. $df = \boldsymbol{\nabla}^T f \, d\mathbf{x} \leq 0$. This is only possible if $\boldsymbol{\nabla}^T f \, d\mathbf{x} = 0$, and consequently in this case $df = 0$.

Remark 3   It can be shown (not proved here) that the KKT stationary conditions are indeed necessary and *sufficient* conditions for a strong constrained global minimum at $\mathbf{x}^*$, if $f(\mathbf{x})$ and $g_i(\mathbf{x})$ are convex functions. This is not surprising because for a convex function, a local unconstrained minimum is also the global minimum.

Remark 4   Also note that if $p > n$ (see possibility (ii) of the proof) then it does not necessarily follow that $\lambda_i \geq 0$ for $i = 1, 2, \ldots, p$, i.e. the KKT conditions do not necessarily apply.

5.4   Saddle point conditions

Two drawbacks of the Kuhn-Tucker stationary conditions are that in general they only yield necessary conditions and that they apply only if $f(\mathbf{x})$ and the $g_i(\mathbf{x})$ are differentiable. These drawbacks can be removed by formulating the Karush-Kuhn-Tucker conditions in terms of the saddle point properties of $L(\mathbf{x}, \boldsymbol{\lambda})$. This is done in the next two theorems.

Theorem 5.4.1

A point $(\mathbf{x}^*, \boldsymbol{\lambda}^*)$ with $\boldsymbol{\lambda}^* \geq \mathbf{0}$ is a saddle point of the Lagrange function of the primal problem (3.18) *if and only if* the following conditions hold:

1. $\mathbf{x}^*$ minimizes $L(\mathbf{x}, \boldsymbol{\lambda}^*)$ over all $\mathbf{x}$;

2. $g_i(\mathbf{x}^*) \leq 0$, $i = 1, 2, \ldots, m$;

3. $\lambda_i^* g_i(\mathbf{x}^*) = 0$, $i = 1, 2, \ldots, m$.

Proof

If $(\mathbf{x}^*, \boldsymbol{\lambda}^*)$ is a saddle point, then $L(\mathbf{x}^*, \boldsymbol{\lambda}) \leq L(\mathbf{x}^*, \boldsymbol{\lambda}^*) \leq L(\mathbf{x}, \boldsymbol{\lambda}^*)$. First prove that if $(\mathbf{x}^*, \boldsymbol{\lambda}^*)$ is a saddle point with $\boldsymbol{\lambda}^* \geq \mathbf{0}$, then conditions 1. to 3. hold.

The right hand side of the above inequality yields directly that $\mathbf{x}^*$ minimizes $L(\mathbf{x}, \boldsymbol{\lambda}^*)$ over all $\mathbf{x}$ and the first condition is satisfied.

By expanding the left hand side:

$$f(\mathbf{x}^*) + \sum_{i=1}^{m} \lambda_i g_i(\mathbf{x}^*) \leq f(\mathbf{x}^*) + \sum_{i=1}^{m} \lambda_i^* g_i(\mathbf{x}^*)$$

and hence

$$\sum_{i=1}^{r} (\lambda_i - \lambda_i^*) g_i(\mathbf{x}^*) \leq 0 \text{ for all } \boldsymbol{\lambda} \geq \mathbf{0}.$$

Assume that $g_i(\mathbf{x}^*) > 0$. Then a contradiction is obtain for arbitrarily large λ_i, and it follows that the second condition, $g_i(\mathbf{x}^*) \leq 0$, holds.

In particular for $\boldsymbol{\lambda} = \mathbf{0}$ it follows that

$$\sum_{i=1}^{m} -\lambda_i^* g_i(\mathbf{x}^*) \leq 0 \text{ or } \sum_{i=1}^{m} \lambda_i^* g_i(\mathbf{x}^*) \geq 0.$$

But for $\boldsymbol{\lambda}^* \geq \mathbf{0}$ and $g_i(\mathbf{x}^*) \leq 0$ it follows that

$$\sum_{i=1}^{m} \lambda_i^* g_i(\mathbf{x}^*) \leq 0.$$

The only way both inequalities can be satisfied is if

$$\sum_{i=1}^{m} \lambda_i^* g_i(\mathbf{x}^*) = 0$$

and as each individual term is non positive the third condition, $\lambda_i^* g_i(\mathbf{x}^*) = 0$, follows.

Now proceed to prove the converse, i.e. that if the three conditions of the theorem hold, then $L(\mathbf{x}^*, \boldsymbol{\lambda}^*)$ has a saddle point at $(\mathbf{x}^*, \boldsymbol{\lambda}^*)$.

The first condition implies that $L(\mathbf{x}^*, \boldsymbol{\lambda}^*) \leq L(\mathbf{x}, \boldsymbol{\lambda}^*)$ which is half of the definition of a saddle point. The rest is obtained from the expansion:

$$L(\mathbf{x}^*, \boldsymbol{\lambda}) = f(\mathbf{x}^*) + \sum_{i=1}^{m} \lambda_i g_i(\mathbf{x}^*).$$

Now as $\mathbf{g}(\mathbf{x}^*) \leq \mathbf{0}$ and $\boldsymbol{\lambda} \geq \mathbf{0}$ it follows that $L(\mathbf{x}^*, \boldsymbol{\lambda}) \leq f(\mathbf{x}^*) = L(\mathbf{x}^*, \boldsymbol{\lambda}^*)$ since, from the third condition, $\sum_{i=1}^{m} \lambda_i^* g_i(\mathbf{x}^*) = 0$.

This completes the proof of the converse.   □

Theorem 5.4.2

If the point $(\mathbf{x}^*, \boldsymbol{\lambda}^*)$ is a saddle point, $\boldsymbol{\lambda}^* \geq \mathbf{0}$, of the Lagrange function associated with the primal problem, then $\mathbf{x}^*$ is the solution of the primal problem.

Proof

If $(\mathbf{x}^*, \boldsymbol{\lambda}^*)$ is a saddle point the previous theorem holds and the inequality constraints are satisfied at $\mathbf{x}^*$. All that is required additionally is to show that $f(\mathbf{x}^*) \leq f(\mathbf{x})$ for all $\mathbf{x}$ such that $\mathbf{g}(\mathbf{x}) \leq \mathbf{0}$. From the definition of a saddle point it follows that

$$f(\mathbf{x}^*) + \sum_{i=1}^{m} \lambda_i^* g_i(\mathbf{x}^*) \leq f(\mathbf{x}) + \sum_{i=1}^{m} \lambda_i^* g_i(\mathbf{x})$$

for all $\mathbf{x}$ in the neighbourhood of $\mathbf{x}^*$. As a consequence of condition 3. of the previous theorem, the left hand side is $f(\mathbf{x}^*)$ and for any $\mathbf{x}$ such that $g_i(\mathbf{x}) \leq 0$, it holds that $\sum_{i=1}^{m} \lambda_i^* g_i(\mathbf{x}) \leq 0$ and hence it follows that $f(\mathbf{x}^*) \leq f(\mathbf{x})$ for all $\mathbf{x}$ such that $\mathbf{g}(\mathbf{x}) \leq \mathbf{0}$, with equality at $\mathbf{x} = \mathbf{x}^*$.

This completes the proof.   □

The main advantage of these saddle point theorems is that necessary conditions are provided for solving optimization problems which are neither convex nor differentiable. Any direct search method can be used to minimize $L(\mathbf{x}, \boldsymbol{\lambda}^*)$ over all $\mathbf{x}$. Of course the problem remains that we do not have an a priori value for $\boldsymbol{\lambda}^*$. In practice it is possible to obtain estimates for $\boldsymbol{\lambda}^*$ using iterative techniques, or by solving the so called dual problem.

Theorem 5.4.3

The dual function $h(\boldsymbol{\lambda}) \leq f(\mathbf{x})$ for all $\mathbf{x}$ that satisfy the constraints $\mathbf{g}(\mathbf{x}) \leq \mathbf{0}$ for all $\boldsymbol{\lambda} \in D$. (Hence the dual function yields a lower bound for the function $f(\mathbf{x})$ with $\mathbf{g}(\mathbf{x}) \leq \mathbf{0}$.)

Proof

Let $X = \{\mathbf{x} | \mathbf{g}(\mathbf{x}) \leq \mathbf{0}\}$, then

$$
\begin{aligned}
h(\boldsymbol{\lambda}) &= \min_{\mathbf{x}} L(\mathbf{x}, \boldsymbol{\lambda}), \quad \boldsymbol{\lambda} \in D \\
&\leq \min_{\mathbf{x} \in X} L(\mathbf{x}, \boldsymbol{\lambda}) \\
&\leq f(\mathbf{x}) + \sum_{i=1}^{m} \lambda_i g_i(\mathbf{x}), \quad \mathbf{x} \in X \\
&\leq f(\mathbf{x}), \quad \mathbf{x} \in X, \boldsymbol{\lambda} \in D.
\end{aligned}
$$

$\square$

The largest lower bound is attained at $\max h(\boldsymbol{\lambda})$, $\boldsymbol{\lambda} \in D$.

Theorem 5.4.4 (Duality Theorem)

The point $(\mathbf{x}^*, \boldsymbol{\lambda}^*)$ with $\boldsymbol{\lambda}^* \geq \mathbf{0}$ is a saddle point of the Lagrange function of the primal problem *if and only if*

1. $\mathbf{x}^*$ is a solution of the primal problem;

2. $\boldsymbol{\lambda}^*$ is a solution of the dual problem;

3. $f(\mathbf{x}^*) = h(\boldsymbol{\lambda}^*)$.

Proof

First assume that $L(\mathbf{x}, \boldsymbol{\lambda})$ has a saddle point at $(\mathbf{x}^*, \boldsymbol{\lambda}^*)$ with $\boldsymbol{\lambda}^* \geq \mathbf{0}$. Then from Theorem 5.4.2 it follows that $\mathbf{x}^*$ is a solution of the primal problem and 1. holds.

By definition:
$$h(\boldsymbol{\lambda}) = \min_{\mathbf{x}} L(\mathbf{x}, \boldsymbol{\lambda}).$$

From Theorem 5.4.1 it follows that $\mathbf{x}^*$ minimizes $L(\mathbf{x}, \boldsymbol{\lambda}^*)$ over all $\mathbf{x}$, thus

$$h(\boldsymbol{\lambda}^*) = f(\mathbf{x}^*) + \sum_{i=1}^{m} \lambda_i^* g_i(\mathbf{x}^*)$$

and as $\lambda_i^* g_i(\mathbf{x}^*) = 0$ it follows that

$$h(\boldsymbol{\lambda}^*) = f(\mathbf{x}^*)$$

which is condition 3.

Also by Theorem 5.4.1 $\mathbf{g}(\mathbf{x}^*) \leq \mathbf{0}$, and it has already been shown in Theorem 5.4.3 that

$$h(\boldsymbol{\lambda}) \leq f(\mathbf{x}) \text{ for all } \mathbf{x} \in \{\mathbf{x} | \mathbf{g}(\mathbf{x}) \leq \mathbf{0}\}$$

and thus in particular $h(\boldsymbol{\lambda}) \leq f(\mathbf{x}^*) = h(\boldsymbol{\lambda}^*)$. Consequently $h(\boldsymbol{\lambda}^*) = \max_{\boldsymbol{\lambda} \in D} h(\boldsymbol{\lambda})$ and condition 2. holds.

Conversely, prove that if conditions 1. to 3. hold, then $L(\mathbf{x}, \boldsymbol{\lambda})$ has a saddle point at $(\mathbf{x}^*, \boldsymbol{\lambda}^*)$, with $\boldsymbol{\lambda}^* \geq \mathbf{0}$, or equivalently that the conditions of Theorem 5.4.1 hold.

As $\mathbf{x}^*$ is a solution of the primal problem, the necessary conditions

$$\mathbf{g}(\mathbf{x}^*) \leq \mathbf{0}$$

hold, which is condition 2. of Theorem 5.4.1.

Also, as $\boldsymbol{\lambda}^*$ is a solution of the dual problem, $\boldsymbol{\lambda}^* \geq \mathbf{0}$. It is now shown that $\mathbf{x}^*$ minimizes $L(\mathbf{x}, \boldsymbol{\lambda}^*)$.

Make the contradictory assumption, i.e. that there exists a point $\hat{\mathbf{x}} \neq \mathbf{x}^*$ such that

$$L(\hat{\mathbf{x}}, \boldsymbol{\lambda}^*) < L(\mathbf{x}^*, \boldsymbol{\lambda}^*).$$

By definition:

$$\begin{aligned} h(\boldsymbol{\lambda}^*) = L(\hat{\mathbf{x}}, \boldsymbol{\lambda}^*) &= f(\hat{\mathbf{x}}) + \sum_{i=1}^{m} \lambda_i^* g_i(\hat{\mathbf{x}}) \\ &< f(\mathbf{x}^*) + \sum_{i=1}^{m} \lambda_i^* g_i(\mathbf{x}^*) \end{aligned}$$

but from condition 3.: $h(\boldsymbol{\lambda}^*) = f(\mathbf{x}^*)$ and consequently

$$\sum_{i=1}^{m} \lambda_i^* g_i(\mathbf{x}^*) > 0$$

which contradicts the fact that $\boldsymbol{\lambda}^* \geq \mathbf{0}$ and $\mathbf{g}(\mathbf{x}^*) \leq \mathbf{0}$; hence $\widehat{\mathbf{x}} = \mathbf{x}^*$ and $\mathbf{x}^*$ minimizes $L(\mathbf{x}, \boldsymbol{\lambda}^*)$ and condition 1. of Theorem 5.4.1 holds.

Also, as

$$h(\boldsymbol{\lambda}^*) = f(\mathbf{x}^*) + \sum_{i=1}^{m} \lambda_i^* g_i(\mathbf{x}^*)$$

and $h(\boldsymbol{\lambda}^*) = f(\mathbf{x}^*)$ by condition 3., it follows that:

$$\sum_{i=1}^{m} \lambda_i^* g_i(\mathbf{x}^*) = 0.$$

As each individual term is non positive the third condition of Theorem 5.4.1 holds: $\lambda_i^* g_i(\mathbf{x}^*) = 0$.

As the three conditions of Theorem 5.4.1 are satisfied it is concluded that $(\mathbf{x}^*, \boldsymbol{\lambda}^*)$ is a saddle point of $L(\mathbf{x}, \boldsymbol{\lambda})$ and the converse is proved. □

5.5   Conjugate gradient methods

Let $\mathbf{u}$ and $\mathbf{v}$ be two non-zero vectors in $\mathbb{R}^n$. Then they are mutually orthogonal if $\mathbf{u}^T \mathbf{v} = (\mathbf{u}, \mathbf{v}) = 0$. Let $\mathbf{A}$ be an $n \times n$ symmetric positive-definite matrix. Then $\mathbf{u}$ and $\mathbf{v}$ are mutually conjugate with respect to $\mathbf{A}$ if $\mathbf{u}$ and $\mathbf{A}\mathbf{v}$ are mutually orthogonal, i.e.

$$\mathbf{u}^T \mathbf{A}\mathbf{v} = (\mathbf{u}, \mathbf{A}\mathbf{v}) = 0. \tag{5.22}$$

Let $\mathbf{A}$ be a square matrix. $\mathbf{A}$ has an eigenvalue λ and an associated eigenvector $\mathbf{x}$ if for $\mathbf{x} \neq \mathbf{0}$, $\mathbf{A}\mathbf{x} = \lambda \mathbf{x}$. It can be shown that if $\mathbf{A}$ is positive-definite and symmetric and $\mathbf{x}$ and $\mathbf{y}$ are distinct eigenvectors, then they are mutually orthogonal, i.e.

$$(\mathbf{x}, \mathbf{y}) = 0 = (\mathbf{y}, \mathbf{x}).$$

Since $(\mathbf{y}, \mathbf{A}\mathbf{x}) = (\mathbf{y}, \lambda\mathbf{x}) = \lambda(\mathbf{y}, \mathbf{x}) = 0$, it follows that the eigenvectors of a positive-definite matrix $\mathbf{A}$ are mutually conjugate with respect to

A. Hence, given positive-definite matrix **A** , there exists at least one pair of mutually conjugate directions with respect to this matrix. It is now shown that a set of mutually conjugate vectors in $\mathbb{R}^n$ forms a basis and thus spans $\mathbb{R}^n$.

Theorem 5.5.1

Let $\mathbf{u}^i$, $i = 1, 2, \ldots, n$ be a set of vectors in $\mathbb{R}^n$ which are mutually conjugate with respect to a given symmetric positive-definite matrix **A**. Then for each $\mathbf{x} \in \mathbb{R}^n$ it holds that

$$\mathbf{x} = \sum_{i=1}^{n} \lambda_i \mathbf{u}^i \text{ where } \lambda_i = \frac{(\mathbf{u}^i, \mathbf{Ax})}{(\mathbf{u}^i, \mathbf{Au}^i)}.$$

Proof

Consider the linear combination $\sum_{i=1}^{n} \alpha_i \mathbf{u}^i = \mathbf{0}$. Then

$$\mathbf{A}\left(\sum_{i=1}^{n} \alpha_i \mathbf{u}^i\right) = \sum_{i=1}^{n} \alpha_i \mathbf{Au}^i = \mathbf{0}.$$

Since the vectors $\mathbf{u}^i$ are mutually conjugate with respect to **A** it follows that

$$(\mathbf{u}^k, \mathbf{A}(\sum_{i=1}^{n} \alpha_i \mathbf{u}^i)) = \alpha_k (\mathbf{u}^k, \mathbf{Au}^k) = 0.$$

Since **A** is positive-definite and $\mathbf{u}^k \neq \mathbf{0}$ it follows that $(\mathbf{u}^k, \mathbf{Au}^k) \neq 0$, and thus $\alpha_k = 0$, $k = 1, 2, \ldots, n$. The set $\mathbf{u}^i$, $i = 1, 2, \ldots, n$ thus forms a linear independent set of vectors in $\mathbb{R}^n$ which may be used as a basis. Thus for any $\mathbf{x}$ in $\mathbb{R}^n$ there exists a unique set λ_i, $i = 1, 2, \ldots, n$ such that

$$\mathbf{x} = \sum_{i=1}^{n} \lambda_i \mathbf{u}^i. \tag{5.23}$$

Now since the $\mathbf{u}^i$ are mutually conjugate with respect to **A** it follows that $(\mathbf{u}^i, \mathbf{Ax}) = (\lambda_i \mathbf{u}^i, \mathbf{Au}^i)$ giving

$$\lambda_i = \frac{(\mathbf{u}^i, \mathbf{Ax})}{(\mathbf{u}^i, \mathbf{Au}^i)} \tag{5.24}$$

which completes the proof. $\square$

The following lemma is required in order to show that the Fletcher-Reeves directions given by $\mathbf{u}^1 = -\boldsymbol{\nabla} f(\mathbf{x}^0)$ and formulae (2.14) and (2.15) are mutually conjugate. For convenience here and in what follows, we also use the notation $\mathbf{g}^k \equiv \nabla f(\mathbf{x}^k)$ for the gradient vector. Here $\mathbf{g}$ should not be confused with the inequality constraint vector function used elsewhere.

Lemma 5.5.2

Let $\mathbf{u}^1$, $\mathbf{u}^2$, ..., $\mathbf{u}^n$ be mutually conjugate directions with respect to A along an optimal descent path applied to $f(\mathbf{x})$ given by (2.9). Then

$$(\mathbf{u}^k, \mathbf{g}^i) = 0, \quad k = 1, 2, \ldots, i; \quad 1 \le i \le n.$$

Proof

For optimal decrease at step k it is required that

$$(\mathbf{u}^k, \nabla f(\mathbf{x}^k)) = 0, \quad k = 1, 2, \ldots, i.$$

Also

$$\begin{aligned}
\nabla f(\mathbf{x}^i) &= \mathbf{A}\mathbf{x}^i + \mathbf{b} \\
&= \left(\mathbf{A}\mathbf{x}^k + \mathbf{A} \sum_{j=k+1}^{i} \lambda_j \mathbf{u}^j \right) + \mathbf{b} \\
&= \nabla f(\mathbf{x}^k) + \sum_{j=k+1}^{i} \lambda_j (\mathbf{A}\mathbf{u}^j)
\end{aligned}$$

and thus

$$(\mathbf{u}^k, \nabla f(\mathbf{x}^i)) = (\mathbf{u}^k, \nabla f(\mathbf{x}^k)) + \sum_{j=k+1}^{i} \lambda_j (\mathbf{u}^k, \mathbf{A}\mathbf{u}^j) = 0$$

which completes the proof.   □

Theorem 5.5.3

The directions $\mathbf{u}^i$, $i = 1, 2, \ldots, n$ of the Fletcher-Reeves algorithm given in Section 2.3.2.4 are mutually conjugate with respect to $\mathbf{A}$ of $f(\mathbf{x})$ given by (2.9).

Proof

The proof is by induction.

First, $\mathbf{u}^1$ and $\mathbf{u}^2$ are mutually conjugate:

$$
\begin{aligned}
(\mathbf{u}^2, \mathbf{A}\mathbf{u}^1) &= -\left((\mathbf{g}^1 + \beta_1 \mathbf{g}^0), \mathbf{A}(\mathbf{x}^1 - \mathbf{x}^0)\tfrac{1}{\lambda_1}\right) \\
&= -\left((\mathbf{g}^1 + \beta_1 \mathbf{g}^0), (\mathbf{g}^1 - \mathbf{g}^0)\tfrac{1}{\lambda_1}\right) \\
&= -\tfrac{1}{\lambda_1}\left(\|\mathbf{g}^1\|^2 - \beta_1\|\mathbf{g}^0\|^2\right) \\
&= 0.
\end{aligned}
$$

Now assume that $\mathbf{u}^1, \mathbf{u}^2, \ldots, \mathbf{u}^i$ are mutually conjugate, i.e.

$$(\mathbf{u}^k, \mathbf{A}\mathbf{u}^j) = 0, \quad k \neq j, \ k, j \leq i.$$

It is now required to prove that $(\mathbf{u}^k, \mathbf{A}\mathbf{u}^{i+1}) = 0$ for $k = 1, 2, \ldots, i$.

First consider $-(\mathbf{g}^k, \mathbf{g}^i)$ for $k = 1, 2, \ldots, i-1$:

$$
\begin{aligned}
-(\mathbf{g}^k, \mathbf{g}^i) &= (-\mathbf{g}^k + \beta_k \mathbf{u}^k, \mathbf{g}^i) \quad \text{from Lemma 5.5.2} \\
&= (\mathbf{u}^{k+1}, \mathbf{g}^i) = 0 \quad \text{also from Lemma 5.5.2.}
\end{aligned}
$$

Hence

$$(\mathbf{g}^k, \mathbf{g}^i) = 0, \quad k = 1, 2, \ldots, i-1. \tag{5.25}$$

Now consider $(\mathbf{u}^k, \mathbf{A}\mathbf{u}^{i+1})$ for $k = 1, 2, \ldots, i-1$:

$$
\begin{aligned}
(\mathbf{u}^k, \mathbf{A}\mathbf{u}^{i+1}) &= -(\mathbf{u}^k, \mathbf{A}\mathbf{g}^i - \beta_i \mathbf{A}\mathbf{u}^i) \\
&= -(\mathbf{u}^k, \mathbf{A}\mathbf{g}^i) \quad \text{from the induction assumption} \\
&= -(\mathbf{g}^i, \mathbf{A}\mathbf{u}^k) \\
&= -\left(\mathbf{g}^i, \mathbf{A}\frac{\mathbf{x}^k - \mathbf{x}^{k-1}}{\lambda_k}\right) \\
&= -\tfrac{1}{\lambda_k}(\mathbf{g}^i, \mathbf{g}^k - \mathbf{g}^{k-1}) = 0 \quad \text{from (5.25).}
\end{aligned}
$$

Hence

$$(\mathbf{u}^k, \mathbf{A}\mathbf{u}^{i+1}) = 0, \quad k = 1, 2, \ldots, i-1. \tag{5.26}$$

All that remains is to prove that $(\mathbf{u}^i, \mathbf{A}\mathbf{u}^{i+1}) = 0$ which implies a β_i such that

$$(\mathbf{u}^i, \mathbf{A}(-\mathbf{g}^i + \beta_i \mathbf{u}^i)) = 0,$$

i.e.

$$-(\mathbf{u}^i, \mathbf{A}\mathbf{g}^i) + \beta_i(\mathbf{u}^i, \mathbf{A}\mathbf{u}^i) = 0$$

or

$$\beta_i = \frac{(\mathbf{g}^i, \mathbf{Au}^i)}{(\mathbf{u}^i, \mathbf{Au}^i)}. \tag{5.27}$$

Now

$$
\begin{aligned}
(\mathbf{g}^i, \mathbf{Au}^i) &= \tfrac{1}{\lambda_i}(\mathbf{g}^i, (\mathbf{Ax}^i - \mathbf{x}^{i-1})) \\
&= \tfrac{1}{\lambda_i}(\mathbf{g}^i, \mathbf{g}^i - \mathbf{g}^{i-1}) \\
&= \tfrac{1}{\lambda_i}\|\mathbf{g}^i\|^2
\end{aligned}
$$

and

$$
\begin{aligned}
(\mathbf{u}^i, \mathbf{Au}^i) &= \tfrac{1}{\lambda_i}(\mathbf{u}^i, \mathbf{A}(\mathbf{x}^i - \mathbf{x}^{i-1})) \\
&= \tfrac{1}{\lambda_i}(\mathbf{u}^i, \mathbf{g}^i - \mathbf{g}^{i-1}) \\
&= -\tfrac{1}{\lambda_i}(\mathbf{u}^i, \mathbf{g}^{i-1}) \quad \text{from Lemma 5.5.2} \\
&= -\tfrac{1}{\lambda_i}(-\mathbf{g}^{i-1} + \beta_{i-1}\mathbf{u}^{i-1}, \mathbf{g}^{i-1}) \\
&= \tfrac{1}{\lambda_i}\|\mathbf{g}^{i-1}\|^2.
\end{aligned}
$$

Thus from (5.27) it is required that $\beta_i = \dfrac{\|\mathbf{g}^i\|^2}{\|\mathbf{g}^{i-1}\|^2}$ which agrees with the value prescribed by the Fletcher-Reeves algorithm. Consequently

$$(\mathbf{u}^i, \mathbf{Au}^{i+1}) = 0.$$

Combining this result with (5.26) completes the proof. □

5.6 DFP method

The following two theorems concern the DFP algorithm. The first theorem shows that it is a descent method.

Theorem 5.6.1

If $\mathbf{G}_i$ is positive-definite and $\mathbf{G}_{i+1}$ is calculated using (2.18) then $\mathbf{G}_{i+1}$ is also positive-definite.

Proof

If $\mathbf{G}_i$ is positive-definite and symmetric then there exists a matrix $\mathbf{F}_i$ such that $\mathbf{F}_i^T \mathbf{F}_i = \mathbf{G}_i$, and

$$(\mathbf{x}, \mathbf{G}_{i+1}\mathbf{x}) = (\mathbf{x}, \mathbf{G}_i\mathbf{x}) + \frac{(\mathbf{v}^j, \mathbf{x})^2}{(\mathbf{y}^j, \mathbf{v}^j)} - \frac{(\mathbf{x}, \mathbf{G}_i\mathbf{y}^j)^2}{(\mathbf{y}^j, \mathbf{G}_i\mathbf{y}^j)}, \; j = i+1$$

$$= \frac{(\mathbf{p}^i, \mathbf{p}^i)(\mathbf{q}^i, \mathbf{q}^i) - (\mathbf{p}^i, \mathbf{q}^i)^2}{(\mathbf{q}^i, \mathbf{q}^i)} + \frac{(\mathbf{v}^j, \mathbf{x})^2}{(\mathbf{y}^j, \mathbf{v}^j)}$$

where $\mathbf{p}^i = \mathbf{F}_i\mathbf{x}$ and $\mathbf{q}^i = \mathbf{F}_i\mathbf{y}^j$.

If $\mathbf{x} \neq \theta\mathbf{y}^j$ for some scalar θ, it follows from the Schwartz inequality that the first term is strictly positive.

For the second term it holds for the denominator that

$$\begin{aligned}
(\mathbf{y}^j, \mathbf{v}^j) &= (\mathbf{g}^j - \mathbf{g}^{j-1}, \mathbf{v}^j) \\
&= -(\mathbf{g}^{j-1}, \mathbf{v}^j) \\
&= \lambda_j(\mathbf{g}^{j-1}, \mathbf{G}_{j-1}\mathbf{g}^{j-1}) \\
&= \lambda_{i+1}(\mathbf{g}^i, \mathbf{G}_i\mathbf{g}^i) > 0 \quad (\lambda_{i+1} > 0 \text{ and } \mathbf{G}_i \text{ positive-definite})
\end{aligned}$$

and hence if $\mathbf{x} \neq \theta\mathbf{y}^j$ the second term is non-negative and the right hand side is strictly positive.

Else, if $\mathbf{x} = \theta\mathbf{y}^j$, the first term is zero and we only have to consider the second term:

$$\begin{aligned}
\frac{(\mathbf{v}^j, \mathbf{x})^2}{(\mathbf{y}^j, \mathbf{v}^j)} &= \frac{\theta^2(\mathbf{y}^j, \mathbf{v}^j)^2}{(\mathbf{y}^j, \mathbf{v}^j)} \\
&= \theta^2(\mathbf{y}^j, \mathbf{v}^j) \\
&= \lambda_{i+1}\theta^2(\mathbf{g}^i, \mathbf{G}_i\mathbf{g}^i) > 0.
\end{aligned}$$

This completes the proof.   □

The theorem above is important as it guarantees descent.

For any search direction $\mathbf{u}^{k+1}$:

$$\frac{df}{d\lambda}(\mathbf{x}^k + \lambda\mathbf{u}^{k+1}) = \mathbf{g}^{kT}\mathbf{u}^{k+1}.$$

For the DFP method, $\mathbf{u}^{k+1} = -\mathbf{G}_k\mathbf{g}^k$ and hence at $\mathbf{x}^k(\lambda = 0)$:

$$\frac{df}{d\lambda} = -\mathbf{g}^{kT}\mathbf{G}_k\mathbf{g}^k.$$

Consequently, if $\mathbf{G}_k$ is positive-definite, descent is guaranteed at $\mathbf{x}^k$, for $\mathbf{g}^k \neq \mathbf{0}$.

Theorem 5.6.2

If the DFP method is applied to a quadratic function of the form given by (2.9), then the following holds:

(i) $(\mathbf{v}^i, \mathbf{A}\mathbf{v}^j) = 0$, $1 \leq i < j \leq k$, $k = 2, 3, \ldots, n$

(ii) $\mathbf{G}_k \mathbf{A}\mathbf{v}^i = \mathbf{v}^i$, $1 \leq i \leq k$, $k = 1, 2, \ldots, n$

where the vectors that occur in the DFP algorithm (see Section 2.4.2.1) are defined by:

$$
\begin{aligned}
\mathbf{x}^{i+1} &= \mathbf{x}^i + \lambda_{i+1}\mathbf{u}^{i+1}, \quad i = 0, 1, 2, \ldots \\
\mathbf{u}^{i+1} &= -\mathbf{G}_i \mathbf{g}^i \\
\mathbf{v}^{i+1} &= \lambda_{i+1}\mathbf{u}^{i+1} \\
\mathbf{y}^{i+1} &= \mathbf{g}^{i+1} - \mathbf{g}^i.
\end{aligned}
$$

Proof

The proof is by induction. The most important part of the proof, the induction step, is presented. (Prove the initial step yourself. Property (ii) will hold for $k = 1$ if $\mathbf{G}_1\mathbf{A}\mathbf{v}^1 = \mathbf{v}^1$. Show this by direct substitution. The first case $(\mathbf{v}^1, \mathbf{A}\mathbf{v}^2) = 0$ in (i) corresponds to $k = 2$ and follows from the fact that (ii) holds for $k = 1$. That (ii) also holds for $k = 2$, follows from the second part of the induction proof given below.)

Assume that (i) and (ii) hold for k. Then it is required to show that they also hold for $k + 1$.

First part: proof that (i) is true for $k + 1$. For the quadratic function:

$$
\mathbf{g}^k = \mathbf{b} + \mathbf{A}\mathbf{x}^k = \mathbf{b} + \mathbf{A}\mathbf{x}^i + \mathbf{A}\sum_{j=i+1}^{k} \mathbf{v}^j.
$$

Now consider:

$$
(\mathbf{v}^i, \mathbf{g}^k) = (\mathbf{v}^i, \mathbf{g}^i) + \sum_{j=i+1}^{k} (\mathbf{v}^i, \mathbf{A}\mathbf{v}^j) = 0, \quad 1 \leq i \leq k.
$$

The first term on the right is zero from the optimal descent property and the second term as a consequence of the induction assumption that (i) holds for k.

Consequently, with $\mathbf{v}^{k+1} = \lambda_{k+1}\mathbf{u}^{k+1} = -\lambda_{k+1}\mathbf{G}_k\mathbf{g}^k$ it follows that

$$(\mathbf{v}^i, \mathbf{A}\mathbf{v}^{k+1}) = -\lambda_{k+1}(\mathbf{v}^i, \mathbf{A}\mathbf{G}_k\mathbf{g}^k) = -\lambda_{k+1}(\mathbf{G}_k\mathbf{A}\mathbf{v}^i, \mathbf{g}^k), \quad \lambda_{k+1} > 0.$$

Hence
$$(\mathbf{v}^i, \mathbf{A}\mathbf{v}^{k+1}) = -\lambda_{k+1}(\mathbf{v}^i, \mathbf{g}^k) = 0$$

as a consequence of the induction assumption (ii) and the result above.

Hence with $(\mathbf{v}^i, \mathbf{A}\mathbf{v}^{k+1}) = 0$, property (i) holds for $k + 1$.

Second part: proof that (ii) holds for $k + 1$. Furthermore

$$
\begin{aligned}
(\mathbf{y}^{k+1}, \mathbf{G}_k\mathbf{A}\mathbf{v}^i) &= (\mathbf{y}^{k+1}, \mathbf{v}^i) \quad \text{for } i \le k \text{ from assumption (ii)} \\
&= (\mathbf{g}^{k+1} - \mathbf{g}^k, \mathbf{v}^i) \\
&= (\mathbf{A}(\mathbf{x}^{k+1} - \mathbf{x}^k), \mathbf{v}^i) \\
&= (\mathbf{A}\mathbf{v}^{k+1}, \mathbf{v}^i) \\
&= (\mathbf{v}^i, \mathbf{A}\mathbf{v}^{k+1}) = 0 \text{ from the first part.}
\end{aligned}
$$

Using the update formula, it follows that

$$
\begin{aligned}
\mathbf{G}_{k+1}\mathbf{A}\mathbf{v}^i &= \mathbf{G}_k\mathbf{A}\mathbf{v}^i + \frac{\mathbf{v}^{k+1}(\mathbf{v}^{k+1})^T\mathbf{A}\mathbf{v}^i}{(\mathbf{v}^{k+1})^T\mathbf{v}^{k+1}} - \frac{(\mathbf{G}_k\mathbf{y}^{k+1})(\mathbf{G}_k\mathbf{y}^{k+1})^T\mathbf{A}\mathbf{v}^i}{(\mathbf{y}^{k+1})^T\mathbf{G}_k\mathbf{y}^{k+1}} \\
&= \mathbf{G}_k\mathbf{A}\mathbf{v}^i \quad \text{(because } (\mathbf{v}^i, \mathbf{A}\mathbf{v}^{k+1}) = 0, \ (\mathbf{y}^{k+1}, \mathbf{G}_k\mathbf{A}\mathbf{v}^i) = 0) \\
&= \mathbf{v}^i \quad \text{for } i \le k \quad \text{from assumption (ii).}
\end{aligned}
$$

It is still required to show that $\mathbf{G}_{k+1}\mathbf{A}\mathbf{v}^{k+1} = \mathbf{v}^{k+1}$. This can be done, as for the initial step where it was shown that $\mathbf{G}_1\mathbf{A}\mathbf{v}^1 = \mathbf{v}^1$, by direct substitution. $\qquad\square$

Thus it was shown that $\mathbf{v}^k$, $k = 1, \ldots, n$ are mutually conjugate with respect to $\mathbf{A}$ and therefore they are linearly independent and form a basis for $\mathbb{R}^n$. Consequently the DFP method is quadratically terminating with $\mathbf{g}^n = \mathbf{0}$. Property (ii) also implies that $\mathbf{G}_n = \mathbf{A}^{-1}$.

A final interesting result is the following theorem.

Theorem 5.6.3

If the DFP method is applied to $f(\mathbf{x})$ given by (2.9), then

$$\mathbf{A}^{-1} = \sum_{i=1}^{n} \mathbf{A}_i.$$

Proof

In the previous proof it was shown that the $\mathbf{v}^i$ vectors , $i = 1, 2, \ldots, n$, are mutually conjugate with respect to $\mathbf{A}$. They therefore form a basis in $\mathbb{R}^n$. Also

$$\mathbf{A}_i = \frac{\mathbf{v}^i \mathbf{v}^{iT}}{\mathbf{v}^{iT} \mathbf{y}^i} = \frac{\mathbf{v}^i \mathbf{v}^{iT}}{\mathbf{v}^{iT}(\mathbf{g}^i - \mathbf{g}^{i-1})} = \frac{\mathbf{v}^i \mathbf{v}^{iT}}{\mathbf{v}^{iT} \mathbf{A} \mathbf{v}^i}.$$

Let

$$\mathbf{B} = \sum_{i=1}^{n} \mathbf{A}_i = \sum_{i=1}^{n} \frac{\mathbf{v}^i \mathbf{v}^{iT}}{\mathbf{v}^{iT} \mathbf{A} \mathbf{v}^i}.$$

Then

$$\mathbf{B}\mathbf{A}\mathbf{x} = \sum_{i=1}^{n} \frac{\mathbf{v}^i \mathbf{v}^{iT}}{\mathbf{v}^{iT} \mathbf{A} \mathbf{v}^i} \mathbf{A}\mathbf{x} = \sum_{i=1}^{n} \frac{\mathbf{v}^{iT} \mathbf{A} \mathbf{x}}{\mathbf{v}^{iT} \mathbf{A} \mathbf{v}^i} \mathbf{v}^i = \mathbf{x}$$

as a consequence of Theorem 5.5.1.

This result holds for arbitrary $\mathbf{x} \in \mathbb{R}^n$ and hence $\mathbf{B}\mathbf{A} = \mathbf{I}$ where $\mathbf{I}$ is the identity matrix, and it follows that

$$\mathbf{B} = \sum_{i=1}^{n} \mathbf{A}_i = \mathbf{A}^{-1}.$$

$\square$

Part II
Gradient-based algorithms

Chapter 6

NEW GRADIENT-BASED TRAJECTORY AND APPROXIMATION METHODS

6.1 Introduction

6.1.1 Why new algorithms?

In spite of the mathematical sophistication of classical *gradient-based* algorithms, certain *inhibiting difficulties remain* when these algorithms are applied to *real-world problems*. This is particularly true in the field of engineering, where unique difficulties occur that have prevented the general application of gradient-based mathematical optimization techniques to design problems.

Optimization difficulties that arise are:

(i) the *functions* are often *very expensive to evaluate*, requiring, for example, the time-consuming finite element analysis of a structure, the simulation of the dynamics of a multi-body system, or a

© Springer International Publishing AG, part of Springer Nature 2018   197
J.A. Snyman and D.N. Wilke, *Practical Mathematical Optimization,*
Springer Optimization and Its Applications 133,
https://doi.org/10.1007/978-3-319-77586-9_6

computational fluid dynamics (CFD) simulation,

(ii) the *existence of noise*, numerical or experimental, in the functions,

(iii) the presence of *discontinuities* in the functions,

(iv) *multiple local minima*, requiring global optimization techniques,

(v) the existence of regions in the design space where the *functions are not defined*, and

(vi) the occurrence of an *extremely large number* of design *variables*, disqualifying, for example, the SQP method.

6.1.2   Research at the University of Pretoria

All the above difficulties have been addressed in research done at the University of Pretoria over the past twenty years. This research has led to, amongst others, the development of the new optimization algorithms and methods listed in the subsections below.

6.1.2.1   Unconstrained optimization

(i) The *leap-frog* dynamic *trajectory* method: LFOP (Snyman 1982, 1983),

(ii) a *conjugate-gradient method* with Euler-trapezium steps in which a novel gradient-only line search method is used: ETOP (Snyman 1985), and

(iii) a *steepest-descent method* applied to successive spherical quadratic approximations: SQSD (Snyman and Hay 2001).

6.1.2.2   Direct constrained optimization

(i) The leap-frog method for constrained optimization, LFOPC (Snyman 2000), and

(ii) the conjugate-gradient method with Euler-trapezium steps and gradient-only line searches, applied to penalty function formulations of constrained problems: ETOPC (Snyman 2005).

6.1.2.3 Approximation methods

(i) A *feasible descent cone method* applied to successive spherical *quadratic sub-problems*: FDC-SAM (Stander and Snyman 1993; Snyman and Stander 1994, 1996; De Klerk and Snyman 1994), and

(ii) the *leap-frog method* (LFOPC) applied to successive spherical *quadratic sub-problems*: Dynamic-Q (Snyman et al. 1994; Snyman and Hay 2002).

6.1.2.4 Methods for global unconstrained optimization

(i) A *multi-start global minimization* algorithm with *dynamic search trajectories*: SF-GLOB (Snyman and Fatti 1987), and

(ii) a *modified bouncing ball* trajectory method for global optimization: MBB (Groenwold and Snyman 2002).

All of the above methods developed at the University of Pretoria are *gradient-based*, and have the common and unique property, for gradient-based methods, that *no explicit objective function line searches are required*.

In this chapter the LFOP/C unconstrained and constrained algorithms are discussed in detail. This is followed by the presentation of the SQSD method, which serves as an introduction to the Dynamic-Q approximation method. Next the ETOP/C algorithms are introduced, with special reference to their ability to deal with the presence of severe noise in the objective function, through the use of a gradient-only line search technique. Finally the SF-GLOB and MBB stochastic global optimization algorithms, which use dynamic search trajectories, are presented and discussed.

6.2 The dynamic trajectory optimization method

The dynamic trajectory method for unconstrained minimization (Snyman 1982, 1983) is also known as the "leap-frog" method. It has been modified (Snyman 2000) to handle constraints via a penalty function formulation of the constrained problem. The outstanding characteristics of the basic method are:

(i) it uses *only* function *gradient* information ∇f,

(ii) *no* explicit *line searches* are performed,

(iii) it is extremely *robust*, handling steep valleys, and discontinuities and noise in the objective function and its gradient vector, with relative ease,

(iv) the algorithm seeks relatively low local minima and can therefore be used as the basic component in a methodology for *global optimization*, and

(v) when applied to smooth and near quadratic functions, it is not as efficient as classical methods.

6.2.1 Basic dynamic model

Assume a particle of *unit mass* in a n-dimensional *conservative* force field with *potential energy* at $\mathbf{x}$ given by $f(\mathbf{x})$, then at $\mathbf{x}$ the force (acceleration $\mathbf{a}$) on the particle is given by:

$$\mathbf{a} = \ddot{\mathbf{x}} = -\nabla f(\mathbf{x}) \tag{6.1}$$

from which it follows that for the motion of the particle over the time interval $[0, t]$:

$$\tfrac{1}{2}\|\dot{\mathbf{x}}(t)\|^2 - \tfrac{1}{2}\|\dot{\mathbf{x}}(0)\|^2 = f(\mathbf{x}(0)) - f(\mathbf{x}(t)) \tag{6.2}$$

or

$$T(t) - T(0) = f(0) - f(t) \tag{6.3}$$

where $T(t)$ represents the kinetic energy of the particle at time t. Thus it follows that

$$f(t) + T(t) = \text{constant} \tag{6.4}$$

i.e. conservation of energy along the trajectory. Note that along the particle trajectory the change in the function f, $\Delta f = -\Delta T$, and therefore, as long as T increases, f decreases. This is the underlying principle on which the dynamic leap-frog optimization algorithm is based.

6.2.2 Basic algorithm for unconstrained problems (LFOP)

The basic elements of the LFOP method are as listed in Algorithm 6.1. A detailed flow chart of the basic LFOP algorithm for unconstrained problems is given in Figure 6.1.

Algorithm 6.1 LFOP algorithm

---

1. Given $f(\mathbf{x})$ and starting point $\mathbf{x}(0) = \mathbf{x}^0$, compute the dynamic trajectory of the particle by solving the *initial value problem*:

$$\ddot{\mathbf{x}}(t) = -\boldsymbol{\nabla} f(\mathbf{x}(t))$$
$$\text{with } \dot{\mathbf{x}}(0) = \mathbf{0}; \text{ and } \mathbf{x}(0) = \mathbf{x}^0. \qquad (6.5)$$

2. Monitor $\mathbf{v}(t) = \dot{\mathbf{x}}(t)$, clearly as long as $T = \frac{1}{2}\|\mathbf{v}(t)\|^2$ increases, $f(\mathbf{x}(t))$ decreases as desired.

3. When $\|\mathbf{v}(t)\|$ decreases, i.e. when the particle moves uphill, apply some *interfering strategy* to gradually extract energy from the particle so as to increase the likelihood of its descent, but not so that descent occurs immediately.

4. In practice the numerical integration of the initial value problem (6.5) is done by the *"leap-frog" method*: compute for $k = 0, 1, 2, \ldots$, and given time step Δt:

$$\mathbf{x}^{k+1} = \mathbf{x}^k + \mathbf{v}^k \Delta t$$
$$\mathbf{v}^{k+1} = \mathbf{v}^k + \mathbf{a}^{k+1} \Delta t$$

where

$$\mathbf{a}^k = -\boldsymbol{\nabla} f(\mathbf{x}^k) \text{ and } \mathbf{v}^0 = \frac{1}{2}\mathbf{a}^0 \Delta t,$$

to ensure an initial step if $\mathbf{a}^0 \neq \mathbf{0}$.

A typical *interfering strategy* is to continue the trajectory when $\|\mathbf{v}^{k+1}\| \geq \|\mathbf{v}^k\|$, otherwise set $\mathbf{v}^k = \frac{1}{4}(\mathbf{v}^{k+1} + \mathbf{v}^k)$, $\mathbf{x}^{k+1} = \frac{1}{2}(\mathbf{x}^{k+1} + \mathbf{x}^k)$ to compute the new $\mathbf{v}^{k+1}$ and then continue.

In addition, Snyman (1982, 1983) introduced additional *heuristics* to determine a suitable initial time step Δt, to allow for the magnification and reduction of Δt, and to control the magnitude of the step $\Delta \mathbf{x} = \mathbf{x}^{k+1} - \mathbf{x}^k$ by setting a step size limit δ along the computed trajectory. The recommended magnitude of δ is $\delta \approx \frac{1}{10}\sqrt{n} \times$ (maximum variable range).

6.2.3 Modification for constrained problems (LFOPC)

The code LFOPC (Snyman 2000) applies the unconstrained optimization algorithm LFOP to a penalty function formulation of the constrained problem (see Section 3.1) in 3 phases (see Algorithm 6.2).

Algorithm 6.2 LFOPC algorithm

Phase 0:
Given some $\mathbf{x}^0$, then with overall penalty parameter $\rho = \rho_0$, apply LFOP to the penalty function $P(\mathbf{x}, \rho_0)$ to give $\mathbf{x}^*(\rho_0)$.
Phase 1:
With $\mathbf{x}^0 := \mathbf{x}^*(\rho_0)$ and $\rho := \rho_1$, where $\rho_1 \gg \rho_0$, apply LFOP to $P(\mathbf{x}, \rho_1)$ to give $\mathbf{x}^*(\rho_1)$ and identify the set of *active* constraints I_a, such that $g_{i_a}(\mathbf{x}^*(\rho_1)) > 0$ for $i_a \in I_a$.
Phase 2:
With $\mathbf{x}^0 := \mathbf{x}^*(\rho_1)$ use LFOP to minimize

$$P_a(\mathbf{x}, \rho_1) = \sum_{i=1}^{r} \rho_1 h_i^2(\mathbf{x}) + \sum_{i_a \in I_a} \rho_1 g_{i_a}^2(\mathbf{x})$$

to give $\mathbf{x}^*$.

For engineering problems (with convergence tolerance $\varepsilon_x = 10^{-4}$) the choice $\rho_0 = 10$ and $\rho_1 = 100$ is recommended. For extreme accuracy ($\varepsilon_x = 10^{-8}$), use $\rho_0 = 100$ and $\rho_1 = 10^4$.

6.2.3.1 Example

Minimize $f(\mathbf{x}) = x_1^2 + 2x_2^2$ such that $g(\mathbf{x}) = -x_1 - x_2 + 1 \leq 0$ with starting point $\mathbf{x}^0 = [3, 1]^T$ by means of the LFOPC algorithm. Use $\rho_0 = 1.0$ and $\rho_1 = 10.0$. The computed solution is depicted in Figure 6.2.

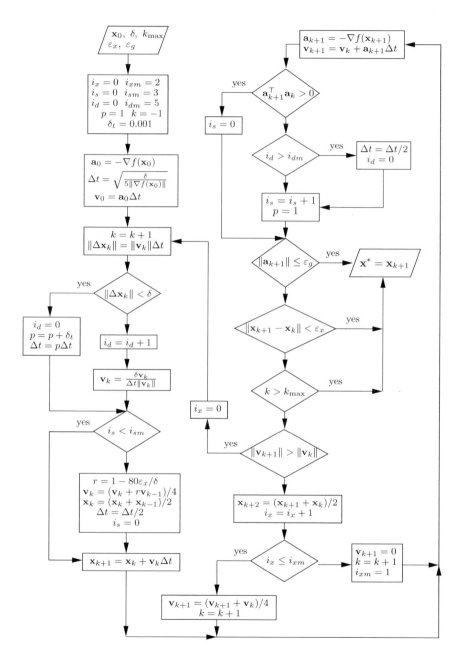

Figure 6.1: Flowchart of the LFOP unconstrained minimization algorithm

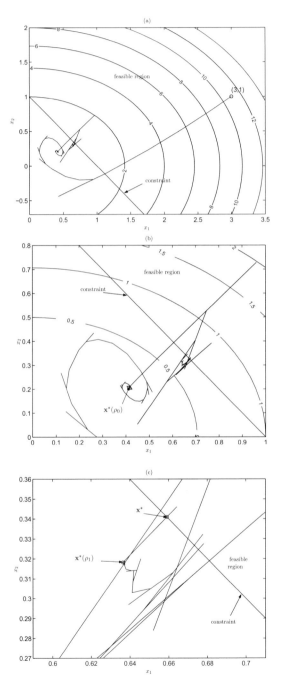

Figure 6.2: The (a) complete LFOPC trajectory for example problem 6.2.3.1, with $\mathbf{x}^0 = [3, 1]^T$, and magnified views of the final part of the trajectory shown in (b) and (c), giving $\mathbf{x}^* \approx [0.659, 0.341]^T$

6.3 The spherical quadratic steepest descent method

6.3.1 Introduction

In this section an extremely simple gradient only algorithm (Snyman and Hay 2001) is proposed that, in terms of storage requirement (only 3 n-vectors need be stored) and computational efficiency, may be considered as an alternative to the conjugate gradient methods. The method effectively applies the steepest descent (SD) method to successive simple spherical quadratic approximations of the objective function in such a way that no explicit line searches are performed in solving the minimization problem. It is shown that the method is convergent when applied to general positive-definite quadratic functions. The method is tested by its application to some standard and other test problems. On the evidence presented the new method, called the SQSD algorithm, appears to be reliable and stable and performs very well when applied to extremely ill-conditioned problems.

6.3.2 Classical steepest descent method revisited

Consider the following unconstrained optimization problem:

$$\min f(\mathbf{x}), \ \mathbf{x} \in \mathbb{R}^n \tag{6.6}$$

where f is a scalar objective function defined on $\mathbb{R}^n$, the n-dimensional real Euclidean space, and $\mathbf{x}$ is a vector of n real components $x_1, x_2, \ldots, x_n$. It is assumed that f is differentiable so that the gradient vector $\nabla f(\mathbf{x})$ exists everywhere in $\mathbb{R}^n$. The solution is denoted by $\mathbf{x}^*$.

The steepest descent (SD) algorithm for solving problem (6.6) may then be stated as follows:

It can be shown that if the steepest descent method is applied to a general positive-definite quadratic function of the form $f(\mathbf{x}) = \frac{1}{2}\mathbf{x}^T \mathbf{A}\mathbf{x} + \mathbf{b}^T\mathbf{x} + c$, then the sequence $\{f(\mathbf{x}^k)\} \to f(\mathbf{x}^*)$. Depending, however, on the starting point $\mathbf{x}^0$ and the condition number of $\mathbf{A}$ associated with the quadratic form, the rate of convergence may become extremely slow.

Algorithm 6.3 SD algorithm

Initialization: Specify convergence tolerances ε_g and ε_x, select starting point $\mathbf{x}^0$. Set $k := 1$ and go to main procedure.
Main procedure:

1. If $\left\| \boldsymbol{\nabla} f(\mathbf{x}^{k-1}) \right\| < \varepsilon_g$, then set $\mathbf{x}^* \cong \mathbf{x}^c = \mathbf{x}^{k-1}$ and stop; otherwise set $\mathbf{u}^k := -\boldsymbol{\nabla} f(\mathbf{x}^{k-1})$.

2. Let λ_k be such that $f(\mathbf{x}^{k-1} + \lambda_k \mathbf{u}^k) = \min_\lambda f(\mathbf{x}^{k-1} + \lambda \mathbf{u}^k)$ subject to $\lambda \geq 0$ {line search step}.

3. Set $\mathbf{x}^k := \mathbf{x}^{k-1} + \lambda_k \mathbf{u}^k$; if $\left\| \mathbf{x}^k - \mathbf{x}^{k-1} \right\| < \varepsilon_x$, then $\mathbf{x}^* \cong \mathbf{x}^c = \mathbf{x}^k$ and stop; otherwise set $k := k + 1$ and go to Step 1.

It is proposed here that for general functions $f(\mathbf{x})$, better overall performance of the steepest descent method may be obtained by applying it successively to a sequence of very simple quadratic approximations of $f(\mathbf{x})$. The proposed modification, named here the spherical quadratic steepest descent (SQSD) method, remains a first order method since only gradient information is used with no attempt being made to construct the Hessian of the function. The storage requirements therefore remain minimal, making it ideally suitable for problems with a large number of variables. Another significant characteristic is that the method requires no explicit line searches.

6.3.3   The SQSD algorithm

In the SQSD approach, given an initial approximate solution $\mathbf{x}^0$, a sequence of spherically quadratic optimization subproblems $P[k], k = 0, 1, 2, \ldots$ is solved, generating a sequence of approximate solutions $\mathbf{x}^{k+1}$. More specifically, at each point $\mathbf{x}^k$ the constructed approximate subproblem is $P[k]$:

$$\min_{\mathbf{x}} \tilde{f}_k(\mathbf{x}) \tag{6.7}$$

where the approximate objective function $\tilde{f}_k(\mathbf{x})$ is given by

$$\tilde{f}_k(\mathbf{x}) = f(\mathbf{x}^k) + \boldsymbol{\nabla}^T f(\mathbf{x}^k)(\mathbf{x} - \mathbf{x}^k) + \frac{1}{2}(\mathbf{x} - \mathbf{x}^k)^T \mathbf{C}_k(\mathbf{x} - \mathbf{x}^k) \tag{6.8}$$

and $\mathbf{C}_k = \mathrm{diag}(c_k, c_k, \ldots, c_k) = c_k\mathbf{I}$. The solution to this problem will be denoted by $\mathbf{x}^{*k}$, and for the construction of the next subproblem $P[k+1]$, $\mathbf{x}^{k+1} := \mathbf{x}^{*k}$.

For the first subproblem the curvature c_0 is set to $c_0 := \|\nabla f(\mathbf{x}^0)\|/\rho$, where $\rho > 0$ is some arbitrarily specified step limit. Thereafter, for $k \geq 1$, c_k is chosen such that $\tilde{f}(\mathbf{x}^k)$ interpolates $f(\mathbf{x})$ at both $\mathbf{x}^k$ and $\mathbf{x}^{k-1}$. The latter conditions imply that for $k = 1, 2, \ldots$

$$c_k := \frac{2\left[f(\mathbf{x}^{k-1}) - f(\mathbf{x}^k) - \nabla^T f(\mathbf{x}^k)(\mathbf{x}^{k-1} - \mathbf{x}^k)\right]}{\|\mathbf{x}^{k-1} - \mathbf{x}^k\|^2}. \tag{6.9}$$

Clearly the identical curvature entries along the diagonal of the Hessian, mean that the level surfaces of the quadratic approximation $\tilde{f}_k(\mathbf{x})$, are indeed concentric hyper-spheres. The approximate subproblems $P[k]$ are therefore aptly referred to as spherical quadratic approximations.

It is now proposed that for a large class of problems the sequence $\mathbf{x}^0, \mathbf{x}^1, \ldots$ will tend to the solution of the original problem (6.6), i.e.

$$\lim_{k \to \infty} \mathbf{x}^k = \mathbf{x}^*. \tag{6.10}$$

For subproblems $P[k]$ that are convex, i.e. $c_k > 0$, the solution occurs where $\nabla \tilde{f}_k(\mathbf{x}) = \mathbf{0}$, that is where

$$\nabla f(\mathbf{x}^k) + c_k\mathbf{I}(\mathbf{x} - \mathbf{x}^k) = \mathbf{0}. \tag{6.11}$$

The solution to the subproblem, $\mathbf{x}^{*k}$ is therefore given by

$$\mathbf{x}^{*k} = \mathbf{x}^k - \frac{\nabla f(\mathbf{x}^k)}{c_k}. \tag{6.12}$$

Clearly the solution to the spherical quadratic subproblem lies along a line through $\mathbf{x}^k$ in the direction of steepest descent. The SQSD method may formally be stated in the form given in Algorithm 6.4.

Step size control is introduced in Algorithm 6.4 through the specification of a step limit ρ and the test for $\|\mathbf{x}^k - \mathbf{x}^{k-1}\| > \rho$ in Step 2 of the main procedure. Note that the choice of c_0 ensures that for $P[0]$ the solution $\mathbf{x}^1$ lies at a distance ρ from $\mathbf{x}^0$ in the direction of steepest descent. Also the test in Step 3 that $c_k < 0$, and setting $c_k := 10^{-60}$ where this condition is true ensures that the approximate objective function is always positive-definite.

Algorithm 6.4 SQSD algorithm

Initialization: Specify convergence tolerances ε_g and ε_x, step limit $\rho > 0$ and select starting point $\mathbf{x}^0$. Set $c_0 := \left\| \boldsymbol{\nabla} f(\mathbf{x}^0) \right\| / \rho$. Set $k := 1$ and go to main procedure.

Main procedure:

1. If $\left\| \boldsymbol{\nabla} f(\mathbf{x}^{k-1}) \right\| < \varepsilon_g$, then $\mathbf{x}^* \cong \mathbf{x}^c = \mathbf{x}^{k-1}$ and stop; otherwise set

$$\mathbf{x}^k := \mathbf{x}^{k-1} - \frac{\boldsymbol{\nabla} f(\mathbf{x}^{k-1})}{c_{k-1}}.$$

2. If $\left\| \mathbf{x}^k - \mathbf{x}^{k-1} \right\| > \rho$, then set

$$\mathbf{x}^k := \mathbf{x}^k - \rho \frac{\boldsymbol{\nabla} f(\mathbf{x}^{k-1})}{\left\| \boldsymbol{\nabla} f(\mathbf{x}^{k-1}) \right\|};$$

if $\left\| \mathbf{x}^k - \mathbf{x}^{k-1} \right\| < \varepsilon_x$, then $\mathbf{x}^* \cong \mathbf{x}^c = \mathbf{x}^k$ and stop.

3. Set

$$c_k := \frac{2 \left[f(\mathbf{x}^{k-1}) - f(\mathbf{x}^k) - \boldsymbol{\nabla}^T f(\mathbf{x}^k)(\mathbf{x}^{k-1} - \mathbf{x}^k) \right]}{\left\| \mathbf{x}^{k-1} - \mathbf{x}^k \right\|^2};$$

if $c_k < 0$ set $c_k := 10^{-60}$.

4. Set $k := k + 1$ and go to Step 1 for next iteration.

6.3.4 Convergence of the SQSD method

An analysis of the convergence rate of the SQSD method, when applied to a general positive-definite quadratic function, affords insight into the convergence behavior of the method when applied to more general functions. This is so because for a large class of continuously differentiable functions, the behavior close to local minima is quadratic. For quadratic functions the following theorem may be proved.

6.3.4.1 Theorem

The SQSD algorithm (without step size control) is convergent when applied to the general quadratic function of the form $f(\mathbf{x}) = \frac{1}{2}\mathbf{x}^T\mathbf{A}\mathbf{x} + \mathbf{b}^T\mathbf{x}$, where $\mathbf{A}$ is a $n \times n$ positive-definite matrix and $\mathbf{b} \in \mathbb{R}^n$.

Proof. Begin by considering the bivariate quadratic function, $f(\mathbf{x}) = x_1^2 + \gamma x_2^2$, $\gamma \geq 1$ and with $\mathbf{x}^0 = [\alpha, \beta]^T$. Assume $c_0 > 0$ given, and for convenience in what follows set $c_0 = 1/\delta, \delta > 0$. Also employ the notation $f_k = f(\mathbf{x}^k)$.

Application of the first step of the SQSD algorithm yields

$$\mathbf{x}^1 = \mathbf{x}^0 - \frac{\nabla f_0}{c_0} = [\alpha(1 - 2\delta), \beta(1 - 2\gamma\delta)]^T \tag{6.13}$$

and it follows that

$$\left\|\mathbf{x}^1 - \mathbf{x}^0\right\|^2 = 4\delta^2(\alpha^2 + \gamma^2\beta^2) \tag{6.14}$$

and

$$\nabla f_1 = [2\alpha(1 - 2\delta), 2\gamma\beta(1 - 2\gamma\delta)]^T. \tag{6.15}$$

For the next iteration the curvature is given by

$$c_1 = \frac{2[f_0 - f_1 - \nabla^T f_1(\mathbf{x}^0 - \mathbf{x}^1)]}{\left\|\mathbf{x}^0 - \mathbf{x}^1\right\|^2}. \tag{6.16}$$

Utilizing the information contained in (6.13)–(6.15), the various entries in expression (6.16) are known, and after substitution c_1 simplifies to

$$c_1 = \frac{2(\alpha^2 + \gamma^3\beta^2)}{\alpha^2 + \gamma^2\beta^2}. \tag{6.17}$$

In the next iteration, Step 1 gives

$$\mathbf{x}^2 = \mathbf{x}^1 - \frac{\nabla f_1}{c_1}. \tag{6.18}$$

And after the necessary substitutions for $\mathbf{x}^1$, ∇f_1 and c_1, given by (6.13), (6.15) and (6.17) respectively, (6.18) reduces to

$$\mathbf{x}^2 = [\alpha(1 - 2\delta)\mu_1, \beta(1 - 2\gamma\delta)\omega_1]^T \tag{6.19}$$

where

$$\mu_1 = 1 - \frac{1 + \gamma^2\beta^2/\alpha^2}{1 + \gamma^3\beta^2/\alpha^2} \tag{6.20}$$

and

$$\omega_1 = 1 - \frac{\gamma + \gamma^3\beta^2/\alpha^2}{1 + \gamma^3\beta^2/\alpha^2}. \tag{6.21}$$

Clearly if $\gamma = 1$, then $\mu_1 = 0$ and $\omega_1 = 0$. Thus by (6.19) $\mathbf{x}^2 = \mathbf{0}$ and convergence to the solution is achieved within the second iteration.

Now for $\gamma > 1$, and for any choice of α and β, it follows from (6.20) that

$$0 \le \mu_1 < 1 \tag{6.22}$$

which implies from (6.19) that for the first component of $\mathbf{x}^2$:

$$\left|x_1^{(2)}\right| = |\alpha(1 - 2\delta)\mu_1| < |\alpha(1 - 2\delta)| = \left|x_1^{(1)}\right| \tag{6.23}$$

or introducing α notation (with $\alpha_0 = \alpha$), that

$$|\alpha_2| = |\mu_1\alpha_1| < |\alpha_1|. \tag{6.24}$$

{Note: because $c_0 = 1/\delta > 0$ is chosen arbitrarily, it cannot be said that $|\alpha_1| < |\alpha_0|$. However α_1 is finite.}

The above argument, culminating in result (6.24), is for the two iterations $\mathbf{x}^0 \to \mathbf{x}^1 \to \mathbf{x}^2$. Repeating the argument for the sequence of overlapping pairs of iterations $\mathbf{x}^1 \to \mathbf{x}^2 \to \mathbf{x}^3$; $\mathbf{x}^2 \to \mathbf{x}^3 \to \mathbf{x}^4$;..., it follows similarly that $|\alpha_3| = |\mu_2\alpha_2| < |\alpha_2|$; $|\alpha_4| = |\mu_3\alpha_3| < |\alpha_3|$;..., since $0 \le \mu_2 < 1$; $0 \le \mu_3 < 1$;..., where the value of μ_k is determined by (corresponding to equation (6.20) for μ_1):

$$\mu_k = 1 - \frac{1 + \gamma^2\beta_{k-1}^2/\alpha_{k-1}^2}{1 + \gamma^3\beta_{k-1}^2/\alpha_{k-1}^2}. \tag{6.25}$$

Thus in general

$$0 \le \mu_k < 1 \tag{6.26}$$

and

$$|\alpha_{k+1}| = |\mu_k \alpha_k| < |\alpha_k|. \tag{6.27}$$

For large positive integer m it follows that

$$|\alpha_m| = |\mu_{m-1} \alpha_{m-1}| = |\mu_{m-1} \mu_{m-2} \alpha_{m-2}| = |\mu_{m-1} \mu_{m-2} \cdots \mu_1 \alpha_1| \tag{6.28}$$

and clearly for $\gamma > 0$, because of (6.26)

$$\lim_{m \to \infty} |\alpha_m| = 0. \tag{6.29}$$

Now for the second component of $\mathbf{x}^2$ in (6.19), the expression for ω_1, given by (6.21), may be simplified to

$$\omega_1 = \frac{1 - \gamma}{1 + \gamma^3 \beta^2 / \alpha^2}. \tag{6.30}$$

Also for the second component:

$$x_2^{(2)} = \beta(1 - 2\gamma\delta)\omega_1 = \omega_1 x_2^{(1)} \tag{6.31}$$

or introducing β notation

$$\beta_2 = \omega_1 \beta_1. \tag{6.32}$$

The above argument is for $\mathbf{x}^0 \to \mathbf{x}^1 \to \mathbf{x}^2$ and again, repeating it for the sequence of overlapping pairs of iterations, it follows more generally for $k = 1, 2, \ldots$, that

$$\beta_{k+1} = \omega_k \beta_k \tag{6.33}$$

where ω_k is given by

$$\omega_k = \frac{1 - \gamma}{1 + \gamma^3 \beta_{k-1}^2 / \alpha_{k-1}^2}. \tag{6.34}$$

Since by (6.29), $|\alpha_m| \to 0$, it follows that if $|\beta_m| \to 0$ as $m \to \infty$, the theorem is proved for the bivariate case. Make the assumption that $|\beta_m|$

does not tend to zero, then there exists a finite positive number ε such that

$$|\beta_k| \geq \varepsilon \tag{6.35}$$

for all k. This allows the following argument:

$$|\omega_k| = \left| \frac{1 - \gamma}{1 + \gamma^3 \beta_{k-1}^2 / \alpha_{k-1}^2} \right| \leq \left| \frac{1 - \gamma}{1 + \gamma^3 \varepsilon^2 / \alpha_{k-1}^2} \right| = \left| \frac{(1 - \gamma)\alpha_{k-1}^2}{\alpha_{k-1}^2 + \gamma^3 \varepsilon^2} \right|. \tag{6.36}$$

Clearly since by (6.29) $|\alpha_m| \to 0$ as $m \to \infty$, (6.36) implies that also $|\omega_m| \to 0$. This result taken together with (6.33) means that $|\beta_m| \to 0$ which contradicts the assumption above. With this result the theorem is proved for the bivariate case.

Although the algebra becomes more complicated, the above argument can clearly be extended to prove convergence for the multivariate case, where

$$f(\mathbf{x}) = \sum_{i=1}^{n} \gamma_i x_i^2, \ \gamma_1 = 1 \leq \gamma_2 \leq \gamma_3 \leq \cdots \leq \gamma_n. \tag{6.37}$$

Finally since the general quadratic function

$$f(\mathbf{x}) = \frac{1}{2}\mathbf{x}^T \mathbf{A} \mathbf{x} + \mathbf{b}^T \mathbf{x}, \ \mathbf{A} \ \text{positive} - \text{definite} \tag{6.38}$$

may be transformed to the form (6.37), convergence of the SQSD method is also ensured in the general case. $\qquad\square$

It is important to note that, the above analysis does not prove that $\|\mathbf{x}^k - \mathbf{x}^*\|$ is monotonically decreasing with k, neither does it necessarily follow that monotonic descent in the corresponding objective function values $f(\mathbf{x}^k)$, is guaranteed. Indeed, extensive numerical experimentation with quadratic functions show that, although the SQSD trajectory rapidly approaches the minimum, relatively large spike increases in $f(\mathbf{x}^k)$ may occur after which the trajectory quickly recovers on its path to $\mathbf{x}^*$. This happens especially if the function is highly elliptical (poorly scaled).

6.3.5   Numerical results and conclusion

The SQSD method is now demonstrated by its application to some test problems. For comparison purposes the results are also given for the standard SD method and both the Fletcher-Reeves (FR) and Polak-Ribiere (PR) conjugate gradient methods. The latter two methods are implemented using the CG+ FORTRAN conjugate gradient program of Gilbert and Nocedal (1992). The CG+ implementation uses the line search routine of Moré and Thuente (1994). The function and gradient values are evaluated together in a single subroutine. The SD method is applied using CG+ with the search direction modified to the steepest descent direction. The FORTRAN programs were run on a 266 MHz Pentium 2 computer using double precision computations.

The standard (refs. Rao 1996; Snyman 1985; Himmelblau 1972; Manevich 1999) and other test problems used are listed in Section 6.3.6 and the results are given in Tables 6.1 and 6.2. The convergence tolerances applied throughout are $\varepsilon_g = 10^{-5}$ and $\varepsilon_x = 10^{-8}$, except for the extended homogeneous quadratic function with $n = 50000$ (Problem 12) and the extremely ill-conditioned Manevich functions (Problems 14). For these problems the extreme tolerances $\varepsilon_g \cong 0 (= 10^{-75})$ and $\varepsilon_x = 10^{-12}$, are prescribed in an effort to ensure very high accuracy in the approximation $\mathbf{x}^c$ to $\mathbf{x}^*$. For each method the number of function-cum-gradient-vector evaluations (N^{fg}) are given. For the SQSD method the number of iterations is the same as N^{fg}. For the other methods the number of iterations (N^{it}) required for convergence, and which corresponds to the number of line searches executed, are also listed separately. In addition the relative error (E^r) in optimum function value, defined by

$$E^r = \left| \frac{f(\mathbf{x}^*) - f(\mathbf{x}^c)}{1 + |f(\mathbf{x}^*)|} \right| \tag{6.39}$$

where $\mathbf{x}^c$ is the approximation to $\mathbf{x}^*$ at convergence, is also listed. For the Manevich problems, with $n \geq 40$, for which the other (SD, FR and PR) algorithms fail to converge after the indicated number of steps, the infinite norm of the error in the solution vector (I^∞), defined by $\|\mathbf{x}^* - \mathbf{x}^c\|_\infty$ is also tabulated. These entries, given instead of the relative error in function value (E^r), are made in italics.

Inspection of the results shows that the SQSD algorithm is consistently competitive with the other three methods and performs notably well for large problems. Of all the methods the SQSD method appears to be the most reliable one in solving each of the posed problems. As expected, because line searches are eliminated and consecutive search directions are

Prob. #	n	SQSD			Steepest Descent		
		ρ	N^{fg}	E^r	N^{fg}	N^{it}	E^r/I^∞
1	3	1	12	3.E-14	41	20	6.E-12
2	2	1	31	1.E-14	266	131	9.E-11
3	2	1	33	3.E-08	2316	1157	4.E-08
4	2	0.3	97	1.E-15	> 20000		3.E-09
5(a)	3	1	11	1.E-12	60	29	6.E-08
5(b)	3	1	17	1.E-12	49	23	6.E-08
6	4	1	119	9.E-09	> 20000		2.E-06
7	3	1	37	1.E-12	156	77	3.E-11
8	2	10	39	1.E-22	12050*	6023*	26*
9	2	0.3	113	5.E-14	6065	3027	2.E-10
10	2	1	43	1.E-12	1309	652	1.E-10
11	4	2	267	2.E-11	16701	8348	4.E-11
12	20	1.E+04	58	1.E-11	276	137	1.E-11
	200	1.E+04	146	4.E-12	2717	1357	1.E-11
	2000	1.E+04	456	2.E-10	> 20000		2.E-08
	20000	1.E+04	1318	6.E-09	> 10000		8.E+01
	50000	1.E+10	4073	3.E-16	> 10000		5.E+02
13	10	0.3	788	2.E-10	> 20000		4.E-07
	100	1	2580	1.E-12	> 20000		3.E+01
	300	1.73	6618	1.E-10	> 20000		2.E+02
	600	2.45	13347	1.E-11	> 20000		5.E+02
	1000	3.16	20717	2.E-10	> 30000		9.E+02
14	20	1	3651	2.E-27	> 20000		9.E-01
		10	3301	9.E-30			
	40	1	13302	5.E-27	> 30000		1.E+00
		10	15109	2.E-33			
	60	1	19016	7.E-39	> 30000		1.E+00
		10	16023	6.E-39			
	100	1	39690	1.E-49	> 50000		1.E+00
		10	38929	3.E-53			
	200	1	73517	5.E-81	> 100000		1.E+00
		10	76621	4.E-81			

* Convergence to a local minimum with $f(\mathbf{x}^c) = 48.9$.

Table 6.1: Performance of the SQSD and SD optimization algorithms when applied to the test problems listed in Section 6.3.6

Prob. #	n	Fletcher-Reeves			Polak-Ribiere		
		N^{fg}	N^{it}	E^r/I^∞	N^{fg}	N^{it}	E^r/I^∞
1	3	7	3	0$	7	3	0$
2	2	30	11	2.E-11	22	8	2.E-12
3	2	45	18	2.E-08	36	14	6.E-11
4	2	180	78	1.E-11	66	18	1.E-14
5(a)	3	18	7	6.E-08	18	8	6.E-08
5(b)	3	65	31	6.E-08	26	11	6.E-08
6	4	1573	783	8.E-10	166	68	3.E-09
7	3	132	62	4.E-12	57	26	1.E-12
8	2	72*	27*	26*	24*	11*	26*
9	2	56	18	5.E-11	50	17	1.E-15
10	2	127	60	6.E-12	30	11	1.E-11
11	4	193	91	1.E-12	99	39	9.E-14
12	20	42	20	9.E-32	42	20	4.E-31
	200	163	80	5.E-13	163	80	5.E-13
	2000	530	263	2.E-13	530	263	2.E-13
	20000	1652	825	4.E-13	1652	825	4.E-13
	50000	3225	1161	1.E-20	3225	1611	1.E-20
13	10	> 20000		2.E-02	548	263	4.E-12
	100	> 20000		8.E+01	1571	776	2.E-12
	300	> 20000		3.E+02	3253	1605	2.E-12
	600	> 20000		6.E+02	5550	2765	2.E-12
	1000	> 30000		1.E+03	8735	4358	2.E-12
14	20	187	75	8.E-24	1088	507	2.E-22
	40	> 30000		*1.E+00*	> 30000		*1.E+00*
	60	> 30000		*1.E+00*	> 30000		*1.E+00*
	100	> 50000		*1.E+00*	> 50000		*1.E+00*
	200	> 100000		*1.E+00*	> 100000		*1.E+00*

* Convergence to a local minimum with $f(\mathbf{x}^c) = 48.9$; $ Solution to machine accuracy.

Table 6.2: Performance of the FR and PR algorithms when applied to the test problems listed in Section 6.3.6

no longer forced to be orthogonal, the new method completely overshadows the standard SD method. What is much more gratifying, however, is the performance of the SQSD method relative to the well-established and well-researched conjugate gradient algorithms. Overall the new method appears to be very competitive with respect to computational efficiency and, on the evidence presented, remarkably stable.

In the implementation of the SQSD method to highly non-quadratic and non-convex functions, some care must however be taken in ensuring that the chosen step limit parameter ρ, is not too large. A too large value may result in excessive oscillations occurring before convergence. Therefore a relatively small value, $\rho = 0.3$, was used for the Rosenbrock problem with $n = 2$ (Problem 4). For the extended Rosenbrock functions of larger dimensionality (Problems 13), correspondingly larger step limit values ($\rho = \sqrt{n}/10$) were used with success.

For quadratic functions, as is evident from the convergence analysis of Section 6.3.4, no step limit is required for convergence. This is borne out in practice by the results for the extended homogeneous quadratic functions (Problems 12), where the very large value $\rho = 10^4$ was used throughout, with the even more extreme value of $\rho = 10^{10}$ for $n = 50000$. The specification of a step limit in the quadratic case also appears to have little effect on the convergence rate, as can be seen from the results for the ill-conditioned Manevich functions (Problems 14), that are given for both $\rho = 1$ and $\rho = 10$. Here convergence is obtained to at least 11 significant figures accuracy ($\|\mathbf{x}^* - \mathbf{x}^c\|_\infty < 10^{-11}$) for each of the variables, despite the occurrence of extreme condition numbers, such as 10^{60} for the Manevich problem with $n = 200$.

The successful application of the new method to the ill-conditioned Manevich problems, and the analysis of the convergence behavior for quadratic functions, indicate that the SQSD algorithm represents a powerful approach to solving quadratic problems with large numbers of variables. In particular, the SQSD method can be seen as an *unconditionally convergent*, *stable* and *economic* alternative iterative method for solving large systems of linear equations, ill-conditioned or not, through the minimization of the sum of the squares of the residuals of the equations.

6.3.6 Test functions used for SQSD

Minimize $f(\mathbf{x})$:

1. $f(\mathbf{x}) = x_1^2 + 2x_2^2 + 3x_3^2 - 2x_1 - 4x_2 - 6x_3 + 6$, $\mathbf{x}^0 = [3,3,3]^T$, $\mathbf{x}^* = [1,1,1]^T$, $f(\mathbf{x}^*) = 0.0$.

2. $f(\mathbf{x}) = x_1^4 - 2x_1^2 x_2 + x_1^2 + x_2^2 - 2x_1 + 1$, $\mathbf{x}^0 = [3,3]^T$, $\mathbf{x}^* = [1,1]^T$, $f(\mathbf{x}^*) = 0.0$.

3. $f(\mathbf{x}) = x_1^4 - 8x_1^3 + 25x_1^2 + 4x_2^2 - 4x_1 x_2 - 32x_1 + 16$, $\mathbf{x}^0 = [3,3]^T$, $\mathbf{x}^* = [2,1]^T$, $f(\mathbf{x}^*) = 0.0$.

4. $f(\mathbf{x}) = 100(x_2 - x_1^2)^2 + (1 - x_1)^2$, $\mathbf{x}^0 = [-1.2,1]^T$, $\mathbf{x}^* = [1,1]^T$, $f(\mathbf{x}^*) = 0.0$ (Rosenbrock's parabolic valley, Rao 1996).

5. $f(\mathbf{x}) = x_1^4 + x_1^3 - x_1 + x_2^4 - x_2^2 + x_2 + x_3^2 - x_3 + x_1 x_2 x_3$, (Zlobec's function, Snyman 1985):

   (a) $\mathbf{x}^0 = [1,-1,1]^T$ and

   (b) $\mathbf{x}^0 = [0,0,0]^T$, $\mathbf{x}^* = [0.57085597, -0.93955591, 0.76817555]^T$, $f(\mathbf{x}^*) = -1.91177218907$.

6. $f(\mathbf{x}) = (x_1 + 10x_2)^2 + 5(x_3 - x_4)^2 + (x_2 - 2x_3)^4 + 10(x_1 - x_4)^4$, $\mathbf{x}^0 = [3,-1,0,1]^T$, $\mathbf{x}^* = [0,0,0,0]^T$, $f(\mathbf{x}^*) = 0.0$ (Powell's quartic function, Rao 1996).

7. $f(\mathbf{x}) = -\left\{ \frac{1}{1+(x_1-x_2)^2} + \sin\left(\frac{1}{2}\pi x_2 x_3\right) + \exp\left[-\left(\frac{x_1+x_3}{x_2} - 2\right)^2\right]\right\}$, $\mathbf{x}^0 = [0,1,2]^T$, $\mathbf{x}^* = [1,1,1]^T$, $f(\mathbf{x}^*) = -3.0$ (Rao 1996).

8. $f(\mathbf{x}) = \{-13 + x_1 + [(5 - x_2)x_2 - 2]x_2\}^2 + \{-29 + x_1 + [(x_2+1)x_2 - 14]x_2\}^2$, $\mathbf{x}^0 = [1/2,-2]^T$, $\mathbf{x}^* = [5,4]^T$, $f(\mathbf{x}^*) = 0.0$ (Freudenstein and Roth function, Rao 1996).

9. $f(\mathbf{x}) = 100(x_2 - x_1^3)^2 + (1 - x_1)^2$, $\mathbf{x}^0 = [-1.2,1]^T$, $\mathbf{x}^* = [1,1]^T$, $f(\mathbf{x}^*) = 0.0$ (cubic valley, Himmelblau 1972).

10. $f(\mathbf{x}) = [1.5 - x_1(1 - x_2)]^2 + [2.25 - x_1(1 - x_2^2)]^2 + [2.625 - x_1(1 - x_2^3)]^2$, $\mathbf{x}^0 = [1,1]^T$, $\mathbf{x}^* = [3,1/2]^T$, $f(\mathbf{x}^*) = 0.0$ (Beale's function, Rao 1996).

11. $f(\mathbf{x}) = [10(x_2 - x_1^2)]^2 + (1 - x_1)^2 + 90(x_4 - x_3^2)^2 + (1 - x_3)^2 + 10(x_2 + x_4 - 2)^2 + 0.1(x_2 - x_4)^2$, $\mathbf{x}^0 = [-3, 1, -3, -1]^T$, $\mathbf{x}^* = [1, 1, 1, 1]^T$, $f(\mathbf{x}^*) = 0.0$ (Wood's function, Rao 1996).

12. $f(\mathbf{x}) = \sum_{i=1}^{n} i x_i^2$, $\mathbf{x}^0 = [3, 3, \ldots, 3]^T$, $\mathbf{x}^* = [0, 0, \ldots, 0]^T$, $f(\mathbf{x}^*) = 0.0$ (extended homogeneous quadratic functions).

13. $f(\mathbf{x}) = \sum_{i=1}^{n-1} [100(x_{i+1} - x_i^2)^2 + (1 - x_i)^2]$, $\mathbf{x}^0 = [-1.2, 1, -1.2, 1, \ldots]^T$, $\mathbf{x}^* = [1, 1, \ldots, 1]^T$, $f(\mathbf{x}^*) = 0.0$ (extended Rosenbrock functions, Rao 1996).

14. $f(\mathbf{x}) = \sum_{i=1}^{n} (1 - x_i)^2 / 2^{i-1}$, $\mathbf{x}^0 = [0, 0, \ldots, 0]^T$, $\mathbf{x}^* = [1, 1, \ldots, 1]^T$, $f(\mathbf{x}^*) = 0.0$ (extended Manevich functions, Manevich 1999).

6.4 The Dynamic-Q optimization algorithm

6.4.1 Introduction

An efficient *constrained* optimization method is presented in this section. The method, called the Dynamic-Q method (Snyman and Hay 2002), consists of applying the dynamic trajectory LFOPC optimization algorithm (see Section 6.2) to successive quadratic approximations of the actual optimization problem. This method may be considered as an extension of the unconstrained SQSD method, presented in Section 6.3, to one capable of handling general constrained optimization problems.

Due to its efficiency with respect to the number of function evaluations required for convergence, the Dynamic-Q method is primarily intended for optimization problems where function evaluations are expensive. Such problems occur frequently in engineering applications where time consuming numerical simulations may be used for function evaluations. Amongst others, these numerical analyses may take the form of a computational fluid dynamics (CFD) simulation, a structural analysis by means of the finite element method (FEM) or a dynamic simulation of a multibody system. Because these simulations are usually expensive to perform, and because the relevant functions may not be known analytically, standard classical optimization methods are normally not suited to these types of problems. Also, as will be shown, the storage requirements of the Dynamic-Q method are minimal. No Hessian information is required. The method is therefore particularly suitable for problems where the number of variables n is large.

6.4.2 The Dynamic-Q method

Consider the general nonlinear optimization problem:

$$\min_{\mathbf{x}} f(\mathbf{x}); \; \mathbf{x} = [x_1, x_2, \ldots, x_n]^T \in \mathbb{R}^n$$

$$\text{subject to} \tag{6.40}$$

$$g_j(\mathbf{x}) \leq \mathbf{0}; \; j = 1, 2, \ldots, p$$

$$h_k(\mathbf{x}) = \mathbf{0}; \; k = 1, 2, \ldots, q$$

where $f(\mathbf{x})$, $g_j(\mathbf{x})$ and $h_k(\mathbf{x})$ are scalar functions of $\mathbf{x}$.

In the Dynamic-Q approach, successive subproblems $P[i]$, $i = 0, 1, 2, \ldots$ are generated, at successive approximations $\mathbf{x}^i$ to the solution $\mathbf{x}^*$, by constructing *spherically quadratic* approximations $\tilde{f}(\mathbf{x})$, $\tilde{g}_j(\mathbf{x})$ and $\tilde{h}_k(\mathbf{x})$ to $f(\mathbf{x})$, $g_j(\mathbf{x})$ and $h_k(\mathbf{x})$. These approximation functions, evaluated at a point $\mathbf{x}^i$, are given by

$$
\begin{aligned}
\tilde{f}(\mathbf{x}) &= f(\mathbf{x}^i) + \boldsymbol{\nabla}^T f(\mathbf{x}^i)(\mathbf{x} - \mathbf{x}^i) + \frac{1}{2}(\mathbf{x} - \mathbf{x}^i)^T \mathbf{A}(\mathbf{x} - \mathbf{x}^i) \\
\tilde{g}_j(\mathbf{x}) &= g_j(\mathbf{x}^i) + \boldsymbol{\nabla}^T g_j(\mathbf{x}^i)(\mathbf{x} - \mathbf{x}^i) \\
&\quad + \frac{1}{2}(\mathbf{x} - \mathbf{x}^i)^T \mathbf{B}_j(\mathbf{x} - \mathbf{x}^i), \; j = 1, \ldots, p \\
\tilde{h}_k(\mathbf{x}) &= h_k(\mathbf{x}^i) + \boldsymbol{\nabla}^T h_k(\mathbf{x}^i)(\mathbf{x} - \mathbf{x}^i) \\
&\quad + \frac{1}{2}(\mathbf{x} - \mathbf{x}^i)^T \mathbf{C}_k(\mathbf{x} - \mathbf{x}^i), \; k = 1, \ldots, q
\end{aligned}
\tag{6.41}
$$

with the Hessian matrices $\mathbf{A}$, $\mathbf{B}_j$ and $\mathbf{C}_k$ taking on the simple forms

$$
\begin{aligned}
\mathbf{A} &= \operatorname{diag}(a, a, \ldots, a) = a\mathbf{I} \\
\mathbf{B}_j &= b_j \mathbf{I} \\
\mathbf{C}_k &= c_k \mathbf{I}.
\end{aligned}
\tag{6.42}
$$

Clearly the identical entries along the diagonal of the Hessian matrices indicate that the approximate subproblems $P[i]$ are indeed spherically quadratic.

For the first subproblem ($i = 0$) a linear approximation is formed by setting the curvatures a, b_j and c_k to zero. Thereafter a, b_j and c_k are chosen so that the approximating functions (6.41) interpolate their corresponding actual functions at both $\mathbf{x}^i$ and $\mathbf{x}^{i-1}$. These conditions imply that for $i = 1, 2, 3, \ldots$

$$
a = \frac{2\left[f(\mathbf{x}^{i-1}) - f(\mathbf{x}^i) - \boldsymbol{\nabla}^T f(\mathbf{x}^i)(\mathbf{x}^{i-1} - \mathbf{x}^i)\right]}{\|\mathbf{x}^{i-1} - \mathbf{x}^i\|^2} \tag{6.43}
$$

$$
b_j = \frac{2\left[g_j(\mathbf{x}^{i-1}) - g_j(\mathbf{x}^i) - \boldsymbol{\nabla}^T g_j(\mathbf{x}^i)(\mathbf{x}^{i-1} - \mathbf{x}^i)\right]}{\|\mathbf{x}^{i-1} - \mathbf{x}^i\|^2}, \quad j = 1, \ldots, p
$$

$$
c_k = \frac{2\left[h_k(\mathbf{x}^{i-1}) - h_k(\mathbf{x}^i) - \boldsymbol{\nabla}^T h_k(\mathbf{x}^i)(\mathbf{x}^{i-1} - \mathbf{x}^i)\right]}{\|\mathbf{x}^{i-1} - \mathbf{x}^i\|^2}, \quad k = 1, \ldots, q.
$$

If the gradient vectors $\boldsymbol{\nabla}^T f$, $\boldsymbol{\nabla}^T g_j$ and $\boldsymbol{\nabla}^T h_k$ are not known analytically, they may be approximated from functional data by means of first-order forward finite differences.

The particular choice of spherically quadratic approximations in the Dynamic-Q algorithm has implications on the computational and storage requirements of the method. Since the second derivatives of the objective function and constraints are approximated using function and gradient data, the $O(n^2)$ calculations and storage locations, which would usually be required for these second derivatives, are not needed. The computational and storage resources for the Dynamic-Q method are thus reduced to $O(n)$. At most, $4 + p + q + r + s$ n-vectors need be stored (where p, q, r and s are respectively the number of inequality and equality constraints and the number of lower and upper limits of the variables). These savings become significant when the number of variables becomes large. For this reason it is expected that the Dynamic-Q method is well suited, for example, to engineering problems such as structural optimization problems where a large number of variables are present.

In many optimization problems, additional bound constraints of the form $\hat{k}_i \leq x_i \leq \check{k}_i$ occur. Constants $\hat{k}_i$ and $\check{k}_i$ respectively represent lower and upper bounds for variable x_i. Since these constraints are of a simple form (having zero curvature), they need not be approximated in the Dynamic-Q method and are instead explicitly treated as special linear inequality constraints. Constraints corresponding to lower and upper limits are respectively of the form

$$\hat{g}_l(\mathbf{x}) = \hat{k}_{vl} - x_{vl} \leq 0, \ l = 1, 2, \ldots, r \leq n \tag{6.44}$$
$$\check{g}_m(\mathbf{x}) = x_{wm} - \check{k}_{wm} \leq 0, \ m = 1, 2, \ldots, s \leq n$$

where $vl \in \hat{I} = (v1, v2, \ldots, vr)$ the set of r subscripts corresponding to the set of variables for which respective lower bounds $\hat{k}_{vl}$ are prescribed, and $wm \in \check{I} = (w1, w2, \ldots, ws)$ the set of s subscripts corresponding to the set of variables for which respective upper bounds $\check{k}_{wm}$ are prescribed. The subscripts vl and wm are used since there will, in general, not be n lower and upper limits, i.e. usually $r \neq n$ and $s \neq n$.

In order to obtain convergence to the solution in a controlled and stable manner, move limits are placed on the variables. For each approximate subproblem $P[i]$ this move limit takes the form of an additional single inequality constraint

$$g_\rho(\mathbf{x}) = \left\| \mathbf{x} - \mathbf{x}^i \right\|^2 - \rho^2 \leq 0 \tag{6.45}$$

where ρ is an appropriately chosen step limit and $\mathbf{x}^i$ is the solution to the previous subproblem.

The approximate subproblem, constructed at $\mathbf{x}^i$, to the optimization problem (6.40) (plus bound constraints (6.44) and move limit (6.45)), thus becomes $P[i]$:

$$\min_{\mathbf{x}} \tilde{f}(\mathbf{x}), \ \mathbf{x} = [x_1, x_2, \ldots, x_n]^T \in \mathbb{R}^n$$

subject to

$$\begin{aligned}
\tilde{g}_j(\mathbf{x}) &\leq 0, \ j = 1, 2, \ldots, p \\
\tilde{h}_k(\mathbf{x}) &= 0, \ k = 1, 2, \ldots, q \\
\hat{g}_l(\mathbf{x}) &\leq 0, \ l = 1, 2, \ldots, r \\
\check{g}_m(\mathbf{x}) &\leq 0, \ m = 1, 2, \ldots, s \\
g_\rho(\mathbf{x}) &= \left\| \mathbf{x} - \mathbf{x}^i \right\|^2 - \rho^2 \leq 0
\end{aligned} \tag{6.46}$$

with solution $\mathbf{x}^{*i}$. The Dynamic-Q algorithm is given by Algorithm 6.5. In the Dynamic-Q method the subproblems generated are solved using the dynamic trajectory, or "leap-frog" (LFOPC) method of Snyman (1982, 1983) for unconstrained optimization applied to penalty function formulations (Snyman et al. 1994; Snyman 2000) of the constrained

---

Algorithm 6.5 Dynamic-Q algorithm

Initialization: Select starting point $\mathbf{x}^0$ and move limit ρ. Set $i := 0$.
Main procedure:

1. Evaluate $f(\mathbf{x}^i)$, $g_j(\mathbf{x}^i)$ and $h_k(\mathbf{x}^i)$ as well as $\boldsymbol{\nabla} f(\mathbf{x}^i)$, $\boldsymbol{\nabla} g_j(\mathbf{x}^i)$ and $\boldsymbol{\nabla} h_k(\mathbf{x}^i)$. If termination criteria are satisfied set $\mathbf{x}^* = \mathbf{x}^i$ and stop.

2. Construct a local approximation $P[i]$ to the optimization problem at $\mathbf{x}^i$ using expressions (6.41) to (6.43).

3. Solve the approximated subproblem $P[i]$ (given by (6.46)) using the constrained optimizer LFOPC with $\mathbf{x}^0 := \mathbf{x}^i$ (see Section 6.2) to give $\mathbf{x}^{*i}$.

4. Set $i := i + 1$, $\mathbf{x}^i := \mathbf{x}^{*(i-1)}$ and return to Step 1.

---

problem. A brief description of the LFOPC algorithm is given in Section 6.2.

The LFOPC algorithm possesses a number of outstanding characteristics, which makes it highly suitable for implementation in the Dynamic-Q methodology. The algorithm requires only gradient information and no explicit line searches or function evaluations are performed. These properties, together with the influence of the fundamental physical principles underlying the method, ensure that the algorithm is extremely robust. This has been proven over many years of testing (Snyman 2000). A further desirable characteristic related to its robustness, and the main reason for its application in solving the subproblems in the Dynamic-Q algorithm, is that if there is no feasible solution to the problem, the LFOPC algorithm will still find the best possible compromised solution. The Dynamic-Q algorithm thus usually converges to a solution from an infeasible remote point without the need to use line searches between subproblems, as is the case with SQP. The LFOPC algorithm used by Dynamic-Q is identical to that presented in Snyman (2000) except for a minor change to LFOP which is advisable should the subproblems become effectively unconstrained.

6.4.3 Numerical results and conclusion

The Dynamic-Q method requires very few parameter settings by the user. Other than convergence criteria and specification of a maximum number of iterations, the only parameter required is the step limit ρ. The algorithm is not very sensitive to the choice of this parameter, however, ρ should be chosen of the same order of magnitude as the diameter of the region of interest. For the problems listed in Table 6.3 a step limit of $\rho = 1$ was used except for problems 72 and 106 where step limits $\rho = \sqrt{10}$ and $\rho = 100$ were used respectively.

Given specified positive tolerances ε_x, ε_f and ε_c, then at step i termination of the algorithm occurs if the normalized step size

$$\frac{\left\| \mathbf{x}^i - \mathbf{x}^{i-1} \right\|}{1 + \left\| \mathbf{x}^i \right\|} < \varepsilon_x \tag{6.47}$$

or if the normalized change in function value

$$\frac{\left| f^i - f_{\text{best}} \right|}{1 + \left| f_{\text{best}} \right|} < \varepsilon_f \tag{6.48}$$

where f_{best} is the lowest previous feasible function value and the current $\mathbf{x}^i$ is feasible. The point $\mathbf{x}^i$ is considered feasible if the absolute value of the violation of each constraint is less than ε_c. This particular function termination criterion is used since the Dynamic-Q algorithm may at times exhibit oscillatory behavior near the solution.

In Table 6.3, for the same starting points, the performance of the Dynamic-Q method on some standard test problems is compared to results obtained for Powell's SQP method as reported by Hock and Schittkowski (1981). The problem numbers given correspond to the problem numbers in Hock and Schittkowski's book. For each problem, the actual function value f_{act} is given, as well as, for each method, the calculated function value f^* at convergence, the relative function error

$$E^r = \frac{\left| f_{\text{act}} - f^* \right|}{1 + \left| f_{\text{act}} \right|} \tag{6.49}$$

and the number of function-gradient evaluations (N^{fg}) required for convergence. In some cases it was not possible to calculate the relative

Prob. #	n	f_{act}	SQP			Dynamic-Q		
			N^{fg}	f^*	E^r	N^{fg}	f^*	E^r
2	2	5.04E-02	16~	2.84E+01	2.70E+01	7*	4.94E+00	<1.00E-08
10	2	-1.00E+00	12	-1.00E+00	5.00E-08	13	-1.00E+00	<1.00E-08
12	2	-3.00E+01	12	-3.00E+01	<1.00E-08	9	-3.00E+01	<1.00E-08
13	2	1.00E+00	45	1.00E+00	5.00E-08	50$	9.59E-01	2.07E-02
14	2	1.39E+00	6	1.39E+00	8.07E-09	5	1.39E+00	7.86E-07
15	2	3.07E+02	5	3.07E+02	<1.00E-08	15*	3.60E+02	5.55E-07
16	2	2.50E-01	6*	2.31E+01	<1.00E-08	5*	2.31E+01	<1.00E-08
17	2	1.00E+00	12	1.00E+00	<1.00E-08	16	1.00E+00	<1.00E-08
20	2	3.82E+01	20	3.82E+01	4.83E-09	4*	4.02E+01	<1.00E-08
22	2	1.00E+00	9	1.00E+00	<1.00E-08	3	1.00E+00	<1.00E-08
23	2	2.00E+00	7	2.00E+00	<1.00E-08	5	2.00E+00	<1.00E-08
24	2	-1.00E+00	5	-1.00E+00	<1.00E-08	4	-1.00E+00	1.00E-08
26	3	0.00E+00	19	4.05E-08	4.05E-08	27	1.79E-07	1.79E-07
27	3	4.00E-02	25	4.00E-02	1.73E-08	28	4.00E-02	9.62E-10
28	3	0.00E+00	5	2.98E-21	2.98E-21	12	7.56E-10	7.56E-10
29	3	-2.26E+01	13	-2.26E+01	8.59E-11	11	-2.26E+01	8.59E-11
30	3	1.00E+00	14	1.00E+00	<1.00E-08	5	1.00E+00	<1.00E-08
31	3	6.00E+00	10	6.00E+00	<1.00E-08	10	6.00E+00	1.43E-08
32	3	1.00E+00	3	1.00E+00	<1.00E-08	4	1.00E+00	<1.00E-08
33	3	-4.59E+00	5*	-4.00E+00	<1.00E-08	3*	-4.00E+00	<1.00E-08
36	3	-3.30E+03	4	-3.30E+03	<1.00E-08	15	-3.30E+03	<1.00E-08
45	5	1.00E+00	8	1.00E+00	<1.00E-08	7	1.00E+00	1.00E-08
52	5	5.33E+00	8	5.33E+00	5.62E-09	12	5.33E+00	1.02E-08
55	6	6.33E+00	1~	6.00E+00	4.54E-02	2*	6.66E+00	1.30E-09
56	7	-3.46E+00	11	-3.46E+00	<1.00E-08	20	-3.46E+00	6.73E-08
60	3	3.26E-02	9	3.26E-02	3.17E-08	11	3.26E-02	1.21E-09
61	3	-1.44E+02	10	-1.44E+02	1.52E-08	10	-1.44E+02	1.52E-08
63	3	9.62E+02	9	9.62E+02	2.18E-09	6	9.62E+02	2.18E-09
65	3	9.54E-01	11~	2.80E+00	9.47E-01	9	9.54E-01	2.90E-08
71	4	1.70E+01	5	1.70E+01	1.67E-08	6	1.70E+01	1.67E-08
72	4	7.28E+02	35	7.28E+02	1.37E-08	30	7.28E+02	1.37E-08
76	4	-4.68E+00	6	-4.68E+00	3.34E-09	8	-4.68E+00	3.34E-09
78	5	-2.92E+00	9	-2.92E+00	2.55E-09	6	-2.92E+00	2.55E-09
80	5	5.39E-02	7	5.39E-02	7.59E-10	6	5.39E-02	7.59E-10
81	5	5.39E-02	8	5.39E-02	1.71E-09	12	5.39E-02	1.90E-10
100	7	6.80E+02	20	6.80E+02	<1.00E-08	16	6.80E+02	1.46E-10
104	8	3.95E+00	19	3.95E+00	8.00E-09	42	3.95E+00	5.26E-08
106	8	7.05E+03	44	7.05E+03	1.18E-05	79	7.05E+03	1.18E-05
108	9	-8.66E-01	9*	-6.97E-01	1.32E-02	26	-8.66E-01	3.32E-09
118	15	6.65E+02	~	~	~	38	6.65E+02	3.00E-08
Svan	21	2.80E+02	150	2.80E+02	9.96E-05	93	2.80E+02	1.59E-06

* Converges to a local minimum - listed E^r relative to function value at local minimum;
~ Fails; $ Terminates on maximum number of steps.

Table 6.3: Performance of the Dynamic-Q and SQP optimization algorithms

function error due to rounding off of the solutions reported by Hock and Schittkowski. In these cases the calculated solutions were correct to at least eight significant digits. For the Dynamic-Q algorithm, convergence tolerances of $\varepsilon_f = 10^{-8}$ on the function value, $\varepsilon_x = 10^{-5}$ on the step size and $\varepsilon_c = 10^{-6}$ for constraint feasibility, were used. These were chosen to allow for comparison with the reported SQP results.

The result for the 12-corner polytope problem of Svanberg (1999) is also given. For this problem the results given in the SQP columns are for Svanberg's Method of Moving Asymptotes (MMA). The recorded number of function evaluations for this method is approximate since the results given correspond to 50 outer iterations of the MMA, each requiring about 3 function evaluations.

A robust and efficient method for nonlinear optimization, with minimal storage requirements compared to those of the SQP method, has been proposed and tested. The particular methodology proposed is made possible by the special properties of the LFOPC optimization algorithm (Snyman 2000), which is used to solve the quadratic subproblems. Comparison of the results for Dynamic-Q with the results for the SQP method show that equally accurate results are obtained with comparable number of function evaluations.

6.5 A gradient-only line search method for conjugate gradient methods

6.5.1 Introduction

Many engineering design optimization problems involve numerical computer analyses via, for example, FEM codes, CFD simulations or the computer modeling of the dynamics of multi-body mechanical systems. The computed objective function is therefore often the result of a complex sequence of calculations involving other computed or measured quantities. This may result in the presence of numerical noise in the objective function so that it exhibits non-smooth trends as design parameters are varied. It is well known that this presence of numerical noise in the design optimization problem inhibits the use of classical and traditional gradient-based optimization methods that employ line searches, such as for example, the conjugate gradient methods. The numerical noise may prevent or slow down convergence during optimization. It may also promote convergence to spurious local optima. The computational expense of the analyses, coupled to the convergence difficulties created by the numerical noise, is in many cases a significant obstacle to performing multidisciplinary design optimization.

In addition to the anticipated difficulties when applying the conjugate gradient methods to noisy optimization problems, it is also known that standard implementations of conjugate gradient methods, in which conventional line search techniques have been used, are less robust than one would expect from their theoretical quadratic termination property. Therefore the conjugate gradient method would, under normal circumstances, not be preferred to quasi-Newton methods (Fletcher 1987). In particular severe numerical difficulties arise when standard line searches are used in solving constrained problems through the minimization of associated penalty functions. However, there is one particular advantage of conjugate gradient methods, namely the particular simple form that requires no matrix operations in determining the successive search directions. Thus, conjugate gradient methods may be the only methods which are applicable to large problems with thousands of variables (Fletcher 1987), and are therefore well worth further investigation.

In this section a new implementation (ETOPC) of the conjugate gradient method (both for the Fletcher-Reeves and Polak-Ribiere versions (see Fletcher 1987) is presented for solving constrained problems. The essential novelty in this implementation is the use of a gradient-only line search technique originally proposed by the author (Snyman 1985), and used in the ETOP algorithm for unconstrained minimization. It will be shown that this implementation of the conjugate gradient method, not only easily overcomes the accuracy problem when applied to the minimization of penalty functions, but also economically handles the problem of severe numerical noise superimposed on an otherwise smooth underlying objective function.

6.5.2 Formulation of optimization problem

Consider again the general constrained optimization problem:

$$\min_{\mathbf{x}} f(\mathbf{x}), \ \mathbf{x} = [x_1, x_2, x_3, \ldots, x_n]^T \in R^n \tag{6.50}$$

subject to the inequality and equality constraints:

$$g_j(\mathbf{x}) \leq 0, \ j = 1, 2, \ldots, m \tag{6.51}$$
$$h_j(\mathbf{x}) = 0, \ j = 1, 2, \ldots, r$$

where the objective function $f(\mathbf{x})$, and the constraint functions $g_j(\mathbf{x})$ and $h_j(\mathbf{x})$, are scalar functions of the real column vector $\mathbf{x}$. The optimum solution is denoted by $\mathbf{x}^*$, with corresponding optimum function value $f(\mathbf{x}^*)$.

The most straightforward way of handling the constraints is via the unconstrained minimization of the penalty function:

$$P(\mathbf{x}) = f(\mathbf{x}) + \sum_{j=1}^{r} \rho_j h_j^2(\mathbf{x}) + \sum_{j=1}^{m} \beta_j g_j^2(\mathbf{x}) \tag{6.52}$$

where $\rho_j \gg 0$, $\beta_j = 0$ if $g_j(\mathbf{x}) \leq 0$, and $\beta_j = \mu_j \gg 0$ if $g_j(\mathbf{x}) > 0$.

Usually $\rho_j = \mu_j = \mu \gg 0$ for all j, with the corresponding penalty function being denoted by $P(\mathbf{x}, \mu)$.

Central to the application of the conjugate gradient method to penalty function formulated problems presented here, is the use of an unconventional line search method for unconstrained minimization, proposed by the author, in which no function values are explicitly required (Snyman 1985). Originally this gradient-only line search method was applied to the conjugate gradient method in solving a few very simple unconstrained problems. For somewhat obscure reasons, given in the original paper (Snyman 1985) and briefly hinted to in this section, the combined method (novel line search plus conjugate gradient method) was called the ETOP (Euler-Trapezium Optimizer) algorithm. For this historical reason, and to avoid confusion, this acronym will be retained here to denote the combined method for unconstrained minimization. In subsequent unreported numerical experiments, the author was successful in solving a number of more challenging practical constrained optimization problems via penalty function formulations of the constrained problem, with ETOP being used in the unconstrained minimization of the sequence of penalty functions. ETOP, applied in this way to constrained problems, was referred to as the ETOPC algorithm. Accordingly this acronym will also be used here.

6.5.3   Gradient-only line search

The line search method used here, and originally proposed by the author (Snyman 1985) uses no explicit function values. Instead the line search is implicitly done by using only two gradient vector evaluations at two points along the search direction and assuming that the function is near-quadratic along this line. The essentials of the gradient-only line search, for the case where the function $f(\mathbf{x})$ is unconstrained, are as follows. Given the current design point $\mathbf{x}^k$ at iteration k and next search direction $\mathbf{v}^{k+1}$, then compute

$$\mathbf{x}^{k+1} = \mathbf{x}^k + \mathbf{v}^{k+1}\tau \qquad (6.53)$$

where τ is some suitably chosen positive parameter. The step taken in (6.53) may be seen as an "Euler step". With this step given by

$$\Delta\mathbf{x}^k = \mathbf{x}^{k+1} - \mathbf{x}^k = \mathbf{v}^{k+1}\tau \qquad (6.54)$$

the line search in the direction $\mathbf{v}^{k+1}$ is equivalent to finding $\mathbf{x}^{*k+1}$ defined by

$$f(\mathbf{x}^{*k+1}) = \min_{\lambda} f(\mathbf{x}^k + \lambda\Delta\mathbf{x}^k). \qquad (6.55)$$

These steps are depicted in Figure 6.3.

It was indicated in Snyman (1985) that for the step $\mathbf{x}^{k+1} = \mathbf{x}^k + \mathbf{v}^{k+1}\tau$ the change in function value Δf_k, in the unconstrained case, can be approximated without explicitly evaluating the function $f(\mathbf{x})$. Here a more formal argument is presented via the following lemma.

6.5.3.1   Lemma 1

For a general quadratic function, the change in function value, for the step $\Delta\mathbf{x}^k = \mathbf{x}^{k+1} - \mathbf{x}^k = \mathbf{v}^{k+1}\tau$ is given by:

$$\Delta f_k = -\langle \mathbf{v}^{k+1}, \frac{1}{2}(\mathbf{a}^k + \mathbf{a}^{k+1})\tau \rangle \qquad (6.56)$$

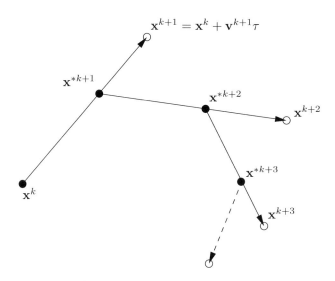

Figure 6.3: Successive steps in line search procedure

where $\mathbf{a}^k = -\nabla f(\mathbf{x}^k)$ and $\langle \ , \ \rangle$ denotes the scalar product.

Proof:

In general, by Taylor's theorem:

$$f(\mathbf{x}^{k+1}) - f(\mathbf{x}^k) = \langle \mathbf{x}^{k+1} - \mathbf{x}^k, \nabla f(\mathbf{x}^k) \rangle + \frac{1}{2}\langle \Delta \mathbf{x}^k, \mathbf{H}(\mathbf{x}^a)\Delta \mathbf{x}^k \rangle$$

and

$$f(\mathbf{x}^{k+1}) - f(\mathbf{x}^k) = \langle \mathbf{x}^{k+1} - \mathbf{x}^k, \nabla f(\mathbf{x}^{k+1}) \rangle - \frac{1}{2}\langle \Delta \mathbf{x}^k, \mathbf{H}(\mathbf{x}^b)\Delta \mathbf{x}^k \rangle$$

where $\mathbf{x}^a = \mathbf{x}^k + \theta_0 \Delta \mathbf{x}^k$, $\mathbf{x}^b = \mathbf{x}^k + \theta_1 \Delta \mathbf{x}^k$ and both θ_0 and θ_1 in the interval $[0, 1]$, and where $\mathbf{H}(\mathbf{x})$ denotes the Hessian matrix of the general function $f(\mathbf{x})$. Adding the above two expressions gives:

$$\begin{aligned} f(\mathbf{x}^{k+1}) - f(\mathbf{x}^k) &= \frac{1}{2}\langle \mathbf{x}^{k+1} - \mathbf{x}^k, \nabla f(\mathbf{x}^k) + \nabla f(\mathbf{x}^{k+1}) \rangle \\ &+ \frac{1}{4}\langle \Delta \mathbf{x}^k, [\mathbf{H}(\mathbf{x}^a) - \mathbf{H}(\mathbf{x}^b)]\Delta \mathbf{x}^k \rangle. \end{aligned}$$

If $f(\mathbf{x})$ is quadratic then $\mathbf{H}(\mathbf{x})$ is constant and it follows that

$$\Delta f_k = f(\mathbf{x}^{k+1}) - f(\mathbf{x}^k) = -\langle \mathbf{v}^{k+1}, \frac{1}{2}(\mathbf{a}^k + \mathbf{a}^{k+1})\tau \rangle$$

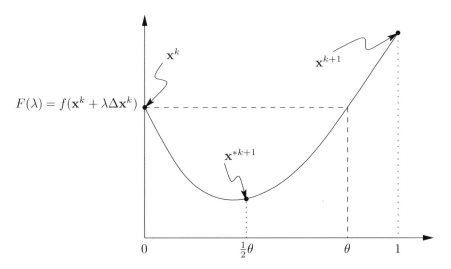

Figure 6.4: Approximation of minimizer $\mathbf{x}^{*k+1}$ in the direction $\mathbf{v}^{k+1}$

where $\mathbf{a}^k = \boldsymbol{\nabla} f(\mathbf{x}^k)$, which completes the proof.   □

By using expression (6.56) the position of the minimizer $\mathbf{x}^{*k+1}$ (see Figure 6.4), in the direction $\mathbf{v}^{k+1}$, can also be approximated without any explicit function evaluation. This conclusion follows formally from the second lemma given below. Note that in (6.56) the second quantity in the scalar product corresponds to an average vector given by the "trapezium rule". This observation together with the remark following equation (6.53), gave rise to the name "Euler-trapezium optimizer (ETOP)" when applying this line search technique in the conjugate gradient method.

6.5.3.2 Lemma 2

For $f(\mathbf{x})$ a positive-definite quadratic function the point $\mathbf{x}^{*k+1}$ defined by $f(\mathbf{x}^{*k+1}) = \min_\lambda f(\mathbf{x}^k + \lambda \Delta \mathbf{x}^k)$ is given by

$$\mathbf{x}^{*k+1} = \mathbf{x}^k + \frac{1}{2}\theta \Delta \mathbf{x}^k \tag{6.57}$$

where

$$\theta = \rho / (\langle \mathbf{v}^{k+1}, \frac{1}{2}(\mathbf{a}^k + \mathbf{a}^{k+1})\tau \rangle + \rho) \text{ and } \rho = -\langle \Delta \mathbf{x}^k, \mathbf{a}^k \rangle. \tag{6.58}$$

Proof:

First determine θ such that

$$f(\mathbf{x}^k + \theta \Delta \mathbf{x}^k) = f(\mathbf{x}^k).$$

By Taylor's expansion:

$$f(\mathbf{x}^{k+1}) - f(\mathbf{x}^k) = \rho + \frac{1}{2}\langle \Delta \mathbf{x}^k, \mathbf{H}\Delta \mathbf{x}^k \rangle, \text{ i.e., } \frac{1}{2}\langle \Delta \mathbf{x}^k, \mathbf{H}\Delta \mathbf{x}^k \rangle = \Delta f_k - \rho$$

which gives for the step $\theta \Delta \mathbf{x}$:

$$f(\mathbf{x}^k + \theta \Delta \mathbf{x}^k) - f(\mathbf{x}^k) = \theta\rho + \frac{1}{2}\theta^2 \langle \Delta \mathbf{x}^k, \mathbf{H}\Delta \mathbf{x}^k \rangle = \theta\rho + \theta^2(\Delta f_k - \rho).$$

For both function values to be the same, θ must therefore satisfy:

$$0 = \theta(\rho + \theta(\Delta f_k - \rho))$$

which has the non-trivial solution:

$$\theta = -\rho/(\Delta f_k - \rho).$$

Using the expression for Δf_k given by Lemma 1, it follows that:

$$\theta = \rho/(\langle \mathbf{v}^{k+1}, \frac{1}{2}(\mathbf{a}^k + \mathbf{a}^{k+1})\tau \rangle + \rho)$$

and by the symmetry of quadratic functions that

$$\mathbf{x}^{*k+1} = \mathbf{x}^k + \frac{1}{2}\theta \Delta \mathbf{x}^k.$$

$\square$

Expressions (6.57) and (6.58) may of course also be used in the general non-quadratic case, to determine an approximation to the minimizer $\mathbf{x}^{*k+1}$ in the direction $\mathbf{v}^{k+1}$, when performing successive line searches using the sequence of descent directions, $\mathbf{v}^{k+1}$, $k = 1, 2, \ldots$ Thus in practice, for the next $(k+1)$-th iteration, set $\mathbf{x}^{k+1} := \mathbf{x}^{*k+1}$, and with the next selected search direction $\mathbf{v}^{k+2}$ proceed as above, using expressions (6.57) and (6.58) to find $\mathbf{x}^{*k+2}$ and then set $\mathbf{x}^{k+2} := \mathbf{x}^{*k+2}$. Continue iterations in this way, with only two gradient vector evaluations done per line search, until convergence is obtained.

In summary, explicit function evaluations are unnecessary in the above line search procedure, since the two computed gradients along the search direction allow for the computation of an approximation (6.56) to the change in objective function, which in turn allows for the estimation of the position of the minimum along the search line via expressions (6.57) and (6.58), based on the assumption that the function is near quadratic in the region of the search.

6.5.3.3 Heuristics

Of course in general the objective function may not be quadratic and positive-definite. Additional heuristics are therefore required to ensure descent, and to see to it that the step size (corresponding to the parameter τ between successive gradient evaluations, is neither too small nor too large. The details of these heuristics are as set out below.

(i) In the case of a successful step having been taken, with Δf_k computed via (6.56) negative, i.e. descent, and θ computed via (6.58) positive, i.e. the function is locally strictly convex, as shown in Figure 6.4, τ is increased by a factor of 1.5 for the next search direction.

(ii) It may turn out that although Δf_k computed via (6.56) is negative, that θ computed via (6.58) is also negative. The latter implies that the function along the search direction is locally concave. In this case set $\theta := -\theta$ in computing $\mathbf{x}^{*k+1}$ by (6.57), so as to ensure a step in the descent direction, and also increase τ by the factor 1.5 before computing the step for the next search direction using (6.53).

(iii) It may happen that Δf_k computed by (6.56) is negative and exactly equal to ρ, i.e. $\Delta f_k - \rho = 0$. This implies zero curvature with $\theta = \infty$ and the function is therefore locally linear. In this case enforce the value $\theta = 1$. This results in the setting, by (6.57), of $\mathbf{x}^{*k+1}$ equal to a point halfway between $\mathbf{x}^k$ and $\mathbf{x}^{k+1}$. In this case τ is again increased by the factor of 1.5.

(iv) If both Δf_k and θ are positive, which is the situation depicted in Figure 6.4, then τ is halved before the next step.

(v) In the only outstanding and unlikely case, should it occur, where Δf^k is positive and θ negative, τ is unchanged.

(vi) For usual unconstrained minimization the initial step size parameter selection is $\tau = 0.5$.

The new gradient-only line search method may of course be applied to any line search descent method for the unconstrained minimization of a general multi-variable function. Here its application is restricted to the conjugate gradient method.

6.5.4 Conjugate gradient search directions and SUMT

The search vectors used here correspond to the conjugate gradient directions (Bazaraa et al. 1993). In particular for $k = 0, 1, \ldots$, the search vectors are

$$\mathbf{v}^{k+1} = (-\nabla f(\mathbf{x}^k) + \beta_{k+1}\mathbf{v}^k/\tau)\tau = \mathbf{s}^{k+1}\tau \qquad (6.59)$$

where $\mathbf{s}^{k+1}$ denote the usual conjugate gradient directions, $\beta_1 = 0$ and for $k > 0$:

$$\beta_{k+1} = \|\nabla f(\mathbf{x}^k)\|^2/\|\nabla f(\mathbf{x}^{k-1})\|^2 \qquad (6.60)$$

for the Fletcher-Reeves implementation, and for the Polak-Ribiere version:

$$\beta_{k+1} = \langle \nabla f(\mathbf{x}^k) - \nabla f(\mathbf{x}^{k-1}), \nabla f(\mathbf{x}^k) \rangle/\|\nabla f(\mathbf{x}^{k-1})\|^2. \qquad (6.61)$$

As recommended by Fletcher (1987), the conjugate gradient algorithm is restarted in the direction of steepest descent when $k > n$.

For the constrained problem the unconstrained minimization is of course applied to successive penalty function formulations $P(\mathbf{x})$ of the form shown in (6.52), using the well known Sequential Unconstrained Minimization Technique (SUMT) (Fiacco and McCormick 1968). In SUMT, for $j = 1, 2, \ldots$, until convergence, successive unconstrained minimizations are performed on successive penalty functions $P(\mathbf{x}) = P(\mathbf{x}, \mu^{(j)})$ in which the overall penalty parameter $\mu^{(j)}$ is successively increased: $\mu^{(j+1)} := 10\mu^{(j)}$. The corresponding initial step size parameter is set at $\tau = 0.5/\mu^{(j)}$ for each sub problem j. This application of ETOP to the constrained problem, via the unconstrained minimization of successive penalty functions, is referred to as the ETOPC algorithm. In practice, if analytical expressions for the components of the gradient of the objective function are not available, they may be calculated with sufficient accuracy by finite differences. However, when the presence of severe noise is suspected, the application of the gradient-only search method with conjugate gradient search directions, requires that *central finite difference* approximations of the gradients be used in order to effectively smooth out the noise. In this case relatively excessive perturbations δx_i in x_i must be used, which in practice may typically be of the order of 0.1 times the range of interest!

In the application of ETOPC a limit Δ_m, is in practice set to the maximum allowable magnitude Δ^* of the step $\boldsymbol{\Delta}^* = \mathbf{x}^{*k+1} - \mathbf{x}^k$. If Δ^* is greater than Δ_m, then set

$$\mathbf{x}^{k+1} := \mathbf{x}^k + (\mathbf{x}^{*k+1} - \mathbf{x}^k)\Delta_m/\Delta^* \qquad (6.62)$$

and restart the conjugate gradient procedure, with $\mathbf{x}^0 := \mathbf{x}^{k+1}$, in the direction of steepest descent. If the maximum allowable step is taken n times in succession, then Δ_m is doubled.

6.5.5   Numerical results

The proposed new implementation of the conjugate gradient method (both the Fletcher- Reeves and Polak-Ribiere versions) is tested here using 40 different problems arbitrarily selected from the famous set of test problems of Hock and Schittkowski (1981). The problem numbers (Pr. #) in the tables, correspond to the numbering used in Hock and Schittkowski. The final test problem, (12-poly), is the 12 polytope problem of Svanberg (1995, 1999). The number of variables (n) of the test problems ranges from 2 to 21 and the number of constraints (m plus r) per problem, from 1 to 59. The termination criteria for the ETOPC algorithm are as follows:

(i) Convergence tolerances for successive approximate sub-problems within SUMT: ε_g for convergence on the norm of the gradient vector, i.e. terminate if $\|\boldsymbol{\nabla} P(\mathbf{x}^{*k+1}, \mu)\| < \varepsilon_g$, and ε_x for con-

vergence on average change of design vector: i.e. terminate if
$\frac{1}{2}\|\mathbf{x}^{*k+1} - \mathbf{x}^{*k-1}\| < \varepsilon_x$.

(ii) Termination of the SUMT procedure occurs if the absolute value
of the relative difference between the objective function values at
the solution points of successive SUMT problems is less than ε_f.

6.5.5.1 Results for smooth functions with no noise

For the initial tests no noise is introduced. For high accuracy requirements (relative error in optimum objective function value to be less than
10^{-8}), it is found that the proposed new conjugate gradient implementation performs as robust as, and more economical than, the traditional
penalty function implementation, FMIN, of Kraft and Lootsma reported
in Hock and Schittkowski (1981). The detailed results are as tabulated
in Table 6.4. Unless otherwise indicated the algorithm settings are:
$\varepsilon_x = 10^{-8}$, $\varepsilon_g = 10^{-5}$, $\Delta_m = 1.0$, $\varepsilon_f = 10^{-8}$, $\mu^{(1)} = 1.0$ and $iout = 15$,
where $iout$ denotes the maximum number of SUMT iterations allowed.
The number of gradient vector evaluations required by ETOPC for the
different problems are denoted by nge (note that the number of explicit
function evaluations is zero), and the relative error in function value at
convergence to the point $\mathbf{x}^c$ is denoted by r_f, which is computed from

$$r_f = |f(\mathbf{x}^*) - f(\mathbf{x}^c)|/(|f(\mathbf{x}^*)| + 1). \tag{6.63}$$

For the FMIN algorithm only the number of explicit objective function
evaluations nfe are listed, together with the relative error r_f at convergence. The latter method requires, in addition to the number of function
evaluations listed, a comparable number of gradient vector evaluations,
which is not given here (see Hock and Schittkowski 1981).

6.5.5.2 Results for severe noise introduced in the objective function

Following the successful implementation for the test problems with no
noise, all the tests were rerun, but with severe relative random noise
introduced in the objective function $f(\mathbf{x})$ and all gradient components

Pr. #	n m r	Fletcher-Reeves		Polak-Ribiere		FMIN	
		nge	r_f	nge	r_f	nfe	r_f
1	2 1 -	100	$< 10^{-14}$	103	$< 10^{-13}$	549	$< 10^{-8}$
2	2 1 -	290	$< 10^{-8}$	318	$< 10^{-8}$	382	1×10^{-8}
10	2 1 -	231	$< 10^{-9}$	247	$< 10^{-9}$	289	7×10^{-8}
12	2 1 -	163	$< 10^{-10}$	184	$< 10^{-10}$	117	1×10^{-8}
13[1]	2 3 -	4993[2]	0.028	4996[2]	0.034	1522	0.163
14	2 1 1	214	$< 10^{-10}$	200	$< 10^{-10}$	232	2×10^{-7}
15	2 3 -	699	$< 10^{-9}$	632	$< 10^{-9}$	729	4×10^{-7}
16	2 5 -	334	$< 10^{-9}$	284	$< 10^{-7}$	362	1×10^{-8}
17	2 5 -	218	$< 10^{-9}$	209	$< 10^{-9}$	541	1×10^{-8}
20[3]	2 5 -	362	$< 10^{-9}$	375	$< 10^{-9}$	701	4×10^{-6}
22	2 2 -	155	$< 10^{-9}$	202	$< 10^{-9}$	174	1×10^{-7}
23	2 9 -	257	$< 10^{-9}$	244	$< 10^{-9}$	423	6×10^{-6}
24	2 5 -	95	$< 10^{-11}$	163	2×10^{-6}	280	2×10^{-8}
26	3 - 1	78	$< 10^{-8}$	100	2×10^{-8}	182	1×10^{-8}
27	3 - 1	129	$< 10^{-8}$	115	$< 10^{-8}$	173	1×10^{-8}
28	3 - 1	17	$< 10^{-28}$	17	$< 10^{-28}$	23	$< 10^{-8}$
29	3 1 -	254	$< 10^{-10}$	267	$< 10^{-10}$	159	$< 10^{-8}$
30	3 7 -	115	$< 10^{-10}$	124	$< 10^{-10}$	1199	4×10^{-8}
31	3 7 1	309	$< 10^{-9}$	274	$< 10^{-9}$	576	$< 10^{-8}$
32	3 7 1	205	$< 10^{-10}$	207	$< 10^{-10}$	874	$< 10^{-8}$
33[3]	3 6 -	272[3]	$< 10^{-10}$	180	$< 10^{-10}$	672[3]	3×10^{-7}
36	3 7 -	336	$< 10^{-12}$	351	$< 10^{-10}$	263	2×10^{-6}
45	5 10 -	175	$< 10^{-10}$	150	$< 10^{-10}$	369	$< 10^{-8}$
52	5 - 3	403	$< 10^{-9}$	388	$< 10^{-9}$	374	$< 10^{-8}$
55[3]	6 8 6	506	$< 10^{-9}$	488	$< 10^{-9}$	581[3]	3×10^{-8}
56	7 - 4	316	6×10^{-8}	289	7×10^{-8}	446	$< 10 - 8$
60	3 6 1	198	$< 10^{-10}$	189	$< 10^{-10}$	347	$1 \times 10 - 8$
61	3 - 2	205	$< 10^{-10}$	201	$< 10^{-10}$	217	$< 10 - 8$
63	3 3 2	205	$< 10^{-10}$	208	$< 10^{-10}$	298	$< 10 - 8$
65	3 7 -	179	$< 10^{-8}$	198	$< 10^{-10}$	-	fails
71	4 9 1	493	$< 10^{-9}$	536	$< 10^{-9}$	1846	5×10^{-3}
72[4]	4 10 -	317	$< 10^{-10}$	298	$< 10^{-10}$	1606	5×10^{-2}
76	4 7 -	224	$< 10^{-10}$	227	$< 10^{-10}$	424	$< 10^{-8}$
78	5 - 3	261	$< 10^{-10}$	264	$< 10^{-10}$	278	$< 10^{-8}$
80	5 10 3	192	$< 10^{-11}$	194	$< 10^{-11}$	1032	2×10^{-8}
81[5]	5 10 3	138	$< 10^{-11}$	158	$< 10^{-10}$	1662	5×10^{-7}
106[6]	8 22 -	6060	5×10^{-6}	6496	3×10^{-5}	-	fails
108	9 14 -	600	$< 10^{-10}$	519	$< 10^{-10}$	984	7×10^{-5}
118[7]	15 29 -	1233	$< 10^{-8}$	1358	$< 10^{-8}$	-	fails
12-poly[7]	21 22 -	844	$< 10^{-9}$	1478	$< 10^{-9}$	-	-

[1]Constraint qualification not satisfied. [2]Termination on maximum number of steps.
[3]Convergence to local minimum. [4]$\mu^{(0)} = 1.0$, $\Delta_m = 1.0$. [5]$\mu^{(0)} = 10^2$. [6]$\Delta_m = 10^2$.
[7]Gradients by central finite differences, $\delta x_i = 10^{-6}, \varepsilon_x = 10^{-6}$.

Table 6.4: The respective performances of the new conjugate gradient implementation ETOPC and FMIN for test problems with no noise introduced

computed by central finite differences. The influence of noise is investigated for two cases, namely, for a variation of the superimposed uniformly distributed random noise as large as (i) 5% and (ii) 10% of $(1 + |f(\mathbf{x}^*)|)$, where $\mathbf{x}^*$ is the optimum of the underlying smooth problem. The detailed results are shown in Table 6.5. The results are listed only for the Fletcher-Reeves version. The results for the Polak- Ribiere implementation are almost identical. Unless otherwise indicated the algorithm settings are: $\delta x_i = 1.0$, $\varepsilon_g = 10^{-5}$, $\Delta_m = 1.0$, $\varepsilon_f = 10^{-8}$, $\mu^{(0)} = 1.0$ and $iout = 6$, where $iout$ denotes the maximum number of SUMT iterations allowed. For termination of sub-problem on step size, ε_x was set to $\varepsilon_x := 0.005\sqrt{n}$ for the initial sub-problem. Thereafter it is successively halved for each subsequent sub-problem.

The results obtained are surprisingly good with, in most cases, fast convergence to the neighbourhood of the known optimum of the underlying smooth problem. In 90% of the cases regional convergence was obtained with relative errors $r_x < 0.025$ for 5% noise and $r_x < 0.05$ for 10% noise, where

$$r_x = \|\mathbf{x}^* - \mathbf{x}^c\|/(\|\mathbf{x}^*\| + 1) \tag{6.64}$$

and $\mathbf{x}^c$ denotes the point of convergence. Also in 90% of the test problems the respective relative errors in final objective function values were $r_f < 0.025$ for 5% noise and $r_f < 0.05$ for 10% noise, where r_f is as defined in (6.63).

6.5.6 Conclusion

The ETOPC algorithm performs exceptionally well for a first order method in solving constrained problems where the functions are smooth. For these problems the gradient only penalty function implementation of the conjugate gradient method performs as well, if not better than the best conventional implementations reported in the literature, in producing highly accurate solutions.

In the cases where severe noise is introduced in the objective function, relatively fast convergence to the neighborhood of $\mathbf{x}^*$, the solution of the underlying smooth problem, is obtained. Of interest is the fact that with the reduced accuracy requirement associated with the presence of noise, the number of function evaluations required to obtain sufficiently

Pr. #	nmr	5% noise			10% noise		
		nge	r_f	r_x	nge	r_f	r_x
1	2 1 -	54	0.035	5×10^{-3}	54	0.06	5×10^{-3}
2	2 1 -	80	2×10^{-3}	2×10^{-3}	87	9×10^{-3}	2×10^{-3}
10	2 1 -	120	0.02	0.022	160	0.048	0.023
12	2 1 -	99	0.018	8×10^{-2}	232	0.006	0.011
13	2 3 -	394	0.079	0.044	187	0.189	0.095
14	2 1 1	138	0.025	6×10^{-4}	126	0.041	6×10^{-4}
15	2 3 -	152	6×10^{-5}	2×10^{-5}	154	0.006	8×10^{-5}
16[1]	2 5 -	250	0.128	0.13	175	0.135	0.16
17	2 5 -	84	0.012	7×10^{-6}	77	0.041	3×10^{-4}
20	2 5 -	89	0.009	2×10^{-5}	105	0.001	2×10^{-5}
22	2 2 -	75	0.01	4×10^{-5}	86	0.035	9×10^{-5}
23	2 9 -	103	0.008	9×10^{-4}	100	0.005	7×10^{-4}
24	2 5 -	75	0.0095	4×10^{-5}	137	0.014	3×10^{-5}
26	3 - 1	63	0.019	2×10^{-3}	71	0.04	3×10^{-3}
27	3 - 1	159	0.015	0.014	132	0.022	0.036
28	3 - 1	46	0.018	6×10^{-3}	49	0.009	0.025
29	3 1 -	232	0.013	0.01	251	0.046	0.015
30	3 7 -	52	0.025	4×10^{-3}	72	0.043	6×10^{-3}
31	3 7 1	123	0.015	9×10^{-4}	183	0.031	0.013
32	3 7 1	89	0.006	4×10^{-3}	107	0.031	5×10^{-3}
33	3 6 -	183	0.016	0.035	83	0.026	3×10^{-3}
36[2]	3 7 -	177	0.018	6×10^{-5}	179	0.01	8×10^{-5}
45	5 10 -	122	0.0013	9×10^{-4}	92	0.009	4×10^{-5}
52	5 - 3	239	0.019	0.042	318	0.041	0.071
55	6 8 6	137	0.016	5×10^{-3}	188	0.041	4×10^{-3}
56	7 - 4	166	0.012	0.014	144	0.03	0.038
60	3 6 1	95	0.021	0.071	83	0.018	0.033
61[2]	3 - 2	105	0.019	2×10^{-3}	83	0.026	9×10^{-4}
63[2]	3 3 2	198	0.02	8×10^{-3}	652	0.016	0.06
65	3 7 -	94	4×10^{-3}	3×10^{-3}	106	0.012	0.003
71	4 9 1	164	0.021	0.022	143	0.021	0.035
72	4 10 -	454	0.01	0.025	578	0.005	0.094
76	4 7 -	131	0.022	0.002	148	0.012	0.041
78	5 - 3	87	0.011	0.004	88	0.037	0.002
80	5 10 3	92	0.011	0.025	105	0.005	0.02
81[3]	5 10 3	39	0.017	0.032	47	0.012	0.031
106[4]	8 22 -	6016	0.023	0.088	8504	0.038	0.113
108[2]	9 14 -	113	0.017	0.04	140	0.04	0.025
118[2]	15 29 -	395	0.012	0.041	371	0.049	0.1
12-poly[2]	21 22 -	476	0.012	0.065	607	0.047	0.1

[1]$\delta x_i = 10^{-1}$. [2]$\delta x_i = 10$. [3]$\mu^{(0)} = 10^2$ [4]$\delta x_i = 10^3, \Delta_m = 10^2$.

Table 6.5: Performance of ETOPC for test problems with severe noise introduced

accurate solutions in the case of noise, is on the average much less than that necessary for the high accuracy solutions for smooth functions. As already stated, ETOPC yields in 90% of the cases regional convergence with relative errors $r_x < 0.025$ for 5% noise, and $r_x < 0.05$ for 10% noise. Also in 90% of the test problems the respective relative errors in the final objective function values are $r_f < 0.025$ for 5% noise and $r_f < 0.05$ for 10% noise. In the other 10% of the cases the relative errors are also acceptably small. These accuracies are more than sufficient for multidisciplinary design optimization problems where similar noise may be encountered.

6.6 Global optimization using dynamic search trajectories

6.6.1 Introduction

The problem of globally optimizing a real valued function is inherently intractable (unless hard restrictions are imposed on the objective function) in that no practically useful characterization of the global optimum is available. Indeed the problem of determining an accurate estimate of the global optimum is mathematically ill-posed in the sense that very similar objective functions may have global optima very distant from each other (Schoen 1991). Nevertheless, the need in practice to find a relative low local minimum has resulted in considerable research over the last decade to develop algorithms that attempt to find such a low minimum, e.g. see Törn and Zilinskas (1989).

The general global optimization problem may be formulated as follows. Given a real valued objective function $f(\mathbf{x})$ defined on the set $\mathbf{x} \in D$ in $\mathbb{R}^n$, find the point $\mathbf{x}^*$ and the corresponding function value f^* such that

$$f^* = f(\mathbf{x}^*) = \text{ minimum } \{f(\mathbf{x})|\mathbf{x} \in D\} \tag{6.65}$$

if such a point $\mathbf{x}^*$ exists. If the objective function and/or the feasible domain D are non-convex, then there may be many local minima which are not global.

If D corresponds to all $\mathbb{R}^n$ the optimization problem is *unconstrained*.

Alternatively, simple bounds may be imposed, with D now corresponding to the hyper box (or domain or region of interest) defined by

$$D = \{\mathbf{x} | \boldsymbol{\ell} \leq \mathbf{x} \leq \boldsymbol{u}\} \qquad (6.66)$$

where $\boldsymbol{\ell}$ and $\boldsymbol{u}$ are n-vectors defining the respective lower and upper bounds on $\mathbf{x}$.

From a *mathematical* point of view, Problem (6.65) is essentially *unsolvable*, due to a lack of mathematical conditions characterizing the global optimum, as opposed to the local optimum of a smooth continuous function, which is characterized by the behavior of the problem function (Hessians and gradients) at the minimum (Arora et al. 1995) (viz. the Karush-Kuhn-Tucker conditions). Therefore, the global optimum f^* can only be obtained by an exhaustive search, except if the objective function satisfies certain subsidiary conditions (Griewank 1981), which mostly are of limited practical use (Snyman and Fatti 1987). Typically, the conditions are that f should satisfy a Lipschitz condition with known constant L and that the search area is bounded, e.g. for all $\mathbf{x}, \bar{\mathbf{x}} \in \mathbf{X}$

$$|f(\mathbf{x}) - f(\bar{\mathbf{x}})| \leq L\|\mathbf{x} - \bar{\mathbf{x}}\|. \qquad (6.67)$$

So called space-covering deterministic techniques have been developed (Dixon et al. 1975) under these special conditions. These techniques are expensive, and due to the need to know L, of limited practical use.

Global optimization algorithms are divided into two major classes (Dixon et al. 1975): deterministic and stochastic (from the Greek word *stokhastikos*, i.e. 'governed by the laws of probability'). Deterministic methods can be used to determine the global optimum through exhaustive search. These methods are typically extremely expensive. With the introduction of a stochastic element into deterministic algorithms, the deterministic *guarantee* that the global optimum can be found is relaxed into a *confidence measure*. Stochastic methods can be used to assess the probability of having obtained the global minimum. Stochastic ideas are mostly used for the development of stopping criteria, or to approximate the regions of attraction as used by some methods (Arora et al. 1995).

The stochastic algorithms presented herein, namely the Snyman-Fatti algorithm and the modified bouncing ball algorithm (Groenwold and Snyman 2002), both depend on dynamic search trajectories to minimize the objective function. The respective trajectories, namely the motion of a particle of unit mass in a n-dimensional conservative force field, and the trajectory of a projectile in a conservative gravitational field, are modified to increase the likelihood of convergence to a low local minimum.

6.6.2 The Snyman-Fatti trajectory method

The essentials of the original SF algorithm (Snyman and Fatti 1987) using dynamic search trajectories for unconstrained global minimization will now be discussed. The algorithm is based on the local algorithms presented by Snyman (1982, 1983). For more details concerning the motivation of the method, its detailed construction, convergence theorems, computational aspects and some of the more obscure heuristics employed, the reader is referred to the original paper and also to the more recent review article by Snyman and Kok (2009).

6.6.2.1 Dynamic trajectories

In the SF algorithm successive sample points $\mathbf{x}^j, j = 1, 2, ...,$ are selected at random from the box D defined by (6.66). For *each* sample point $\mathbf{x}^j$, a sequence of trajectories T^i, $i = 1, 2, ...,$ is computed by numerically solving the successive initial value problems:

$$\ddot{\mathbf{x}}(t) = -\nabla f(\mathbf{x}(t))$$

$$\mathbf{x}(0) = \mathbf{x}_0^i \; ; \;\; \dot{\mathbf{x}}(0) = \dot{\mathbf{x}}_0^i.$$

(6.68)

This trajectory represents the motion of a particle of unit mass in a n-dimensional conservative force field, where the function to be minimized represents the potential energy.

Trajectory T^i is terminated when $\mathbf{x}(t)$ reaches a point where $f(\mathbf{x}(t))$ is arbitrarily close to the value $f(\mathbf{x}_0^i)$ while moving "uphill", or more precisely, if $\mathbf{x}(t)$ satisfies the conditions

$$f(\mathbf{x}(t)) > f(\mathbf{x}_0^i) - \epsilon_u$$

$$\text{and} \;\; \dot{\mathbf{x}}(t)^T \nabla f(\mathbf{x}(t)) > 0$$

(6.69)

where ϵ_u is an arbitrary small prescribed positive value.

An argument is presented in Snyman and Fatti (1987) to show that when the level set $\{\mathbf{x}|f(\mathbf{x}) \leq f(\mathbf{x}_0^i)\}$ is bounded and $\nabla f(\mathbf{x}_0^i) \neq \mathbf{0}$, then conditions (6.69) above will be satisfied at some finite point in time.

Each computed step along trajectory T^i is monitored so that at termination the point $\mathbf{x}_m^i$ at which the minimum value was achieved is recorded together with the associated velocity $\dot{\mathbf{x}}_m^i$ and function value f_m^i. The values of $\mathbf{x}_m^i$ and $\dot{\mathbf{x}}_m^i$ are used to determine the initial values for the next trajectory T^{i+1}. From a comparison of the minimum values the best point $\mathbf{x}_b^i$, for the current j over all trajectories to date is also recorded. In more detail the minimization procedure for *a given sample point* $\mathbf{x}^j$, in computing the sequence $\mathbf{x}_b^i$, $i = 1, 2, ...,$ is as follows.

Algorithm 6.6 Minimization Procedure MP1

---

1. For given sample point $\mathbf{x}^j$, set $\mathbf{x}_0^1 := \mathbf{x}^j$ and compute T^1 subject to $\dot{\mathbf{x}}_0^1 := 0$; record $\mathbf{x}_m^1, \dot{\mathbf{x}}_m^1$ and f_m^1 ; set $\mathbf{x}_b^1 := \mathbf{x}_m^1$ and $i := 2$,

2. compute trajectory T^i with $\mathbf{x}_0^i := \frac{1}{2}\left(\mathbf{x}_0^{i-1} + \mathbf{x}_b^{i-1}\right)$ and $\dot{\mathbf{x}}_0^i := \frac{1}{2}\dot{\mathbf{x}}_m^{i-1}$, record $\mathbf{x}_m^i, \dot{\mathbf{x}}_m^i$ and f_m^i,

3. if $f_m^i < f(\mathbf{x}_b^{i-1})$ then $\mathbf{x}_b^i := \mathbf{x}_m^i$; else $\mathbf{x}_b^i := \mathbf{x}_b^{i-1}$,

4. set $i := i + 1$ and go to 2.

In the original paper (Snyman and Fatti 1987) an argument is presented to indicate that under normal conditions on the continuity of f and its derivatives, $\mathbf{x}_b^i$ will converge to a local minimum. Procedure MP1, for a given j, is accordingly terminated at step Algorithm 6.6 above if $\|\nabla f(\mathbf{x}_b^i)\| \leq \epsilon$, for some small prescribed positive value ϵ, and $\mathbf{x}_b^i$ is taken as the local minimizer $\mathbf{x}_f^j$, i.e. set $\mathbf{x}_f^j := \mathbf{x}_b^i$ with corresponding function value $f_f^j := f(\mathbf{x}_f^j)$.

Reflecting on the overall approach outlined above, involving the computation of energy conserving trajectories and the minimization procedure, it should be evident that, in the presence of many local minima, the probability of convergence to a relative low local minimum is increased. This one expects because, with a small value of ϵ_u (see conditions (6.69)), it is likely that the particle will move through a trough associated with a relative high local minimum, and move over a ridge to record a lower function value at a point beyond. Since we assume that the level set associated with the starting point function is bounded, termination of the search trajectory will occur as the particle eventually moves to a region of higher function values.

6.6.3 The modified bouncing ball trajectory method

The essentials of the modified bouncing ball algorithm using dynamic search trajectories for unconstrained global minimization are now presented. The algorithm is in an experimental stage, and details concerning the motivation of the method, its detailed construction, and computational aspects will be presented in future.

6.6.3.1 Dynamic trajectories

In the MBB algorithm successive sample points $\mathbf{x}^j, j = 1, 2, ...$, are selected at random from the box D defined by (6.66). For *each* sample point $\mathbf{x}^j$, a sequence of *trajectory steps* $\Delta\mathbf{x}^i$ and associated *projection points* $\mathbf{x}^{i+1}$, $i = 1, 2, ...$, are computed from the successive analytical relationships (with $\mathbf{x}^1 := \mathbf{x}^j$ and prescribed $V_{0_1} > 0$):

$$\Delta\mathbf{x}^i = V_{0_i} t_i \cos\theta_i \nabla f(\mathbf{x}^i)/\|\nabla f(\mathbf{x}^i)\| \qquad (6.70)$$

where

$$\theta_i = \tan^{-1}(\|\nabla f(\mathbf{x}^i)\|) + \frac{\pi}{2}, \qquad (6.71)$$

$$t_i = \frac{1}{g}\left[V_{0_i}\sin\theta_i + \{(V_{0_i}\sin\theta_i)^2 + 2gh(\mathbf{x}^i)\}^{1/2}\right], \qquad (6.72)$$

$$h(\mathbf{x}^i) = f(\mathbf{x}^i) + k \qquad (6.73)$$

with k a constant chosen such that $h(\mathbf{x}) > 0 \ \forall \ \mathbf{x} \in D$, g a positive constant, and

$$\mathbf{x}^{i+1} = \mathbf{x}^i + \Delta\mathbf{x}^i. \qquad (6.74)$$

For the next step, select $V_{0_{i+1}} < V_{0_i}$. Each step $\Delta\mathbf{x}^i$ represents the ground or horizontal displacement obtained by projecting a particle in a vertical gravitational field (constant g) at an elevation $h(\mathbf{x}^i)$ and speed V_{0_i} at an inclination θ_i. The angle θ_i represents the angle that the outward normal $\mathbf{n}$ to the hypersurface represented by $y = h(\mathbf{x})$ makes, at $\mathbf{x}^i$ in $n+1$ dimensional space, with the horizontal. The time of flight t_i is the time taken to reach the ground corresponding to $y = 0$.

More formally, the minimization trajectory for *a given sample point* $\mathbf{x}^j$ and some initial prescribed speed V_0 is obtained by computing the sequence $\mathbf{x}^i, \ i = 1, 2, ...$, as follows.

Algorithm 6.7 Minimization Procedure MP2

---

1. For given sample point $\mathbf{x}^j$, set $\mathbf{x}^1 := \mathbf{x}^j$ and compute trajectory step $\Delta\mathbf{x}^1$ according to (6.70)–(6.73) and subject to $V_{0_1} := V_0$; record $\mathbf{x}^2 := \mathbf{x}^1 + \Delta\mathbf{x}^1$, set $i := 2$ and $V_{0_2} := \alpha V_{0_1}$ $(\alpha < 1)$.

2. Compute $\Delta\mathbf{x}^i$ according to (6.70)–(6.73) to give $\mathbf{x}^{i+1} := \mathbf{x}^i + \Delta\mathbf{x}^i$, record $\mathbf{x}^{i+1}$ and set $V_{0_{i+1}} := \alpha V_{0_i}$.

3. Set $i := i + 1$ and go to 2.

In the vicinity of a local minimum $\hat{\mathbf{x}}$ the sequence of projection points $\mathbf{x}^i, \ i = 1, 2, ...$, constituting the search trajectory for starting point $\mathbf{x}^j$ will converge since $\Delta\mathbf{x}^i \to 0$ (see (6.70)). In the presence of many local minima, the probability of convergence to a relative low local minimum is increased, since the kinetic energy can only decrease for $\alpha < 1$.

Procedure MP2, for a given j, is successfully terminated if $\|\nabla f(\mathbf{x}^i)\| \le \epsilon$ for some small prescribed positive value ϵ, or when $\alpha V_0^i < \beta V_0^1$, and $\mathbf{x}^i$ is taken as the local minimizer $\mathbf{x}_f^j$ with corresponding function value $f_f^j := h(\mathbf{x}_f^j) - k$.

Clearly, the condition $\alpha V_0^i < \beta V_0^1$ will always occur for $0 < \beta < \alpha$ and $0 < \alpha < 1$.

MP2 can be viewed as a variant of the steepest descent algorithm. However, as opposed to steepest descent, MP2 has (as has MP1) the ability for 'hill-climbing', as is inherent in the physical model on which MP2 is based (viz., the trajectories of a bouncing ball in a conservative gravitational field.) Hence, the behavior of MP2 is quite different from that of steepest descent and furthermore, because of it's physical basis, it tends to seek local minima with relative low function values and is therefore suitable for implementation in global searches, while steepest descent is not.

For the MBB algorithm, convergence to a local minimum is not proven. Instead, the underlying physics of a bouncing ball is exploited. Unsuccessful trajectories are terminated, and do not contribute to the probabilistic stopping criterion (although these points are included in the number of unsuccessful trajectories $\tilde{n}$). In the validation of the algorithm the philosophy adopted here is that the practical demonstration of convergence of a proposed algorithm on a variety of demanding test problems may be as important and convincing as a rigorous mathematical convergence argument.

Indeed, although for the steepest descent method convergence can be proven, in practice it often fails to converge because effectively an infinite number of steps is required for convergence.

6.6.4 Global stopping criterion

The above methods require a termination rule for deciding when to end the sampling and to take the current overall minimum function value $\tilde{f}$, i.e.

$$\tilde{f} = \text{minimum} \left\{ f_f^j, \text{ over all } j \text{ to date} \right\} \quad (6.75)$$

as an approximation of the global minimum value f^*.

Define the *region of convergence* of the dynamic methods for a local minimum $\hat{\mathbf{x}}$ as the set of all points $\mathbf{x}$ which, used as starting points for the above procedures, converge to $\hat{\mathbf{x}}$. One may reasonably expect that in the case where the *regions of attraction* (for the usual gradient-descent methods, see Dixon et al. 1976) of the local minima are more or less equal, that the region of convergence of the global minimum will be relatively increased.

Let R_k denote the region of convergence for the above minimization procedures MP1 and MP2 of local minimum $\hat{\mathbf{x}}^k$ and let α_k be the associated probability that a sample point be selected in R_k. The region of convergence and the associated probability for the global minimum $\mathbf{x}^*$

are denoted by R^* and α^* respectively. The following basic assumption, which is probably true for many functions of practical interest, is now made. BASIC ASSUMPTION:

$$\alpha^* \geq \alpha_k \text{ for all local minima } \hat{\mathbf{x}}^k. \tag{6.76}$$

The following theorem may be proved.

6.6.4.1   Theorem (Snyman and Fatti 1987)

Let r be the number of sample points falling within the region of convergence of the current overall minimum $\tilde{f}$ after $\tilde{n}$ points have been sampled. Then under the above assumption and a statistically non-informative prior distribution the probability that $\tilde{f}$ corresponds to f^* may be obtained from

$$Pr\left[\tilde{f} = f^*\right] \geq q(\tilde{n}, r) = 1 - \frac{(\tilde{n}+1)!(2\tilde{n}-r)!}{(2\tilde{n}+1)!(\tilde{n}-r)!}. \tag{6.77}$$

On the basis of this theorem the *stopping rule* becomes: STOP when $Pr\left[\tilde{f} = f^*\right] \geq q^*$, where q^* is some prescribed desired confidence level, typically chosen as 0.99.

Proof:

We present here an outline of the proof of (6.77), and follow closely the presentation in Snyman and Fatti (1987). (We have since learned that the proof can be shown to be a generalization of the procedure proposed by Zieliński 1981.) Given $\tilde{n}^*$ and α^*, the probability that at least one point, $\tilde{n} \geq 1$, has converged to f^* is

$$Pr[\tilde{n}^* \geq 1 | \tilde{n}, r] = 1 - (1 - \alpha^*)^{\tilde{n}} . \tag{6.78}$$

In the Bayesian approach, we characterize our uncertainty about the value of α^* by specifying a prior probability distribution for it. This distribution is modified using the sample information (namely, $\tilde{n}$ and r) to form a posterior probability distribution. Let $p_*(\alpha^*|\tilde{n}, r)$ be the posterior probability distribution of α^*. Then,

$$
\begin{aligned}
\Pr[\tilde{n}^* \geq 1 | \tilde{n}, r] &= \int_0^1 \left[1 - (1 - \alpha^*)^{\tilde{n}} \right] p_*(\alpha^* | \tilde{n}, r) d\alpha^* \\
&= 1 - \int_0^1 (1 - \alpha^*)^{\tilde{n}} p_*(\alpha^* | \tilde{n}, r) d\alpha^*.
\end{aligned}
\tag{6.79}
$$

Now, although the r sample points converge to the current overall minimum, we do not know whether this minimum corresponds to the global minimum of f^*. Utilizing (6.76), and noting that $(1-\alpha)^{\tilde{n}}$ is a decreasing function of α, the replacement of α^* in the above integral by α yields

$$
\Pr[\tilde{n}^* \geq 1 | \tilde{n}, r] \geq \int_0^1 \left[1 - (1 - \alpha)^{\tilde{n}} \right] p(\alpha | \tilde{n}, r) d\alpha .
\tag{6.80}
$$

Now, using Bayes theorem we obtain

$$
p(\alpha | \tilde{n}, r) = \frac{p(r | \alpha, \tilde{n}) p(\alpha)}{\int_0^1 p(r | \alpha, \tilde{n}) p(\alpha) d\alpha} .
\tag{6.81}
$$

Since the $\tilde{n}$ points are sampled at random and each point has a probability α of converging to the current overall minimum, r has a binomial distribution with parameters α and $\tilde{n}$. Therefore

$$
p(r | \alpha, \tilde{n}) = \binom{\tilde{n}}{r} \alpha^r (1 - \alpha)^{\tilde{n}-r} .
\tag{6.82}
$$

Substituting (6.82) and (6.81) into (6.80) gives:

$$
\Pr[\tilde{n}^* \geq 1 | \tilde{n}, r] \geq 1 - \frac{\int_0^1 \alpha^r (1 - \alpha)^{2\tilde{n}-r} p(\alpha) d\alpha}{\int_0^1 \alpha^r (1 - \alpha)^{\tilde{n}-r} p(\alpha) d\alpha} .
\tag{6.83}
$$

A suitable flexible prior distribution $p(\alpha)$ for α is the beta distribution with parameters a and b. Hence,

$$
p(\alpha) = [1/\boldsymbol{\beta}(a, b)] \alpha^{a-1} (1 - \alpha)^{b-1}, \qquad 0 \leq \alpha \leq 1.
\tag{6.84}
$$

Using this prior distribution gives:

$$
\begin{aligned}
\Pr[\tilde{n}^* \geq 1 | \tilde{n}, r] &\geq 1 - \frac{\Gamma(\tilde{n} + a + b) \, \Gamma(2\tilde{n} - r + b)}{\Gamma(2\tilde{n} + a + b) \, \Gamma(\tilde{n} - r + b)} \\
&= 1 - \frac{(\tilde{n} + a + b - 1)! \, (2\tilde{n} - r + b - 1)!}{(2\tilde{n} + a + b - 1)! \, (\tilde{n} - r + b - 1)!}.
\end{aligned}
$$

Assuming a prior expectation of 1, (viz. $a = b = 1$), we obtain

$$\Pr[\tilde{n}^* \geq 1 | \tilde{n}, r] \;=\; 1 - \frac{(\tilde{n} + 1)! \, (2\tilde{n} - r)!}{(2\tilde{n} + 1)! \, (\tilde{n} - r)!},$$

which is the required result.   □

6.6.5 Numerical results

No.	Name	ID	n	Ref.
1	Griewank G1	G1	2	Törn and Zilinskas; Griewank
2	Griewank G2	G2	10	Törn and Zilinskas; Griewank
3	Goldstein-Price	GP	2	Törn and Zilinskas; Dixon and Szegö
4	Six-hump Camelback	C6	2	Törn and Zilinskas; Branin
5	Shubert, Levi No. 4	SH	2	Lucidl and Piccioni
6	Branin	BR	2	Törn and Zilinskas; Branin and Hoo
7	Rastrigin	RA	2	Törn and Zilinskas
8	Hartman 3	H3	3	Törn and Zilinskas; Dixon and Szegö
9	Hartman 6	H6	6	Törn and Zilinskas; Dixon and Szegö
10	Shekel 5	S5	4	Törn and Zilinskas; Dixon and Szegö
11	Shekel 7	S7	4	Törn and Zilinskas; Dixon and Szegö
12	Shekel 10	S10	4	Törn and Zilinskas; Dixon and Szegö

Table 6.6: The test functions

No.	ID	SF - This Study			SF - Previous		MBB		
		N_f	$(r/\tilde{n})_b$	$(r/\tilde{n})_w$	N_f	$r/\tilde{n}$	N_f	$(r/\tilde{n})_b$	$(r/\tilde{n})_w$
1	G1	4199	6/40	6/75	1606	6/20	2629	5/8	6/23
2	G2	25969	6/84	6/312	26076	6/60	19817	6/24	6/69
3	GP	2092	4/4	5/12	668	4/4	592	4/4	5/10
4	C6	426	4/4	5/9	263	4/4	213	4/4	5/10
5	SH	8491	6/29	6/104	—	—	1057	5/7	6/26
6	BR	3922	4/4	5/12	—	—	286	4/4	5/6
7	RA	4799	6/67	6/117	—	—	1873	4/4	6/42
8	H3	933	4/4	5/8	563	5/6	973	5/9	6/29
9	H6	1025	4/4	5/10	871	5/8	499	4/4	5/9
10	S5	1009	4/4	6/24	1236	6/17	2114	5/8	6/39
11	S7	1057	5/8	6/37	1210	6/17	2129	6/16	6/47
12	S10	845	4/4	6/31	1365	6/20	1623	5/7	6/39

Table 6.7: Numerical results

Method	Test Function					
	BR	C6	GP	RA	SH	H3
TRUST	55	31	103	59	72	58
MBB	25	29	74	168	171	24

Table 6.8: Cost (N_f) using *a priori* stopping condition

The test functions used are tabulated in Table 6.6, and tabulated numerical results are presented in Tables 6.7 and 6.8. In the tables, the reported number of function values N_f are the average of 10 independent (random) starts of each algorithm.

Unless otherwise stated, the following settings were used in the SF algorithm (see Snyman and Fatti 1987): $\gamma = 2.0$, $\alpha = 0.95$, $\epsilon = 10^{-2}$, $\omega = 10^{-2}$, $\delta = 0.0$, $q^* = 0.99$, and $\Delta t = 1.0$. For the MBB algorithm, $\alpha = 0.99$, $\epsilon = 10^{-4}$, and $q^* = 0.99$ were used. For each problem, the initial velocity V_0 was chosen such that Δx^1 was equal to half the 'radius' of the domain D. A local search strategy was implemented with varying α in the vicinity of local minima.

In Table 6.7, $(r/\tilde{n})_b$ and $(r/\tilde{n})_w$ respectively indicate the best and worst $r/\tilde{n}$ ratios (see equation (6.77)), observed during 10 independent optimization runs of both algorithms. The SF results compare well with the previously published results by Snyman and Fatti, who reported values for a single run only. For the Shubert, Branin and Rastrigin functions, the MBB algorithm is superior to the SF algorithm. For the Shekel functions (S5, S7 and S10), the SF algorithm is superior. As a result of the stopping criterion (6.77), the SF and MBB algorithms found the global optimum between 4 and 6 times for each problem.

The results for the trying Griewank functions (Table 6.7) are encouraging. G1 has some 500 local minima in the region of interest, and G2 several thousand. The values used for the parameters are as specified, with $\Delta t = 5.0$ for G1 and G2 in the SF-algorithm. It appears that both the SF and MBB algorithms are highly effective for problems with a large number of local minima in D, and problems with a large number of design variables.

In Table 6.8 the MBB algorithm is compared with the deterministic

TRUST algorithm (Barhen et al. 1997). Since the TRUST algorithm was terminated when the global approximation was within a specified tolerance of the (known) global optimum, a similar criterion was used for the MBB algorithm. The table reveals that the two algorithms compare well. Note however that the highest dimension of the test problems used in Barhen et al. (1997) is 3. It is unclear if the deterministic TRUST algorithm will perform well for problems of large dimension, or problems with a large number of local minima in D.

In conclusion, the numerical results indicate that both the Snyman-Fatti trajectory method and the modified bouncing ball trajectory method are effective in finding the global optimum efficiently. In particular, the results for the trying Griewank functions are encouraging. Both algorithms appear effective for problems with a large number of local minima in the domain, and problems with a large number of design variables. A salient feature of the algorithms is the availability of an apparently effective global stopping criterion.

Chapter 7

SURROGATE MODELS

7.1 Introduction

A Taylor series expansion of a function allows us to approximate a function $f(\mathbf{x})$ at any point $\mathbf{x}$, based solely on information about the function at a single point $\mathbf{x}^i$. Here, information about the function implies zero order, first order, second order and higher order information of the function at $\mathbf{x}^i$. The higher the order of information included in a Taylor series representation of a function, the higher the accuracy of the approximation distant from $\mathbf{x}^i$. However, higher order information for multivariate functions grows exponentially in dimensionality, i.e. the gradient vector constitutes n values, the Hessian matrix is expressed by n^2 values and the 3rd derivative is comprised of n^3 values. In practice, zero order and first order information about a problem is usually computable and convenient to store, while second order information is usually not readily available for engineering problems and needs to be inferred from first order information as described in Section 2.4.

Surrogate modelling offers an alternative approach to Taylor series for constructing approximations of functions. Instead of constructing approximations based on ever higher and higher order information at a single point, surrogate modelling approximates functions using lower order information at numerous points in the domain of interest. The advantage of such an approach is that it is (i) computationally inexpensive to

© Springer International Publishing AG, part of Springer Nature 2018
J.A. Snyman and D.N. Wilke, *Practical Mathematical Optimization*,
Springer Optimization and Its Applications 133,
https://doi.org/10.1007/978-3-319-77586-9_7

approximate zero and first order information of the function at additional points in the domain, and that (ii) lower order information can be computed in parallel on distributed computing platforms. Hence, the approximation functions can be exhaustively optimized, while the computationally demanding evaluations of the actual function can be distributed over multiple cores and computers. It is not surprising that surrogate modelling is the preferred strategy to solve computationally demanding multidisciplinary engineering design problems as highlighted by Forrester et al. (2008). However, as information grows exponentially with the order of information for a multivariate problem, so too does the design space grow exponentially with problem dimensionality. This is referred to as, the *curse of dimensionality*, as phrased by Bellman (1957). This limits surrogate modelling to lower-dimensional problems in the same way that Taylor series approximations are limited to lower order information for higher-dimensional problems.

Formally, a surrogate model approximates a non-linear function $f(\mathbf{x})$, when information about $f(\mathbf{x})$ is known at m discrete locations $\mathbf{x}^i$, $i = 1, \ldots, m$. The information is usually limited to zero order information, i.e. only function values as described by Hardy (1971, 1990); Franke (1982); Dyn et al. (1986); Khuri and Mukhopadhyay (2010). More recently both zero and first order information have been considered more readily in the construction of surrogate models, Hardy (1975, 1990); Morris et al. (1993); Chung and Alonso (2001); Lauridsen et al. (2002). Lastly, Wilke (2016) proposed the construction of surrogate models using only first order information. This allows for smooth surrogate approximations of piecewise smooth discontinuous functions as will be shown at the end of this chapter.

7.2   Radial basis surrogate models

A number of surrogate models are available to approximate a non-linear function $f(\mathbf{x})$. Hardy (1971) pioneered radial basis functions by approximating a non-linear function, $f(\mathbf{x})$, as a linear combination of p chosen non-linear basis functions $\phi_j(\mathbf{x}, \mathbf{x}_c^j)$, $j = 1, \ldots, p$, that are centered around p spatial points $\mathbf{x}_c^j$, which is conveniently expressed by

$$f(\mathbf{x}) \approx \sum_{j=1}^{p} w_j \phi_j(\mathbf{x}, \mathbf{x}_c^j) = \tilde{f}(\mathbf{x}). \qquad (7.1)$$

The non-linear basis functions, $\phi_j(\mathbf{x}, \mathbf{x}_c^j)$, $j = 1, \ldots, p$, are usually chosen to be of identical form but centered around distinct spatial points $\mathbf{x}_c^j$. For examples of radial basis functions refer to Table 7.1. Typically Figures 7.1 (a) and (b) depict three basis functions centered about three points in a two-dimensional design domain.

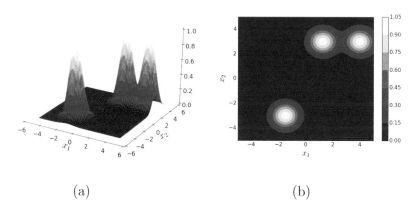

(a)   (b)

Figure 7.1: Illustration of three radial basis functions centered around three spatial locations, namely, $x_1 = -1.5$ and $x_2 = -3$, $x_1 = 1.25$ and $x_2 = 3$, and $x_1 = 4$ and $x_2 = 3$

A significant benefit of this approach is that all the non-linearity of $f(\mathbf{x})$ is approximated by the non-linear basis functions $\phi_j(\mathbf{x})$, while the approximation is linear in the weights w_j that need to be estimated. Hence, w_j can be obtained by merely solving a linear problem.

7.2.1 Zero order only radial basis surrogate models

Solving for the weights w_j, $j = 1, \ldots, p$ from only zero order (zo) information requires the function, $f(\mathbf{x}^j)$, to be evaluated at least for $m \geq p$ distinct designs $\mathbf{x}^j$, $j = 1, \ldots, m$. The *response* surface $\tilde{f}(\mathbf{x})$ is then required to recover the actual function value at the $\mathbf{x}^j$, $j = 1, \ldots, m$ designs, i.e. $\tilde{f}(\mathbf{x}^j) \approx f(\mathbf{x}^j)$. By choosing the number of designs and basis functions to be equal, i.e. $m = p$, we recover an *interpolation* surface, i.e. $\tilde{f}(\mathbf{x}^j) = f(\mathbf{x}^j)$, $j = 1, \ldots, m$,. When we have fewer basis functions than design vectors, i.e. $p < m$, we recover a *regression* surface $\tilde{f}(\mathbf{x}^j) \approx f(\mathbf{x}^j)$, $j = 1, \ldots, m$, that requires a least squares problem to be solved to recover the weights. In general interpolation surfaces are preferred when only a few design vectors are available, while regression surfaces are favoured when the design domain is densely sampled.

In general, by choosing p basis functions and m design vectors, results in p unknowns to be solved from m equations

$$f(\mathbf{x}^1) = \sum_{j=1}^{p} w_j \phi_j(\mathbf{x}^1, \mathbf{x}_c^j) = \tilde{f}(\mathbf{x}^1)$$

$$f(\mathbf{x}^2) = \sum_{j=1}^{p} w_j \phi_j(\mathbf{x}^2, \mathbf{x}_c^j) = \tilde{f}(\mathbf{x}^2)$$

$$\vdots \tag{7.2}$$

$$f(\mathbf{x}^m) = \sum_{j=1}^{p} w_j \phi_j(\mathbf{x}^m, \mathbf{x}_c^j) = \tilde{f}(\mathbf{x}^m).$$

This can be rewritten in block matrix form

$$\mathbf{R}_{zo} \mathbf{w}^{zo} = \mathbf{r}^{zo}, \tag{7.3}$$

with

$$\mathbf{R}_{zo} = \begin{bmatrix} \phi_1(\mathbf{x}^1, \mathbf{x}_c^1) & \phi_2(\mathbf{x}^1, \mathbf{x}_c^2) & \cdots & \phi_m(\mathbf{x}^1, \mathbf{x}_c^p) \\ \vdots & & & \\ \phi_1(\mathbf{x}^m, \mathbf{x}_c^1) & \phi_2(\mathbf{x}^m, \mathbf{x}_c^2) & \cdots & \phi_m(\mathbf{x}^m, \mathbf{x}_c^p) \end{bmatrix},$$

$$\mathbf{w}^{zo} = \begin{bmatrix} w_1^{zo} \\ w_2^{zo} \\ \vdots \\ w_p^{zo} \end{bmatrix}, \mathbf{r}^{zo} = \begin{bmatrix} f(\mathbf{x}^1) \\ \vdots \\ f(\mathbf{x}^m) \end{bmatrix},$$

to obtain an m by p linear system of equations. For $p = m$ the system can be solved directly. However, $p < m$ yields an overdetermined system of equations to be solved in a least squares sense. This is achieved by pre-multiplying (7.3) by $\mathbf{R}_{zo}^T$ to obtain the following $p \times p$ linear system of equations

$$\mathbf{R}_{zo}^T \mathbf{R}_{zo} \mathbf{w}^{zo} = \mathbf{R}_{zo}^T \mathbf{r}^{zo}, \tag{7.4}$$

from which $\mathbf{w}^{zo}$ can be solved for.

For convenience, when choosing $p = m$, the m designs $\mathbf{x}^j$, $j = 1, \ldots, m$ are usually chosen to coincide with the p basis function centers, i.e. $\mathbf{x}_c^j = \mathbf{x}^j$, $j = 1, \ldots, m$.

7.2.2 Combined zero and first order radial basis surrogate models

Solving for the weights w_j, $j = 1, \ldots, p$ from both zero and first order information, i.e. mixed order (mo), requires the function, $f(\mathbf{x}^j)$, and gradient of the function, $\nabla f(\mathbf{x}^j)$, to be evaluated at m distinct designs $\mathbf{x}^j$, $j = 1, \ldots, m$. The response surface $\tilde{f}(\mathbf{x})$ is then required to recover the actual function value and gradient at the $\mathbf{x}^j$, $j = 1, \ldots, m$ designs, i.e. $\tilde{f}(\mathbf{x}^j) \approx f(\mathbf{x}^j)$ and $\nabla \tilde{f}(\mathbf{x}^j) \approx \nabla f(\mathbf{x}^j)$. By choosing m designs for $p = m$ basis functions we recover a regression surface, i.e. $\tilde{f}(\mathbf{x}^j) \approx f(\mathbf{x}^j)$, $j = 1, \ldots, m$, and $\nabla \tilde{f}(\mathbf{x}^j) \approx \nabla f(\mathbf{x}^j)$, $j = 1, \ldots, m$. A least squares problem is then to be solved to recover the weights.

In general, by choosing p basis functions and m design vectors, each of dimension n, we now have p unknowns to be solved for from $m(n + 1)$

equations:

$$
\begin{aligned}
f(\mathbf{x}^1) &= \sum_{j=1}^{p} w_j \phi_j(\mathbf{x}^1, \mathbf{x}_c^j) = \tilde{f}(\mathbf{x}^1), \\
\nabla f(\mathbf{x}^1) &= \sum_{j=1}^{p} w_j \frac{\partial \phi_j(\mathbf{x}^1, \mathbf{x}_c^j)}{\partial \mathbf{x}} = \nabla \tilde{f}(\mathbf{x}^1), \\
f(\mathbf{x}^2) &= \sum_{j=1}^{p} w_j \phi_j(\mathbf{x}^2, \mathbf{x}_c^j) = \tilde{f}(\mathbf{x}^2), \\
\nabla f(\mathbf{x}^2) &= \sum_{j=1}^{p} w_j \frac{\partial \phi_j(\mathbf{x}^2, \mathbf{x}_c^j)}{\partial \mathbf{x}} = \nabla \tilde{f}(\mathbf{x}^2), \\
&\vdots \\
f(\mathbf{x}^m) &= \sum_{j=1}^{p} w_j \phi_j(\mathbf{x}^m, \mathbf{x}_c^j) = \tilde{f}(\mathbf{x}^m), \\
\nabla f(\mathbf{x}^m) &= \sum_{j=1}^{p} w_j \frac{\partial \phi_j(\mathbf{x}^m, \mathbf{x}_c^j)}{\partial \mathbf{x}} = \nabla \tilde{f}(\mathbf{x}^m),
\end{aligned}
\tag{7.5}
$$

which can be rewritten in block matrix form to obtain

$$
\mathbf{R}_{mo}\mathbf{w}^{mo} = \mathbf{r}^{mo},
\tag{7.6}
$$

with

$$
\mathbf{R}_{mo} = \begin{bmatrix}
\phi_1(\mathbf{x}^1, \mathbf{x}_c^1) & \phi_2(\mathbf{x}^1, \mathbf{x}_c^2) & \cdots & \phi_m(\mathbf{x}^1, \mathbf{x}_c^p) \\
\frac{\partial \phi_1(\mathbf{x}^1, \mathbf{x}_c^1)}{\partial \mathbf{x}} & \frac{\partial \phi_2(\mathbf{x}^1, \mathbf{x}_c^2)}{\partial \mathbf{x}} & \cdots & \frac{\partial \phi_m(\mathbf{x}^1, \mathbf{x}_c^p)}{\partial \mathbf{x}} \\
\vdots & & & \\
\phi_1(\mathbf{x}^m, \mathbf{x}_c^1) & \phi_2(\mathbf{x}^m, \mathbf{x}_c^2) & \cdots & \phi_m(\mathbf{x}^m, \mathbf{x}_c^p) \\
\frac{\partial \phi_1(\mathbf{x}^m, \mathbf{x}_c^1)}{\partial \mathbf{x}} & \frac{\partial \phi_2(\mathbf{x}^m, \mathbf{x}_c^2)}{\partial \mathbf{x}} & \cdots & \frac{\partial \phi_m(\mathbf{x}^m, \mathbf{x}_c^p)}{\partial \mathbf{x}}
\end{bmatrix},
$$

$$
\mathbf{w}^{mo} = \begin{bmatrix} w_1 \\ w_2 \\ \vdots \\ w_p \end{bmatrix}, \mathbf{r}^{mo} = \begin{bmatrix} f(\mathbf{x}^1) \\ \nabla f(\mathbf{x}^1) \\ \vdots \\ f(\mathbf{x}^m) \\ \nabla f(\mathbf{x}^m) \end{bmatrix},
$$

to obtain a $m(n + 1)$ by p overdetermined linear system of equations. The overdetermined system of equations can be solved in a least squares sense by pre-multiplying (7.6) by $\mathbf{R}_{mo}^{\mathrm{T}}$ to obtain the following $p \times p$ linear system of equations

$$\mathbf{R}_{mo}^{\mathrm{T}}\mathbf{R}_{mo}\mathbf{w}^{mo} = \mathbf{R}_{mo}^{\mathrm{T}}\mathbf{r}^{mo}, \tag{7.7}$$

from which $\mathbf{w}^{mo}$ can be solved for.

7.2.3   First order only radial basis surrogate models

Solving for the weights w_j, $j = 1, \ldots, p$ from only first order (fo) or gradient only information requires that the gradient $\nabla f(\mathbf{x})$ of the non-linear function to be evaluated at m distinct designs $\mathbf{x}^j$, $j = 1, \ldots, m$. The response surface $\tilde{f}(\mathbf{x})$ is then required to recover the actual gradient at the $\mathbf{x}^j$, $j = 1, \ldots, m$ designs, i.e. $\nabla \tilde{f}(\mathbf{x}^j) \approx \nabla f(\mathbf{x}^j)$. By choosing m designs and $p = m$ basis functions we recover an interpolation surface for the gradient, i.e. $\nabla \tilde{f}(\mathbf{x}^j) = \nabla f(\mathbf{x}^j)$, $j = 1, \ldots, m$, only for univariate functions.

In general, higher-dimensional functions result in regression response surfaces, i.e. $\nabla \tilde{f}(\mathbf{x}^j) \approx \nabla f(\mathbf{x}^j)$, since we only have $p = m$ weights to recover $m \times n = mn$ gradient components, where n is the dimension of the design vector. The resulting mn equations

$$\nabla f(\mathbf{x}^1) = \sum_{j=1}^{p} w_j \frac{\partial \phi_j(\mathbf{x}^1, \mathbf{x}_c^j)}{\partial \mathbf{x}} = \nabla \tilde{f}(\mathbf{x}^1)$$

$$\nabla f(\mathbf{x}^2) = \sum_{j=1}^{p} w_j \frac{\partial \phi_j(\mathbf{x}^2, \mathbf{x}_c^j)}{\partial \mathbf{x}} = \nabla \tilde{f}(\mathbf{x}^2)$$

$$\vdots \tag{7.8}$$

$$\nabla f(\mathbf{x}^m) = \sum_{j=1}^{p} w_j \frac{\partial \phi_j(\mathbf{x}^m, \mathbf{x}_c^j)}{\partial \mathbf{x}} = \nabla \tilde{f}(\mathbf{x}^m),$$

can be rewritten in block matrix form to obtain

$$\mathbf{R}_{fo}\mathbf{w}^{fo} = \mathbf{r}^{fo}, \tag{7.9}$$

with

$$\mathbf{R}_{fo} = \begin{bmatrix} \frac{\partial \phi_1(\mathbf{x}^1, \mathbf{x}_c^1)}{\partial \mathbf{x}} & \frac{\partial \phi_2(\mathbf{x}^1, \mathbf{x}_c^2)}{\partial \mathbf{x}} & \cdots & \frac{\partial \phi_m(\mathbf{x}^1, \mathbf{x}_c^p)}{\partial \mathbf{x}} \\ \vdots & & & \\ \frac{\partial \phi_1(\mathbf{x}^m, \mathbf{x}_c^1)}{\partial \mathbf{x}} & \frac{\partial \phi_2(\mathbf{x}^m, \mathbf{x}_c^2)}{\partial \mathbf{x}} & \cdots & \frac{\partial \phi_m(\mathbf{x}^m, \mathbf{x}_c^p)}{\partial \mathbf{x}} \end{bmatrix},$$

$$\mathbf{w}^{fo} = \begin{bmatrix} w_1 \\ w_2 \\ \vdots \\ w_p \end{bmatrix}, \mathbf{r}^{fo} = \begin{bmatrix} \nabla f(\mathbf{x}^1) \\ \vdots \\ \nabla f(\mathbf{x}^m) \end{bmatrix},$$

giving an mn by m overdetermined linear system of equations. This overdetermined system of equations can be solved in a least squares sense by pre-multiplying (7.9) by $\mathbf{R}_{fo}^T$ to obtain the following $m \times m$ linear system of equations

$$\mathbf{R}_{fo}^T \mathbf{R}_{fo} \mathbf{w}^{fo} = \mathbf{R}_{fo}^T \mathbf{r}^{fo}, \tag{7.10}$$

from which $\mathbf{w}^{fo}$ is to be solved for. However, $\mathbf{R}_{fo}^T \mathbf{R}_{fo}$ is singular as an approximated surrogate from first order only information has infinite representations since any constant added to the surrogate leaves the gradient of the surrogate unchanged. This can be addressed by adding at least one equation that enforces the function value at a design to (7.10). Alternatively the minimum norm solution for the least squares system (7.10) can be computed. For convenience, the m designs $\mathbf{x}^j$, $j = 1, \ldots, m$ are usually chosen to coincide with the chosen $p = m$ basis function centers, i.e. $\mathbf{x}_c^j = \mathbf{x}^j$, $j = 1, \ldots, m$.

7.3   Basis functions

Numerous radial basis functions $\phi_j(r_j)(\mathbf{x})$ have been proposed with the most popular global support basis functions listed in Table 7.1. The shape parameter ϵ is in general unknown and needs to be determined as will be discussed in the next section. Smooth basis functions are preferred when constructing surrogate models for optimization applications, as they are everywhere differentiable. A preferred choice is the Gaussian basis function,

$$\phi_j(\mathbf{x}, \mathbf{x}_c^j) = \phi_j(r_j(\mathbf{x})) = e^{(-\epsilon r_j(\mathbf{x})^2)}. \tag{7.11}$$

Name	Abbreviation	Equation
Gaussian	GA	$\phi_j(r_j(\mathbf{x})) = e^{-\epsilon r_j(\mathbf{x})^2}$
Exponential	EXP	$\phi_j(r_j(\mathbf{x})) = e^{-\epsilon r_j(\mathbf{x})}$
Multiquadric	MQ	$\phi_j(r_j(\mathbf{x})) = \sqrt{1 + (\epsilon r_j(\mathbf{x}))^2}$
Inverse quadratic	IQ	$\phi_j(r_j(\mathbf{x})) = \frac{1}{1+(\epsilon r_j(\mathbf{x}))^2}$
Inverse multiquadric	IMQ	$\phi_j(r_j(\mathbf{x})) = \frac{1}{\sqrt{1+(\epsilon r_j(\mathbf{x}))^2}}$

Table 7.1: Radial basis functions $\phi_j(r_j(\mathbf{x}))$ with $r_j(\mathbf{x}) = \|\mathbf{x} - \mathbf{x}^j\|$

The gradient is given by

$$\nabla \phi_j(r_j(\mathbf{x})) = \frac{\partial \phi_j}{\partial r_j} \frac{\partial r_j}{\partial \mathbf{x}} = \left(-2\epsilon r_j e^{(-\epsilon r_j^2)}\right) \left(\frac{1}{2r_j} 2(\mathbf{x} - \mathbf{x}_c^j)\right), \quad (7.12)$$

and tends to $\mathbf{0}$ as $r_j(\mathbf{x}) \to 0$. However, as (7.12) is prone to numerical instabilities as $r_j(\mathbf{x}) \to 0$, it is required to apply L'Hospital's rule to ensure that numerically $\nabla \phi_j(r_j(\mathbf{x}))$ indeed evaluates to $\mathbf{0}$ at $r_j(\mathbf{x}_c^j)$.

7.3.1 Shape parameter

The shape parameter ϵ needs to be estimated. It determines the radius of the domain over which the basis function has significant influence. For example, consider the Gaussian basis function (7.11), for ϵ chosen large and unit r_j, $-\epsilon r_j^2$ evaluates to a large negative value of which the exponential is close to zero. Consequently, the larger ϵ the smaller the domain over which the basis function has a significant influence. The choice of ϵ is therefore of utmost importance, with smaller ϵ preferred, but choosing ϵ too small will result in severe numerical ill-conditioning for finite precision computing.

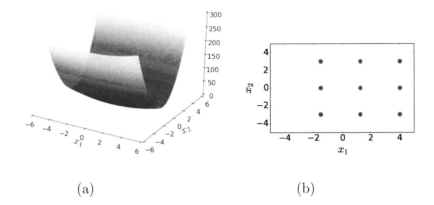

(a)   (b)

Figure 7.2: (a) Two-dimensional quadratic function evaluated at (b) nine points in the two-dimensional design domain

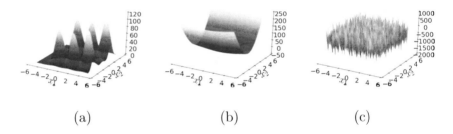

(a)   (b)   (c)

Figure 7.3: Radial basis approximation surfaces constructed using (a) $\epsilon = 10^0$, (b) $\epsilon = 10^{-1}$ and (c) $\epsilon = 10^{-5}$

Consider the construction of zero-order Gaussian radial basis surrogate approximations of the quadratic function, $f(\mathbf{x}) = x_1^2 + 10x_2^2$, depicted in Figure 7.2 (a) that is evaluated at nine points as shown in Figure 7.2 (b).

The construction done for three values of ϵ are depicted in Figures 7.3 (a)–(c). The results clearly indicate the influence of the shape parameter on the constructed radial basis function approximation surfaces. Choosing ϵ too large results in locally compact support that fails to capture smooth trends over large domains as shown in Figure 7.3 (a). Ill-conditioning is evident in Figure 7.3 (c) in which the numerical solution for the weights have broken down resulting in severe noise. Lastly, choosing an appropriate shape parameter gives the ability to capture the actual function using spatially distributed information as illustrated in Figure 7.3 (b).

In general, ϵ is computed using k-fold cross validation (Kohavi (1995)). In k-fold cross validation the m design vectors are randomly partitioned into k equal sized subsets, i.e. each subset containing $\frac{m}{k}$ design vectors. Of the k subsets, the first is retained to test the model (test subset), while the union of the remaining $k - 1$ subsets (the training subsets) is used to construct the response surface. Once, the response surface has been constructed, using only the training subset, the model is used to predict the response at the design vectors in the test subset. Since the responses at the designs in the test subset are known an error can be computed that captures the difference between the predicted and actual response. Here, the sum of the difference squared is usually computed. This process is repeated by choosing the next subset as the test subset, while the union of the remaining $k - 1$ subsets again define the training subset until all subsets have been used as a test subset, i.e. this process is repeated k times. All the errors over the k test subsets are averaged to define the k-fold cross validation error (k-CVE). The k-CVE is then computed for different choices of ϵ to find the ϵ^* that minimizes k-CVE error. Afterwards, ϵ^* is used to construct the surrogate, usually using all m points. A typical choice for k is between 5 and 20, while choosing $k = m$ results in leave-one-out cross validation (LOOCV). This forms the basis for the predicted residual error sum of squares (PRESS) statistic proposed by Allen (1974).

The process of constructing surrogate models using the k-CVE is listed in Algorithm 7.1.

7.4 Numerical examples

To demonstrate the implications and utility of using only zero or only first order information in constructing surrogate models, consider the two test functions depicted in Figure 7.4 (a) and (b). They are the smooth continuous quadratic function

$$f(\mathbf{x}) = \sum_{i=1}^{n} 10^{i-1} x_{i-1}^2, \tag{7.13}$$

Algorithm 7.1 Radial basis surrogate model.

Initialization: Select a radial basis function and associated initial shape parameter ϵ_0. Choose an integer value for k and select the number of design vectors m (a multiple of k) in the design of experiments. Randomly partition the m design vectors into k subsets of equal size. These k subsets are then used to compute the k-fold cross validation error (k-CVE). Set $l = 0$ and perform the following steps:

1. **Design of Experiments:** Identify m design vectors $\mathbf{x}^i$, $i = 1, \ldots, m$.

2. **Evaluate Designs:** For each design, compute the function value $f_i = f(\mathbf{x}^i)$, $i = 1, \ldots m$ and/or gradient vector $\nabla f^i = \nabla f(\mathbf{x}^i)$, $i = 1, \ldots m$.

3. **Trial Surrogate model:** For $\epsilon = \epsilon_l$ construct k trial surrogate models by solving for $\mathbf{w}$ from (7.4), (7.7) or (7.10) using successively, each of the k subsets containing $\frac{m}{k}$ design vectors as the test subset, and the union of the remaining $k - 1$ subsets as training set. For each of the k surrogate models use its corresponding test set and compute the test set error. Then compute the k-CVE as the average of the k test set errors.

4. **Update ϵ_l:**

   (a) Set $l := l + 1$

   (b) Update ϵ_l using a minimization strategy to reduce the k-CVE.

   (c) If ϵ_l has converged within an acceptable tolerance then set $\epsilon^* = \epsilon_l$ and go to Step 5, else go to Step 3.

5. **Construct Surrogate Model:** Construct the surrogate model by solving for $\mathbf{w}$ from (7.4), (7.7) or (7.10) using $\epsilon = \epsilon^*$. Instead of only using the training set of design vectors it is often advised to use all m design vectors to construct the final surrogate model. A typical choice for k is between 5 and 20.

---

depicted in Figure 7.4 (a) with $n = 2$, and the piece-wise smooth step discontinuous quadratic function

$$f(\mathbf{x}) = \sum_{i=2}^{n} 10^{i-2} x_{i-2}^2 + 10^{i-1} x_{i-1}^2 + a(\text{sign}(x_{i-2})) -$$
$$b(\text{sign}(x_{i-1})) + c(\text{sign}(x_{i-2}))(\text{sign}(x_{i-1})) \quad (7.14)$$

depicted in Figure 7.4 (b) with $n = 2$, $a = 100$, $b = 50$ and $c = 133$. It is important to note that the piece-wise smooth step discontinuous quadratic function is the same quadratic function given by (7.13) with step discontinuities imposed that are controlled by the coefficients a, b and c. Hence, the first *associated partial derivatives* w.r.t. x_1 and x_2 for both the smooth quadratic function and step discontinuous quadratic function are the same and respectively depicted in Figures 7.4 (c) and (d). The functions are evaluated at 3×3 designs as depicted in Figure 7.2 (b) to obtain the zero or first order information required to construct the surrogate models. Gaussian radial basis functions are used in the models and $k = m$ is specified for computing the CVE. Only zero order information or only first order information are considered.

The constructed zero order only and first order only (with the exception of enforcing $f = 0$ at $x_1 = 0$, $x_2 = 0$) surrogate models for the smooth quadratic function are depicted in Figures 7.5 (a) and (b). In addition, the first order partial derivatives w.r.t. x_1 and x_2 of the surrogate models are respectively depicted in Figures 7.6 (a) and (b) and Figures 7.6 (c) and (d) over the domain -5 to 5 for both variables.

It is evident that the two constructed surfaces, depicted in Figures 7.5 (a) and (b), are nearly identical. Similarly, the estimated first order partial derivatives for the zero order only and first order only constructed surrogate models w.r.t. x_1 and x_2 are respectively depicted in Figures 7.6 (a) and (b) and Figures 7.6 (c) and (d). The approximated partial derivatives are nearly identical to the actual partial derivatives depicted in Figures 7.4 (c) and (d).

For the piece-wise smooth step discontinuous quadratic function, the constructed zero order only and first order only (with the exception of enforcing $f = 0$ at $x_1 = 0$, $x_2 = 0$) surrogate models are depicted in Figures 7.7 (a) and (b). In addition, the first order partial derivatives w.r.t. x_1 and x_2 of the surrogate models are respectively depicted in Figures 7.8 (a) and (b) and Figures 7.8 (c) and (d) over the domain -5 to 5 for both variables. The approximated associated partial derivatives of the zero order only surrogate model differs significantly from the actual associated partial derivatives depicted in Figures 7.4 (c) and (d). However, the associated partial derivatives of the first order only approximated surrogate model is nearly identical to the actual associated partial derivatives. Similarly, the zero order only constructed surrogate model

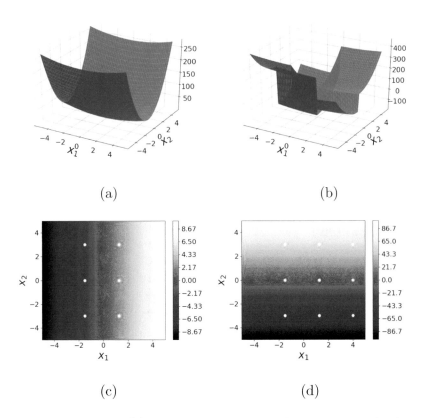

(a)   (b)

(c)   (d)

Figure 7.4: Surface of (a) the smooth quadratic function given by (7.13), and (b) the piece-wise smooth step discontinuous quadratic function given by (7.14). In addition, the first associated partial derivatives w.r.t. x_1 and x_2 for both functions are the same and respectively depicted in (c) and (d), with the full set of identified designs points indicated by white circles

differs significantly from both the actual smooth quadratic function and the piece-wise smooth step discontinuous function respectively depicted in Figures 7.4 (a) and (b). In contrast, the first order only constructed surrogate model is nearly identical to the actual smooth quadratic function, demonstrating that the step discontinuities are effectively ignored when only first order information is considered.

It is evident that the zero order only and first order only constructed surrogate functions differ significantly in approximating both the func-

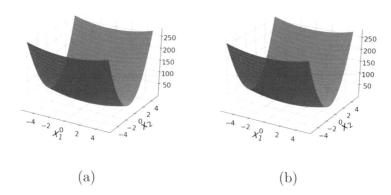

(a)   (b)

Figure 7.5: (a) Zero order only and (b) first order only constructed response surfaces for the smooth quadratic function

tion value and associated gradient of the actual piece-wise smooth step discontinuous function. The zero order constructed surrogate function poorly represents both the function and associated gradients of the piece-wise smooth step discontinuous function. It is only the approximated associated gradients of the first order constructed surrogate function that is consistent with the actual associated gradient of the piece-wise smooth step discontinuous function. The result, when only first order information is considered to construct a surrogate, is a smooth surrogate that is consistent with the first order information of the piece-wise smooth step discontinuous function. As will be pointed out in Chapter 8, this is ideal when the step discontinuities are numerical artefacts that need to be ignored. The effectiveness of first order only constructed surrogates to ignore or filter out step discontinuities is remarkable and an important aspect to consider when constructing surrogates for discontinuous functions. The resulting smooth surrogate function that is approximated from only first order information can then be optimized using conventional gradient based approaches.

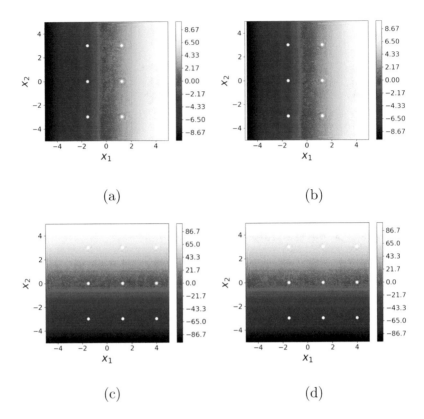

(a)   (b)

(c)   (d)

Figure 7.6: First order partial derivatives w.r.t. x_1 and x_2 respectively computed from the constructed (a),(c) zero order only surrogate model and (b),(d) first order only surrogate model for the smooth quadratic function. The sampled designs are indicated by white circles

7.5   Exercises

The reader is encouraged to employ a convenient computing environment to complete the exercises. With Python being freely available it is recommended to be used as outlined in Chapter 9.

7.4.1 Consider some non-linear function $f(\mathbf{x}, \mathbf{c})$ that is linear w.r.t. some parametrization $\mathbf{c} = [c_0, c_1, \ldots, c_r]$. For example consider the quadratic function

$$f(\mathbf{x}, \mathbf{c}) = c_5 x_1^2 + c_4 x_2^2 + c_3 x_1 x_2 + c_2 x_1 + c_1 x_2 + c_0,$$

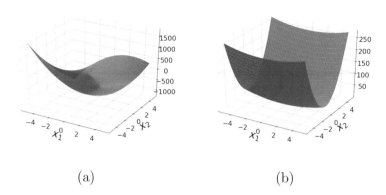

(a) (b)

Figure 7.7: (a) Zero order only and (b) first order only constructed response surfaces for the piece-wise smooth step discontinuous quadratic function

that is non-linear in $\mathbf{x}$ but linear in $\mathbf{c}$. For unknown $\mathbf{c} = \mathbf{c}^*$, given k observations $f(\mathbf{x}^0, \mathbf{c}^*), f(\mathbf{x}^1, \mathbf{c}^*) \dots, f(\mathbf{x}^k, \mathbf{c}^*)$ for known $\mathbf{x}^0, \mathbf{x}^1, \dots, \mathbf{x}^k$ and with $k > r$. Formulate an unconstrained minimization problem, using matrix notation, that estimates $\mathbf{c}^*$ from the information of k observations.

7.4.2 Derive the general first order optimality criterion that solves for $\mathbf{c}^*$ for the formulation presented in Problem **7.4.1**.

7.4.3 Equation 7.15 represents 12 scalar observations $f^i = f(\mathbf{x}^i, \mathbf{c}^*)$, $i = 1, \dots, 12$ that depends on two variables x_1 and x_2.

i	x_1^i	x_2^i	f^i
1	−3	−3	9
2	−1.5	−3	8
3	0	−3	7
4	1.5	−3	6
5	−3	0	0.5
6	−1.5	0	1.0
7	0	0	1.25
8	1.5	0	1.5
9	−3	3	6
10	−1.5	3	7
11	0	3	8
12	1.5	3	9

(7.15)

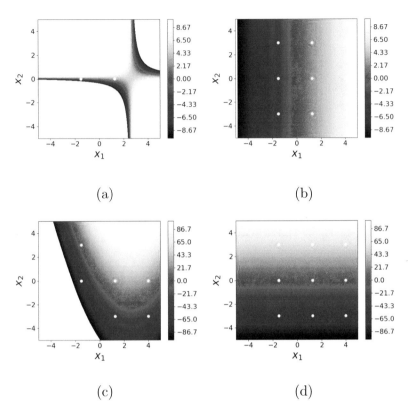

Figure 7.8: First order partial derivatives w.r.t. x_1 and x_2 respectively computed from the constructed (a),(c) zero order only surrogate model and (b),(d) first order only surrogate model for the piece-wise smooth step discontinuous quadratic function. The sampled designs are indicated by white circles and the colorbar scale limited to the minimum and maximum values of the actual associated partial derivatives

Fit the following quadratic function

$$f(\mathbf{x}, \mathbf{c}) = c_5 x_1^2 + c_4 x_2^2 + c_3 x_1 x_2 + c_2 x_1 + c_1 x_2 + c_0, \qquad (7.16)$$

to the data in Equation 7.15 using the optimal criterion derived in Exercise 7.4.2.

7.4.4 For Exercise 7.4.3 present, in a detailed discussion, all the verifications you can envisage to ensure that it was indeed correctly solved.

7.4.5 Verify the solution in Exercise 7.4.3 by outlining and conducting a numerical study that perturbs the optimal solution.

7.4.6 Verify the solution in Exercise 7.4.3 by outlining and conducting a numerical study that solves the formulated problem iteratively using an appropriate minimization algorithm.

7.4.7 Instead of using the f^i given in Equation 7.15, evaluate the function $f(\mathbf{x}) = 13x_1^2 + 11x_2^2 + 7x_1x_2 + 5x_1 + 3$ at the designs listed in Equation 7.15. What do you expect the solution for $c_0, c_1, \ldots, c_5$ to be, and motivate your answer by a detailed discussion of your reasoning.

7.4.8 Conduct a numerical study that investigates the validity of your expectation in Exercise 7.4.7.

7.4.9 Higher order polynomials compute higher order powers of designs of $\mathbf{x}$ that may lead to numerically very large numbers and ultimately to ill-conditioning of the system to be solved. Propose an appropriate scaling of $\mathbf{x}$ that would eliminate such ill-conditioning.

7.4.10 Given two Gaussian representations, $\phi_1(r) = e^{-\epsilon r^2}$ and $\phi_2(r) = e^{-\frac{r^2}{\epsilon}}$. Discuss the implication of increasing ϵ on the size of the effect of the domain covered by $\phi_1(r)$ and $\phi_2(r)$ respectively.

7.4.11 Utilize the function Rbf in the module scipy.interpolate to construct a zero order only radial basis surrogate model of the quadratic function in Exercise 7.4.7 using the Gaussian basis function. Note that Rbf returns a function object that takes the same number of inputs as the dimensionality of the problem, where each entry is a list of values for a coordinate at which to evaluate the constructed RBF function. The basis function can be selected by assigning the string gaussian to the parameter function of the function object, and the shape parameter by assigning a value to the parameter epsilon. Type help(Rbf) for additional information and note the usage of the shape parameter ϵ in the Gaussian formulation.

7.4.12 Optimize the actual quadratic function given in Exercise 7.4.7 using 100 random starting guesses between -1.5 and 1.5 for both x_1 and x_2 using a zero and first order algorithm of your choice.

Repeat the optimization using the same initial starting guesses, but this time optimize the RBF approximation to the quadratic function constructed in Exercise 7.4.11. Critically compare the solutions obtained.

7.4.13 Repeat Exercise 7.4.12 but increase the bounds for the initial starting guesses from -1.5 and 1.5 to (i) -15 and 15, and to (ii) -150 and 150 respectively and critically compare the solutions obtained as well as pointing out the cause of any evident changes in the results.

7.4.14 Repeat Exercise 7.4.12 but use the Exponential basis function given by $\phi(r) = e^{-\epsilon r}$ (Python code `exp(-(r*self.epsilon))`). Critically compare the solutions obtained as well as pointing out the cause of any evident changes in the results.

7.4.15 Given the Gaussian radial basis function $\phi(r(\mathbf{x})) = e^{-\epsilon r(\mathbf{x})^2}$, and the Exponential radial basis function $\phi(r(\mathbf{x})) = e^{-\epsilon r(\mathbf{x})}$. Plot the functions and critically discuss any expected implications of the smoothness of the basis function on the performance of gradient based optimization strategies. Which basis function would you prefer for gradient based optimization strategies?

7.4.16 Consider the Gaussian basis function given in (7.11). Conduct a numerical study that identifies the observed ill-conditioning of $\mathbf{R}_{zo}$ as ϵ gets smaller, and the absence of ill-conditioning as ϵ gets larger.

7.4.17 Given a two-dimensional function defined over the domain -1 and 1 for x_1, and -100 and 100 for x_2. Discuss the potential problems associated with constructing an RBF surrogate model over such a domain. Detail a potential solution strategy.

7.4.18 Conduct a numerical investigation that highlights the potential difficulties associated with Exercise 7.4.17 and that demonstrates the potential benefits of the solution strategy proposed in Exercise 7.4.17.

7.4.19 Write your own Python code to construct a zero order only radial basis surrogate model for a function of n dimensions. Test your code on the quadratic function given in Exercise 7.4.7, and compare your RBF approximation against `scipy.interpolate.Rbf` for equivalent shape parameters.

7.4.20 Approximate the gradient vector of the quadratic function given in Exercise 7.4.7. Use the radial basis surrogate model constructed in Exercise 7.4.19 at 100 random designs over the domain defined by the design of experiments. Compute the average error for each component of the gradient vector at the 100 random designs.

7.4.21 Write your own Python code to construct a combined zero and first order radial basis surrogate model for a function of n dimensions. Test your code on the quadratic function given in Exercise 7.4.7.

Chapter 8

GRADIENT-ONLY SOLUTION STRATEGIES

8.1 Introduction

As outlined in Section 1.1, mathematical optimization is the systematic process of finding a best solution, generally subject to some constraints, to a problem based on a mathematical model. Here the model is constructed in such a way that the solution we seek often corresponds to a quantity that minimizes some multi-dimensional scalar function which is the outcome of the model. Specifically in this chapter we restrict ourselves to unconstrained optimization problems. Thus formally the process now becomes (i) the formulation of the model $f(\mathbf{x})$ and (ii) the minimization of $f(\mathbf{x})$:

$$\underset{\mathbf{x}}{\text{minimize}}\, f(\mathbf{x}), \;\; \mathbf{x} = [x_1, x_2, \ldots, x_n]^T \in \mathbb{R}^n,$$

where $f(\mathbf{x})$ is a scalar function of the real *column vector* $\mathbf{x}$. Care is usually taken during the mathematical modelling and numerical computation of the scalar function $f(\mathbf{x})$ to ensure that it is smooth and twice continuously differentiable. As highlighted in Section 6.5, the presence of numerical noise in the objective function is sometimes an unintended consequence of the complicated numerical nature frequently associated with the computation of the output function of a multi-disciplinary design optimization model. Numerical noise can also be the conse-

© Springer International Publishing AG, part of Springer Nature 2018
J.A. Snyman and D.N. Wilke, *Practical Mathematical Optimization*,
Springer Optimization and Its Applications 133,
https://doi.org/10.1007/978-3-319-77586-9_8

quence of a deliberate computational savings strategy employed by a design engineer.

To elaborate, consider for example the computational cost associated with integrating a system of partial differential equations that is required to construct the objective function for some design optimization problem. For each chosen design vector $\mathbf{x}$, the numerical integration may require the solution of a finite element model. The associated computational cost to solve the finite element model is directly related to the number of elements used to discretize the spatial domain over which the integration needs to be performed. Not unexpectedly, more elements places higher demands on computing resources at the added benefit of reducing the discretization error which results in more accurate solutions. This clear trade-off between time to solution and solution accuracy may be exploited with care in the design optimization process. A concrete demonstration of this is given by the adaptive remeshing strategy proposed by Wilke et al. (2013a). In this strategy the accuracy of each analysis is increased as the optimizer converges towards an optimum when conducting structural shape optimization. A complication of this strategy is however that because of the initial rough meshing the computed piece-wise smooth objective function is discontinuous. This complication requires a significant adaptation in both (i) the formulation and (ii) the solution strategy to solve the successive approximate discontinuous mathematical optimization problem to the real continuous underlying problem.

This chapter is dedicated to explore alternative formulations and solution strategies when specifically dealing with piece-wise smooth discontinuous objective functions (Wilke et al. (2013b)). In essence, this chapter elaborates and formalizes the concepts and ideas introduced and hinted to in Section 6.5, that includes the handling of noisy objective functions and the use of gradient-only optimization strategies.

8.2 Piece-wise smooth step discontinuous functions

Nowadays, the computation of the objective function routinely requires the integration of a system of differential or partial differential equations. The inherent numerical nature of modern mathematical optimization often allows for the same mathematical model to be solved using different numerical methods (Strang (2007)). Applying care usually renders the same optimum when optimizing the computed objective function with different numerical methods. This observation is valid when the resulting discretization error (Strang (2007)) of the various numerical methods are (i) comparable, (ii) small, and (iii) varies smoothly and continuously between designs.

Failure to meet the first two criteria may still result in numerically computed continuous and smooth objective functions but may render inaccurate approximations of the optimum of the mathematical optimization problem. Consequently, the computed optima for the different methods may differ significantly as a direct result of large or non-comparable discretization errors for the various numerical methods used in computing the objective functions. Lastly, failing to conform to the third criterion results in piece-wise smooth discontinuous objective functions, where the size of the discontinuity decreases as the discretization error reduces. Hence, by ensuring that the discretization error is negligible both (i) smoothness of the objective function and (ii) accuracy in the determination of the optimum to the underlying mathematical model is ensured. This may however place an unattainable computational burden on available computing resources.

To illustrate these concepts consider the Lotka-Volterra system of first order, non-linear, differential equations:

$$\frac{dz(t)}{dt} = (1 - \lambda)z(t) - \beta z(t)y(t) \tag{8.1}$$

$$\frac{dy(t)}{dt} = \delta z(t)y(t) - \gamma y(t), \tag{8.2}$$

which was first proposed to model auto-catalytic chemical reactions (Yorke and W.N. Anderson (1973)). Two initial conditions $z(0) = z_0$ and $y(0) = y_0$ completes the formulation. Given that $\beta = 0.3$, $\delta = \gamma = 1$

and $z_0 = y_0 = 0.9$, the entire response of the system depends solely on λ. In addition, given that at $t = 8$ the actual values of the two functions $z(8) = \tilde{z}$ and $y(8) = \tilde{y}$ are known, we can construct an inverse problem in which we aim to find λ such that $z(8) \approx \tilde{z}$ and $y(8) \approx \tilde{y}$. The sum of the errors squared w.r.t. λ (Strang (2007)) is a convenient unconstrained optimization problem that defines this inverse problem, with the sum of the errors squared objective function given by

$$E(\lambda) = (z(t = 8, \lambda) - \tilde{z})^2 + (y(t = 8, \lambda) - \tilde{y})^2. \qquad (8.3)$$

In particular, accept the accurate global optimum to be $\lambda^* = 0.5$ computed using a forward Euler integration scheme (Strang (2007)) for 50 000 equal time steps between 0 and 8 seconds to define $\tilde{z}$ and $\tilde{y}$. Computing the objective function and respective derivative using now only 10 000 equal time steps, values are obtained as depicted in Figures 8.1 (a)–(b) for λ between 0 and 1. Clearly the figures indicate that the accurate optimum λ^* and computed global optimum λ^o in the figures closely coincide with $\lambda^o \approx \lambda^* = 0.5$. There is no apparent difference since the discretization error is negligible whether computing the objective function using 50 000 or 10 000 equally spaced time steps. However when computing the objective function using only 30 equally spaced time steps gives an apparent global optimum $\lambda^o \approx 0.56$ that is significantly different from the accepted global optimum $\lambda^* = 0.5$. This is clear from Figures 8.2 (a)–(b) for which the objective function and its derivative plotted for λ ranging between 0.4 to 0.8. Also of interest is that the computed objective function and its derivative still appears to be continuous and smooth.

Instead of considering only fixed time steps, an integration scheme where the number of time steps varies for different values of λ could be considered. This is reminiscent of adaptive time stepping strategies often employed to solve systems of differential equations (Strang (2007)). As illustration, consider a time stepping strategy that selects a random number of time steps for each λ from a defined distribution around a mean number of time steps, μ. In particular, for $\mu = 30$ consider the number of time steps varying between 27 and 33 with an equal probability. When computing the objective function using between 27 and 33 randomly varying time steps gives an apparent global optimum $\lambda^o \approx 0.56$ that is significantly different from the accepted global optimum $\lambda^* = 0.5$. In addition, the inherent step discontinuous nature of

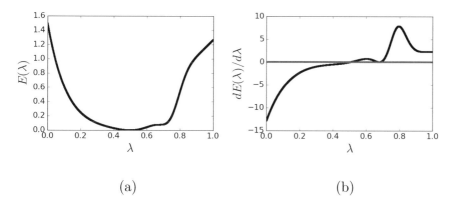

(a) (b)

Figure 8.1: (a) Error function $E(\lambda)$ and corresponding (b) derivative function $\frac{dE(\lambda)}{d\lambda}$ plotted against values of λ ranging from 0 to 1. In the numerical integration done in computing the function values for each plotted value of λ, 10 000 equally spaced time steps, over the time interval 0 to 8 seconds, were used. The figures show that the apparent global optimum $\lambda^o \approx 0.5$ closely coincides with the accepted accurate global optimum $\lambda^* = 0.5$

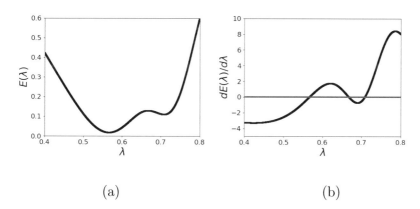

(a) (b)

Figure 8.2: (a) Error function $E(\lambda)$ and corresponding (b) derivative function $\frac{dE(\lambda)}{d\lambda}$ plotted against values of λ ranging from 0.4 to 0.8. In the numerical integration done in computing the function values for each plotted value of λ, only 30 equally spaced time steps, over the time interval 0 to 8 seconds, were used. The figures show that the apparent global optimum $\lambda^o \approx 0.56$ significantly differs from the accepted accurate global optimum $\lambda^* = 0.5$

both the objective function and respective derivative function is evident from Figures 8.3 (a) and (b). The additional complications that the step discontinuities introduce to the minimization of $E(\lambda)$, is at first glance disconcerting.

The usual approach to address numerical step discontinuities is to increase the computational cost associated with each analysis with the aim of reducing the magnitudes of the step discontinuities. This is illustrated in Figures 8.4 (a)–(d), in which (a) and (b) indicate the resulting computed objective function and corresponding derivative function when the number of time steps are randomly selected with an equal probability between 54 and 66 for the computation for each λ, while (c) and (d) depicts the same but with the number of time steps now randomly selected with equal probability between 108 and 132. Clearly the magnitudes of these numerical step discontinuities decrease as the number of randomly selected time steps increases. Thus in practice the number of time steps may be increased until the step discontinuity magnitudes are small enough so as to not cause complications when minimization strategies are applied. In addition, the progression of the apparent global optimum λ^o to the accepted global optimum λ^* is also evident as the number of random selected time steps, for the computation of the objective function for each λ, is increased.

Another source of discontinuities in design optimization problems are discontinuities which are as a result of a sudden change in the physical response of a model as the design vector changes, hereafter referred to as physical discontinuities. Examples of physical discontinuities include the onset or breakdown of shock waves between designs in fluid dynamic applications (Homescu and Navon (2003)), or the inherent chaotic nature of dynamic strain ageing during plastic deformation (Sarkar et al. (2007)) that may abruptly change as the design vectors vary. Physical discontinuities are distinct from the numerically induced step discontinuities discussed above. Numerically induced step discontinuities are due to abrupt changes in the discretization error between design vectors. It is important to distinguish between physical and numerical step discontinuities as the complications that arise from their presence may need to be addressed differently.

Physical discontinuities are consequences of the underlying physics being modelled and need to be considered as they relate to actual information

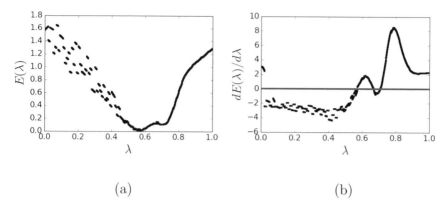

(a) (b)

Figure 8.3: (a) Error function $E(\lambda)$ and corresponding (b) derivative function $\frac{dE(\lambda)}{d\lambda}$ plotted against values of λ ranging from 0 to 1. In the numerical integration over the time interval 0 to 8 seconds done for each plotted λ, the number of equally spaced time steps were randomly selected between 27 and 33. were used. The figures again show that the apparent global optimum $\lambda^o \approx 0.56$ differs significantly from the accepted accurate global optimum $\lambda^* = 0.5$. The piece-wise smooth step discontinuous nature of both the objective and derivative functions is evident

about the nature of the problem under consideration. Hence, physical discontinuities need to be resolved and, indeed, a significant effort has been made to explore their presence and impact in terms of a design optimization problem (Smith and Gilbert (2007)). On the other hand, as demonstrated for the example error function problem, numerical step discontinuities are a consequence of the numerical solution strategy employed to solve an underlying mathematical model, and if significantly present in the computed objective function they may hide the underlying physical nature of the problem under consideration. Thus, ideally we would wish for sufficient computational accuracy so that numerical step discontinuities are effectively eliminated and may be ignored, allowing for the underlying physical trends present in the problem to drive an optimizer towards the solution of an optimization problem. The remainder of this chapter is dedicated to addressing the complications arising from numerical step discontinuities in such a way that piece-wise smooth discontinuous objective functions can still be optimized, efficiently and robustly. The main emphasis will therefore be

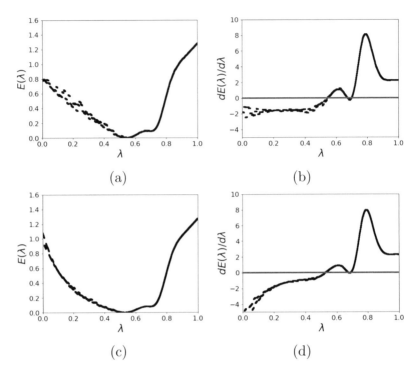

Figure 8.4: (a),(c) Error function $E(\lambda)$ and (b),(d) corresponding derivative function $\frac{dE(\lambda)}{d\lambda}$ plotted against values of λ ranging from 0 to 1 for
respectively 54–66 and 108–132 randomly selected time steps computed
for each λ. The numerical integration is conducted over the time interval 0 to 8 seconds. The figures show that the magnitude of the step
discontinuities decreases as the number of randomly selected time steps
increases. In addition, the apparent global optimum λ^o draws closer
to the accepted accurate global optimum $\lambda^* = 0.5$ as the number of
randomly selected time steps for each λ increases

on computational minimization procedures and strategies that allow for
overcoming the presence of step discontinuities in piece-wise smooth step
discontinuous objective functions. In addition, a gradient-only problem
formulation is included that defines the underlying problem that is consistent with what the optimization procedures and strategies considered
in this chapter actually solves.

8.3   Computational minimization procedures

Consideration is now given to different formulations and approaches to compute the unconstrained minimum of the model function $f(\mathbf{x})$, which represents the first step in the overall mathematical optimization process. Although the minimization is generally viewed as the systematic searching of the design space to find at least a local minimum, alternative approaches and solution strategies are at our disposal that may simplify the complexities arising from the presence of numerical step discontinuities in $f(\mathbf{x})$.

To aid the discussion consider Figures 8.5 (a)–(d) that depict the error function $E(\lambda)$ and corresponding derivative function $\frac{dE(\lambda)}{d\lambda}$ plotted against values of λ ranging from 0.5 to 0.6. The error function and corresponding derivative function in Figures 8.5 (a) and (b) respectively are computed using a fixed number of time steps, while they are computed using a randomly selected number of time steps in Figures 8.5 (c) and (d). In particular the numerical integration done, over the time interval 0 to 8 seconds, for each plotted value of λ uses either 30 equally spaced time steps or a randomly selected number of time steps between 27 and 33. Instead of now focusing on the difference between the apparent global minimum $\lambda^o \approx 0.56$ and accepted global minimum $\lambda^* = 0.5$, attention is specifically given to determine the apparent global minimum.

Consider Figure 8.5 (a) that depicts a smooth objective function. Conventional zero- or first order gradient based search minimization methods covered in the previous chapters can be used to accurately solve this optimization problem. An alternative second order approach may be to apply the so-called optimality criteria, i.e. the necessary and sufficient conditions to find all points with zero first derivative and corresponding positive second derivative to determine candidate points for the global minimum. The indicated horizontal line, in Figure 8.5 (b), determines the level-set of a derivative of magnitude zero. As the gradient is increasing with an increase in λ the second derivative is positive. It is clear that in this case (a) the minimization formulation and application of the optimality criteria would recover the same apparent minimum over the indicated λ domain.

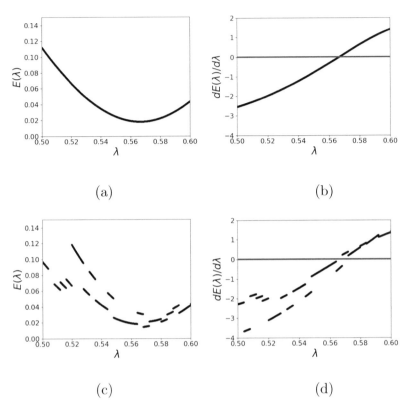

(a)   (b)

(c)   (d)

Figure 8.5: (a),(c) Error function $E(\lambda)$ and (b),(d) corresponding derivative function $\frac{dE(\lambda)}{d\lambda}$ plotted against values of λ ranging from 0.5 to 0.6 using respectively a fixed time step (a)–(b) and randomly selected time step (c)–(d) strategy. In particular the numerical integration done for each plotted value of λ uses respectively either 30 equally spaced time steps or 27–33 randomly selected time steps, over the time interval 0 to 8 seconds

The difference between the two approaches is more evident when multiple local minima occur. In this case the application of a zero to first order minimization approach will clearly terminate at a local minimum (not necessary the global minimum), whereas the application of optimality criteria may reveal all local minima including the global minimum, irrespective of their associated function values, as candidate solutions to the problem. Thus the implication for the two solution approaches changes significantly when applied to the piece-wise smooth step discontinuous function and related derivative function, depicted in Figures 8.5 (c) and (d). It is evident that the application of either minimization strategy now poses additional challenges. The economically computed objective function, as result of the introduction of step discontinuities, has become highly multi-modal with the presence of numerous local minima. In the real world the difficulties in the application of the above two conventional approaches to step discontinuous functions become so daunting that it often leads to the justification and employment of computationally demanding evolutionary approaches (Arnold (2002)), over more efficient gradient based strategies. By inspection of Figure 8.5 (d) it is clear that the second order optimality criteria approach would not suffice as there is no λ for which the derivative is close to 0. It is cautionary noted that the two points with the smallest derivative magnitude is around $\lambda \approx 0.56$ and $\lambda \approx 0.58$, while the apparent global minimum is around $\lambda^o \approx 0.57$. This may have significant implications for solution strategies that aim to find points where the derivative is close to zero.

A third approach is now proposed that consistently interprets the derivative information presented in Figures 8.5 (b) and (d). This third approach, that may be called gradient-only minimization, follows from defining descent for a function along a search direction by directional derivatives that are negative, while ascent is associated with directional derivatives that are positive, with the directional derivative computed as the projection of the gradient vector onto the search direction. Instead of associating a candidate local minimum along a search direction as a point at which the directional derivative is zero, we associate a local minimum with a point where the directional derivative changes in sign from negative to positive. It is important to note that this defines a local minimum and not merely a candidate local minimum, since the second order information is incorporated by requiring the directional derivative to change from negative to positive. By inspection of Figure 8.5 (d) it is clear that by interpreting descent of the function indirectly by requiring the directional derivative to be negative and the minimum by a sign change in the directional derivative, from negative to positive, results in a robust minimization strategy for step discontinuous functions.

8.4   Overview of discontinuities

Before we proceed   with an in depth investigation into gradient-only minimization strategies for piece-wise smooth step discontinuous functions (also known as semi-continuous functions), some background on discontinuities in general is required. To assist the discussion consider Figure 8.6 (a)–(d) that depicts four types of discontinuities for univariate functions $f(x)$ at $x = s$. They are respectively an infinite (or asymptotic) discontinuity, a removable discontinuity, an endpoint discontinuity and lastly a step (or jump) discontinuity.

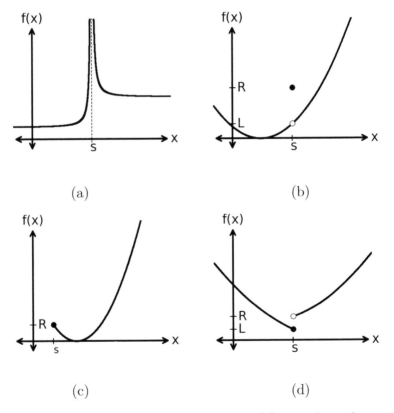

Figure 8.6: Four discontinuity types namely, (a) an infinite (or asymptotic) discontinuity, (b) a removable discontinuity, (c) an endpoint discontinuity and (d) a step (or jump) discontinuity

Figure 8.6 (a) indicates that an infinite discontinuity, at s, is usually the result of vertical asymptotes present at s, with the function unbounded as x approaches s. Therefore, both the left-hand limit, $\lim_{x \to s^-} f(x) = +\infty$, and right-hand limit, $\lim_{x \to s^+} f(x) = +\infty$, indicate an unbounded response of the function. This type of discontinuity is usually the result of the behaviour of an underlying mathematical model and therefore inherent to the characteristics of a problem to be solved i.e. they can be seen as physical discontinuities. Examples include the Stokes phenomenon present in the solution of differential equations (Meyer (1989)) and the asymptotic Ruderman-Kittel-Kasuya-Yosida (RKKY) interaction between magnetic impurities in graphene (Klier et al. (2014)).

In some instances both one-sided limits exist and are equal, given by $\lim_{x \to s^-} f(x) = \lim_{x \to s^+} f(x) = L$, but the function at $x = s$ does not correspond to the value of the limits, that is $f(s) = R \neq L$, as illustrated in Figure 8.6 (b). This is referred to as a removable discontinuity, since the discontinuity can simply be removed by redefining the function $f(s) = \lim_{x \to s^-} f(x) = L$ at $x = s$.

An end-point discontinuity exists in cases where one of the one-sided limits does not exist, as depicted in Figure 8.6 (c). In this case only the right-hand limit exists, that is $\lim_{x \to s^+} f(x) = R$, while the left-hand limit is not defined. This type of discontinuity is often associated with mathematical models that cannot be evaluated over non-physical domains, e.g. non-positive mass. This type of discontinuity is again associated with physical discontinuities.

Consider the step or jump discontinuity, at $x = s$, in Figure 8.6 (d). Here, both one-sided limits exist and are finite. However, the one-sided limits, at $x = s$, have different values as the left-hand limit, $\lim_{x \to s^-} f(x) = L$, differs from the right-hand limit, $\lim_{x \to s^+} f(x) = R \neq L$. As demonstrated in previous sections, numerical inconsistencies as design vectors vary are a significant source of this type of discontinuity. However, not all step discontinuities are necessarily problematic using classical minimization strategies. We distinguish between two types of step discontinuities. Namely those that are *consistent* with the function *trend*, and those that are *inconsistent* with the function *trend*, as shown respectively in Figures 8.7 (a) and (b).

Consistent discontinuities do not hamper descent while inconsistent discontinuities result in a local minimum. However, the related derivative function in both cases indicates only descent as the sign of the derivative along x remains negative over the depicted domain. This is an important distinction as a gradient-only minimization strategy would ignore a local minimum due to an inconsistent discontinuity, thereby improving the robustness of a minimization strategy, as opposed to a classical minimization strategy that may aim to resolve such a local minimum.

Note that other combinations of function trends and step discontinuities result in either local minima or local maxima that are consistently indicated when considering either only the function or only the corresponding derivative of the function. In addition to the discontinuity type, the semi-continuity of f is indicated using a double empty/filled circle convention in Figure 8.7 (a)–(b). First, a filled circle indicates that $f(s)$ is defined as indicated, while an empty circle indicates no function value is defined as indicated. Lower semi-continuity is represented by the filled/empty circle pairs annotated 1, i.e. the filled circle is always associated with the lower function value, in turn, upper semi-continuity is represented by the empty/filled circle pairs annotated 2, i.e. the filled circle is always associated with the higher function value at the discontinuity. This distinction allows for an explicit treatment of the defined function at a discontinuity, which will become evident as we treat derivatives and gradients associated with semi-continuous functions in more detail.

8.5   Derivatives and gradients of step discontinuous functions

Semi-continuous functions are not everywhere differentiable, as the derivative at step discontinuities is not defined. However, computationally the derivatives and gradients are everywhere computable since the analysis is *per se* restricted to the part of the objective function to the left, or right of a step discontinuity along a search direction. Reference to the derivative of a semi-continuous function therefore requires a rigorous treatment of limits and derivatives to allow for a new definition of what is implied with a derivative at a step discontinuity.

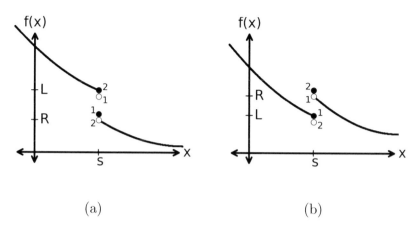

(a) (b)

Figure 8.7: Lower and upper semi-continuous univariate functions for (a) a consistent step discontinuity, and (b) an inconsistent step discontinuity, where the semi-continuity is indicated using a double empty/filled circle convention. The empty/filled circle pairs annotated, 1, are lower semi-continuous, whereas, empty/filled circle pairs annotated, 2, are upper semi-continuous function representations

For a univariate function $f(x)$, the limit at $x = a$ exists if (i) both the left-hand limit, $\lim_{x \to a^-} f(x) = L$, and right-hand limit, $\lim_{x \to a^+} f(x) = \lim_{x \to s^+} f(x) = R$, exist, and (ii) the left-hand and right-hand limits are equal, that is $L = R$. It follows from the limit definition of the derivative

$$f'(x) = \frac{df(x)}{dx} = \lim_{\Delta x \to 0} \frac{f(x + \Delta x) - f(x)}{\Delta x}, \tag{8.4}$$

that the derivative at $x = a$ is only defined if the limit at $x = a$ is defined. However, as treated in Section 8.4, both the left-hand and right-hand limits exist at a step discontinuity but the limits are not equal, implying that the derivative function is not everywhere defined.

We therefore supplement the definition of a derivative to define the *associated derivative*, $f'^A(x)$, that is given by either the left derivative or right derivative whenever the limit and therefore derivative does not exist, otherwise it is defined by the derivative when the limit does exist. Let $f : X \subset \mathbb{R} \to \mathbb{R}$ be a piece-wise smooth real univariate step-discontinuous function that is everywhere defined. The *associated derivative* $f'^A(x)$ for $f(x)$ at a point x is given by the derivative of $f(x)$ at x when $f(x)$ is differentiable at x. The *associated derivative*, f'^A, for

$f(x)$ non-differentiable at x, is given by the left derivative of $f(x)$:

$$f'^-(x) = \frac{df^-(x)}{dx} = \lim_{\Delta x^- \to 0} \frac{f(x + \Delta x) - f(x)}{\Delta x}, \qquad (8.5)$$

when x is associated with the piece-wise continuous section of the function to the left of the discontinuity, otherwise it is given by the right derivative of $f(x)$:

$$f'^+(x) = \frac{df^+(x)}{dx} = \lim_{\Delta x^+ \to 0} \frac{f(x + \Delta x) - f(x)}{\Delta x}. \qquad (8.6)$$

The associated derivative is therefore everywhere defined when piece-wise smooth step discontinuous functions are considered. This implies that $f : (a, b) \subset \mathbb{R} \to \mathbb{R}$ for which $f(x)$ and $f'^A(x)$ are uniquely defined for every $x \in (a, b)$ is said to have a strictly negative *associated derivative* on (a, b) if $f'^A(x) < 0$, $\forall\, x \in (a, b)$, e.g. see Figure 8.7 (a) and (b). Conversely, $f(x)$ is said to have a strictly *positive associated derivative* on (a, b) if $f'^A(x) > 0$, $\forall\, x \in (a, b)$.

Similarly for multi-variate functions we define the *associated gradient* $\nabla_A f(\mathbf{x})$, by letting $f : X \subset \mathbb{R}^n \to \mathbb{R}$ be a piece-wise continuous function that is everywhere defined. The *associated gradient* $\nabla_A f(\mathbf{x})$ for $f(\mathbf{x})$ at a point $\mathbf{x}$ is given by the gradient of $f(\mathbf{x})$ at $\mathbf{x}$ when $f(\mathbf{x})$ is differentiable at $\mathbf{x}$. The *associated gradient* $\nabla_A f(\mathbf{x})$ for $f(\mathbf{x})$ non-differentiable at $\mathbf{x}$ is defined as the vector of partial derivatives with each partial derivative defined by its corresponding *associated derivative*. It follows that the *associated gradient* reduces to the gradient of a function when it is everywhere differentiable.

Similarly, to recognizing the lower and upper semi-continuity of a uni-variate function as highlighted in Figures 8.7 (a) and (b), we now consider the semi-continuous nature of the associated derivative for such functions. The associated derivative can be related to a univariate function or the directional derivative of a multivariate function. For example the *associated directional derivative* along a normalized direction $\mathbf{u} \in \mathbb{R}^n$ is lower semi-continuous at $\mathbf{x} \in X$, if

$$F'^A(\lambda) = \nabla_A^{\mathrm{T}} f(\mathbf{x}) \mathbf{u} \leq \liminf_{\lambda \to 0^\pm} \nabla_A^{\mathrm{T}} f(\mathbf{x} + \lambda \mathbf{u}) \mathbf{u}, \ \lambda \in \mathbb{R}, \qquad (8.7)$$

as depicted in Figure 8.8 (b), with the related function, depicted in Figure 8.8 (a), which is also lower semi-continuous. An upper semi-continuous *associated directional derivative* at $\mathbf{x} \in X$ along a normalized

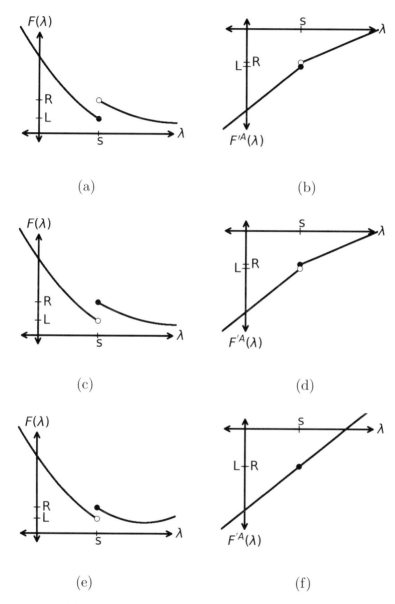

Figure 8.8: (a) Lower semi-continuous function and (b) associated derivative, (c) upper semi-continuous function and (d) associated derivative, and (e) upper semi-continuous function with (f) pseudo-continuous associated derivative

direction $\mathbf{u} \in \mathbb{R}^n$ is defined by

$$F'^A(\lambda) = \nabla_A{}^\mathrm{T} f(\mathbf{x})\mathbf{u} \geq \limsup_{\lambda \to 0^\pm} \nabla_A{}^\mathrm{T} f(\mathbf{x} + \lambda\mathbf{u})\mathbf{u}, \ \lambda \in \mathbb{R}, \qquad (8.8)$$

depicted in Figure 8.8 (d) with the related function depicted in Figure 8.8 (c). Lastly, the *associated directional derivative* along a normalized direction $\mathbf{u} \in \mathbb{R}^n$ is pseudo-continuous at a step discontinuity, $\mathbf{x} \in \mathbb{R}^n$, if it is both upper and lower semi-continuous as demonstrated in Figures 8.8 (e) and (f).

8.5.1   Associated gradients by finite differences

The associated gradient can be computed analytically by direct differentiation of the equations that numerically evaluate the objective function (Strang (2007)). Recall that step discontinuities are due to changes in the discretization of the numerical scheme used to evaluate the objective function as the design vector changes, while the computed analytical sensitivity is associated with a given discretization for a specific design.

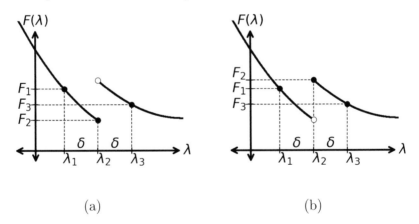

Figure 8.9: Finite difference step, δ, over (a) a lower semi-continuous and (b) an upper semi-continuous function with inconsistent step discontinuities

Without loss of generality, we first limit the discussion to a single directional derivative $F'(\lambda)$ as the gradient vector, $\nabla f(\mathbf{x})$, is comprised of directional derivatives aligned with the Cartesian directions. First, consider the finite difference strategies outlined in Section 2.3.1.6, applied here to an upper or lower step discontinuous function with an inconsistent step discontinuity. For the lower semi-continuous function, $F(\lambda)$, depicted in Figure 8.9 (a) we estimate only the sign of the derivative at λ_2 using the forward (FD), backward (BD) and central difference (CD) schemes, which gives

$$
\begin{aligned}
\left(\frac{dF(\lambda_2)}{d\lambda}\right)_{FD} &\approx \frac{F(\lambda_2+\delta)-F(\lambda_2)}{\delta} = \frac{F_3-F_2}{\lambda_3-\lambda_2} > 0, \\
\left(\frac{dF(\lambda_2)}{d\lambda}\right)_{BD} &\approx \frac{f(\lambda_2)-f(\lambda_2-\delta)}{\delta} = \frac{F_2-F_1}{\lambda_2-\lambda_1} < 0, \\
\left(\frac{dF(\lambda_2)}{d\lambda}\right)_{CD} &\approx \frac{F(\lambda_2+\delta)-F(\lambda_2-\delta)}{2\delta} = \frac{F_3-F_1}{\lambda_3-\lambda_1} < 0,
\end{aligned}
\tag{8.9}
$$

whereas for the upper semi-continuous function in Figure 8.9 (b) the sign of the derivative at λ_2 is estimated as follows:

$$
\begin{aligned}
\left(\frac{dF(\lambda_2)}{d\lambda}\right)_{FD} &\approx \frac{F(\lambda_2+\delta)-F(\lambda_2)}{\delta} = \frac{F_3-F_2}{\lambda_3-\lambda_2} < 0, \\
\left(\frac{dF(\lambda_2)}{d\lambda}\right)_{BD} &\approx \frac{F(\lambda_2)-F(\lambda_2-\delta)}{\delta} = \frac{F_2-F_1}{\lambda_2-\lambda_1} > 0, \\
\left(\frac{dF(\lambda_2)}{d\lambda}\right)_{CD} &\approx \frac{F(\lambda_2+\delta)-F(\lambda_2-\delta)}{2\delta} = \frac{F_3-F_1}{\lambda_3-\lambda_1} < 0.
\end{aligned}
\tag{8.10}
$$

It is evident that finite differences over inconsistent step discontinuities are problematic resulting in inconsistencies not only in the magnitude of the derivative but also the signs of the computed derivatives. Here, the problem is that an actual finite difference step, δ, is taken over the step discontinuity. This then results in inconsistent estimates of the derivative. Although not always practical or possible, the step discontinuity can be removed by forcing the numerical computation scheme to only have smooth variations in the discretization error whilst computing the derivatives.

The complex-step method circumvents the above issues related to conventional finite difference strategies by only taking finite difference steps of δi in the imaginary plane. The implication is that no step over a discontinuity is ever taken to compute the derivative even at a discontinuity (Wilke and Kok (2014)). As discussed in Section 2.3.1.6, the complex-step method has the additional advantages of allowing for much smaller finite difference steps to be taken as it is not susceptible to a subtraction error.

To demonstrate our arguments, consider the following simple piece-wise linear step discontinuous function:

$$f(x) = \begin{cases} x < 1 : & -2x - 0.5 \\ x \geq 1 : & -2x \end{cases} . \tag{8.11}$$

with analytical associated derivative of $f'(x) = -2$. The choice for a piece-wise linear function implies the Taylor series approximation is exact for all schemes. Hence, the truncation error is exactly zero for all finite difference schemes on each section of the piece-wise linear function. Any error that varies as a function of the step size is due to the numerical errors introduced by the subtraction of two numbers or the influence of the discontinuity errors.

Mathematically, the derivative is not defined at $x = 1$. However, the *associated* derivative of this function is continuous and -2 everywhere, including at $x = 1$. Computing the derivative with the complex-step method yields exactly -2 everywhere, including $x = 1$, allowing a full field computation of the derivative of a discontinuous function. The computed sensitivity difference for the forward difference scheme varies between $\approx 10^{-16}$ and $\approx 10^{0}$ as the step size is decreased from 10^{0} to 10^{-20}. In turn, the computed sensitivity difference for both the backward and central difference schemes vary between $\approx 10^{-1}$ and $\approx 10^{15}$ as the step size is decreased from 10^{0} to 10^{-20}. Hence, the difference in magnitude increases as the step size decreases, in addition to the derivative having the wrong sign. Therefore extending the numerical computation of the components of the associated gradient vector $\nabla_A f(\mathbf{x})$, namely $\frac{\partial_A f(\mathbf{x})}{\partial_A x_j}$, $j = 1, \ldots, n$, follows by complex-step differences:

$$\frac{\partial_A f(\mathbf{x})}{\partial_A x_j} \cong \frac{Im[f(\mathbf{x} + i\boldsymbol{\delta}_j)]}{\delta_j}, \tag{8.12}$$

where $\boldsymbol{\delta}_j = [0, 0, \ldots \delta_j, 0, \ldots, 0]^T$, $\delta_j > 0$ in the j-th position.

8.6   Gradient-only line search descent methods

Gradient-only line search descent methods closely follow the structure of line search descent methods treated in Chapter 2. Following the general six-step structure, outlined in Chapter 2, for line search descent methods, we outline gradient-only line search descent methods as follows:

Algorithm 8.1 Gradient-only line search descent framework.

Initialization: Select tolerance $\varepsilon_1 > 0$. Select the maximum number of iterations i_{max} and perform the following steps:

1. Given $\mathbf{x}^0$, set $i := 1$.

2. Select a descent direction $\mathbf{u}^i$ (see descent condition (2.1)).

3. Perform a *one-dimensional gradient-only line search* in direction $\mathbf{u}^i$: i.e. find λ_i that designates a sign change from negative to positive in the directional derivative:

$$F'(\lambda) = \nabla^{\mathrm{T}} f(\mathbf{x}^{i-1} + \lambda \mathbf{u}^i)\mathbf{u}^i,$$

to indirectly compute the minimizer, λ_i.

4. Set $\mathbf{x}^i = \mathbf{x}^{i-1} + \lambda_i \mathbf{u}^i$.

5. **Convergence test:** *if* $\|\mathbf{x}^i - \mathbf{x}^{i-1}\| < \varepsilon_1$ or $i > i_{max}$, then stop and $\mathbf{x}^* \cong \mathbf{x}^i$, *else* go to Step 6.

6. Set $i = i + 1$ and go to Step 2.

We turn our attention to the third step, which requires the minimum along the search direction to be indirectly resolved by finding a sign change in the derivative from negative to positive, as opposed to, by direct minimization of the function.

8.6.1 One-dimensional gradient-only line search

Clearly, in implementing the gradient-only descent algorithms as outlined above, requires the univariate problem along direction $\mathbf{u}^i$ to be solved:

Find λ_i that designates a sign change from negative to positive in the directional derivative:

$$F'(\lambda) = \nabla^{\mathrm{T}} f(\mathbf{x}^{i-1} + \lambda \mathbf{u}^i), \mathbf{u}^i,$$

to indirectly compute the minimizer, λ_i.

The problem of finding a sign change in the directional derivative, $F'(\lambda)$, is closely related to a one-dimensional root finding problem, which is usually conducted in two phases (Wilke et al. (2013b)). First by bracketing a sign change and then secondly by reducing the bracket size to refine the location of the sign change.

8.6.1.1 Gradient-only exact line searches

The bracketing phase only requires two points, as two points uniquely define a sign change from a negative directional derivative to a positive directional derivative. Given some user specified parameter, h, this bracket can be achieved by evaluating the directional derivative at

$$\lambda_i = ih + h, \quad i = 0, 1, 2, \ldots,$$

until a sign change from negative to positive is located. Alternatively, instead of using fixed interval sizes between consecutive points, the interval sizes between consecutive points can be increased:

$$\lambda_i = \sum_{k=0}^{i} ha^k, \quad i = 0, 1, 2, \ldots,$$

where h is an initial interval step size and a the interval growth parameter. Choosing $a \approx 1.618$ recovers the bracketing strategy often used for the popular golden section method used in line search descent (Arora (2004)).

Once an interval has been bracketed, with lower bound λ_L^0 and upper bound λ_U^0, that isolates a sign change in the directional derivative from negative to positive, that is $F'^A(\lambda_L^0) < 0$ and $F'^A(\lambda_U^0) > 0$, the interval is reduced to isolate the sign change within a specified tolerance ϵ.

The interval reduction can be done using a standard bisection approach by evaluating the directional derivative in the middle of successively bracketed intervals. Thus starting with $k = 1$ set middle value for $\lambda_M^{k-1} = \frac{\lambda_U^{k-1} + \lambda_L^{k-1}}{2}$ to give $F'^A(\lambda_M^{k-1})$.

If $F'^A(\lambda_M^{k-1}) < 0$ then

1. set $\lambda_L^k = \lambda_M^{k-1}$, and

2. set $\lambda_U^k = \lambda_U^{k-1}$,

while if, $F'^A(\lambda_M^{k-1}) > 0$, then

1. set $\lambda_L^k = \lambda_L^{k-1}$, and

2. set $\lambda_U^k = \lambda_M^{k-1}$.

This process is repeated, for $k = 1, 2, \ldots$, until $\frac{\lambda_U^k - \lambda_L^k}{2} < \epsilon$.

Note that the gradient-only bi-section interval is reduced by 50% every iteration, while the most efficient line search descent interval strategy, namely the golden section method, only reduces the interval by 38.2% at every iteration. Since the bracketed interval is efficiently reduced using interval bi-section, it is preferable to opt for a bracketing strategy for which the interval between successive points increases.

8.6.2 Conditions for sufficient improvement

Similar to the conditions for sufficient improvement discussed in Section 2.3.1.5, there are gradient-only conditions that can be utilized to indicate sufficient improvement and that can be used as a termination criteria for a line search strategy. Consider the following conditions that may be imposed on the step $\lambda_i \mathbf{u}^i$ in the direction $\mathbf{u}^i$ during the line search:

1. Predefined:

$$\lambda_i = d_i, \text{ where } d_i \text{ is prescribed at step } i$$

2. Descent:

$$\mathbf{u}^{i^T} \nabla f(\mathbf{x}^i + \lambda_i \mathbf{u}^i) \leq 0,$$

3. Curvature:

$$c_1 \mathbf{u}^{i^T} \nabla f(\mathbf{x}^i) \leq \mathbf{u}^{i^T} \nabla f(\mathbf{x}^i + \lambda_i \mathbf{u}^i),$$

4. Strong curvature:

$$|\mathbf{u}^{i^T} \nabla f(\mathbf{x}^i + \lambda_i \mathbf{u}^i)| \leq c_2 |\mathbf{u}^{i^T} \nabla f(\mathbf{x}^i)|,$$

5. Upper curvature:

$$\mathbf{u}^{i^T}\nabla f(\mathbf{x}^i + \lambda_i \mathbf{u}^i) \leq c_3 |\mathbf{u}^{i^T}\nabla f(\mathbf{x}^i)|,$$

with c_1, c_2 and c_3 required to be selected as non-negative parameters. These parameters control the degree to which the conditions are enforced. The simplest condition is a predefined strategy in which the step length evolution d_i is chosen *a priori* before the start of the optimization run and only depends on the iteration number i as detailed by Bertsekas (2015). A constant step length is popular amongst subgradient methods originally introduced by Shor et al. (1985). The other three strategies aims to assimilate information about the problem to inform step lengths. That is the step length depends on the gradient at the current point and the current search direction.

The descent condition ensures that the search direction remains a descent direction at the update. The disadvantage of such a condition is that the sign change is only approached from the left side. The curvature condition attempts to rectify this but may result in updates that are too large as any positive directional derivative satisfies this condition. The strong curvature condition in turn limits the largest update step size but do require the magnitude of the directional derivative to diminish as c_2 is reduced. This is sufficient for problems that are smooth in the vicinity of the optimum, whereas, it may be problematic at discontinuous solutions, i.e. it may be possible that no point along a search direction satisfies this condition. The upper curvature condition ensures that an update always exists, however, small step sizes also satisfy this condition. This can be circumvented by combining this condition with another condition that limits the minimum step size, e.g. using an *a priori* step length strategy.

8.7 Gradient-only sequential approximate optimization

In sequential approximate optimization (SAO) methods, the approximation functions used can easily be formulated using truncated second order Taylor expansions following Snyman and Hay (2001) and Groenwold et al. (2007). For the purposes of gradient-only optimization we aim here to approximate the gradient of a function $\nabla f(\mathbf{x})$ around some current iterate $\mathbf{x}^i$ to be given by

$$\nabla \tilde{f}^i(\mathbf{x}) = \nabla f(\mathbf{x}^i) + \mathbf{H}_i(\mathbf{x} - \mathbf{x}^i), \tag{8.13}$$

where, according to Wilke et al. (2010), some approximation of the curvature $\mathbf{H}_i$ using only gradient information at iteration i is required. Approximations for $\mathbf{H}_i$ are usually obtained by requiring the gradient to be recovered at the previous iteration, $i - 1$, where the gradient had been computed. Given the well-known secant equation,

$$\nabla f^i(\mathbf{x}^{i-1}) = \nabla f(\mathbf{x}^i) + \mathbf{H}_i(\mathbf{x}^{i-1} - \mathbf{x}^i),$$
$$\mathbf{H}_i(\mathbf{x}^{i-1} - \mathbf{x}^i) = \nabla f^i(\mathbf{x}^{i-1}) - \nabla f(\mathbf{x}^i),$$
$$\mathbf{H}_i \Delta \mathbf{x}^{i-1} = \Delta \nabla f^{i-1},$$

$$\tag{8.14}$$

where $\mathbf{H}_i$ in general requires n^2 components to be solved or $\frac{n^2-n}{2} + n$ components for a symmetric $\mathbf{H}_i$, where symmetry is guaranteed for twice continuously differentiable functions. Hence, to uniquely solve for $\mathbf{H}_i$ requires n^2 linear equations, but each gradient vector only contributes n equations towards the system of linear equations. Generalizing the secant equation results in the following system of equations from which to solve for $\mathbf{H}_i$,

$$\mathbf{H}_i \left[\Delta \mathbf{x}^{i-1}, \ldots, \Delta \mathbf{x}^{i-k} \right] = \left[\Delta \nabla f^{i-1}, \ldots, \Delta \nabla f^{i-k} \right]. \tag{8.15}$$

Consequently, by requiring $k = n$ unique gradient vectors to be recovered at the n previous iterates results in a linear system of equations that can be uniquely solved to yield $\mathbf{H}_i$. However, choosing $k < n$ results in an underdetermined system of equations to be solved, which can be regularized by requiring a minimum norm solution to $\mathbf{H}_i$. Alternatively, instead of solving for the full $\mathbf{H}_i$, a form for $\mathbf{H}_i$ can be assumed that requires less components to be solved for. Here, different assumptions regarding the form of $\mathbf{H}_i$ results in different assumptions on the curvature of the problem, and ultimately different approximation strategies. Typical forms include a constant diagonal Hessian matrix that implies constant curvature (also known as spherical approximations), general diagonal Hessian matrix implies that changes in curvature are aligned with the Cartesian coordinate axes (subset of separable problems), full non-symmetric Hessian matrix and a full symmetric Hessian matrix that assumes second order continuity.

Once $\mathbf{H}_i$ is approximated the current subproblem i is constructed and solved analytically since the subproblem is continuous by construction; the minimizer of subproblem i follows from setting the gradient of (8.13) equal to $\mathbf{0}$ to give the update

$$\mathbf{x}^{i*} = \mathbf{x}^i - (\mathbf{H}_i)^{-1}\nabla f(\mathbf{x}^i), \tag{8.16}$$

which can be solved from $\mathbf{H}_i(\mathbf{x}^{i*} - \mathbf{x}^i) = -\nabla f(\mathbf{x}^i)$. Solving a linear system may be computationally demanding when large systems are to be considered. Extending on the discussion in Section 2.3.2, this computational burden can be avoided when the inverse Hessian $\mathbf{G}_i = (\mathbf{H}_i)^{-1}$ is directly approximated. This then merely requires a matrix vector product

$$\mathbf{x}^{i*} = \mathbf{x}^i - \mathbf{G}_i\nabla f(\mathbf{x}^i) \tag{8.17}$$

to compute the update.

A general framework of gradient-only sequential approximation algorithms is listed in Algorithm 8.2.

8.7.1 Constant diagonal Hessian matrix approximations

Assuming the curvature can be described by a constant diagonal Hessian matrix results in a spherical approximation of the Hessian, which is approximated by a single scalar. As highlighted by Gould et al. (2005) this allows for a sparse description well suited for high-dimensional optimization problems. Hence, the approximate Hessian or curvature is of the form $\mathbf{H}_i = c_i\mathbf{I}$, with c_i a scalar, and $\mathbf{I}$ the identity matrix. This gives

$$\nabla \tilde{f}^i(\mathbf{x}) = \nabla f(\mathbf{x}^i) + c_i(\mathbf{x} - \mathbf{x}^i), \tag{8.18}$$

with the scalar curvature c_i unknown.

At $\mathbf{x} = \mathbf{x}^i$, the gradient of the function ∇f and the gradient of the approximation function $\nabla \tilde{f}$ match exactly. The approximate Hessian

---

Algorithm 8.2 Gradient-only sequential approximation algorithm.

Initialization: Given $\mathbf{x}^0$, select the real constant $\epsilon > 0$ and initial curvature $c_0 > 0$. Select the maximum number of iterations i_{max}. Set $\mathbf{H}_0 = \mathbf{I}$ or $\mathbf{G}_0 = \mathbf{I}$. Set $i := 0$ and perform the following steps:

1. **Gradient evaluation:** Compute $\nabla f(\mathbf{x}^i)$.

2. **Approximate optimization:** Construct a local approximate subproblem (8.13) at $\mathbf{x}^i$ using an appropriate approximation to $\mathbf{H}_i$ (or $\mathbf{G}_i$) obtained from only gradient information. Solve this subproblem analytically via (8.16) (or (8.17)) to arrive at a new candidate solution $\mathbf{x}^{i*}$.

3. **Move to the new iterate:** Set $\mathbf{x}^{i+1} := \mathbf{x}^{i*}$.

4. **Convergence test:** if $\|\mathbf{x}^{i+1} - \mathbf{x}^i\| \leq \epsilon$, OR $i = i_{max}$, stop.

5. **Initiate an additional outer loop:** Set $i := i + 1$ and go to Step 1.

$\mathbf{H}_i$ of the approximation $\tilde{f}$ is chosen to match additional information. c_i is obtained by matching the gradient vector at $\mathbf{x}^{i-1}$. Since only a single free parameter c_i is available, the n components of the respective gradient vectors are matched in a least square sense. The least squares error is given by

$$E^i = (\nabla \tilde{f}^i(\mathbf{x}^{i-1}) - \nabla f(\mathbf{x}^{i-1}))^{\mathrm{T}}(\nabla \tilde{f}^i(\mathbf{x}^{i-1}) - \nabla f(\mathbf{x}^{i-1})), \qquad (8.19)$$

which, after substitution of (8.18) into (8.19), gives

$$E^i = (\nabla f(\mathbf{x}^i) + c_i(\mathbf{x}^{i-1} - \mathbf{x}^i) - \nabla f(\mathbf{x}^{i-1}))^{\mathrm{T}}$$
$$(\nabla f(\mathbf{x}^i) + c_i(\mathbf{x}^{i-1} - \mathbf{x}^i) - \nabla f(\mathbf{x}^{i-1})). \quad (8.20)$$

Minimization of the least squares error E^i w.r.t. c_i then gives

$$\frac{dE^i}{dc_i} = (\nabla f(\mathbf{x}^i) + c_i(\mathbf{x}^{i-1} - \mathbf{x}^i)$$
$$- \nabla f(\mathbf{x}^{i-1}))^{\mathrm{T}}(\mathbf{x}^{i-1} - \mathbf{x}^i)$$
$$+ (\mathbf{x}^{i-1} - \mathbf{x}^i)^{\mathrm{T}}(\nabla f(\mathbf{x}^i)$$
$$+ c_i(\mathbf{x}^{i-1} - \mathbf{x}^i) - \nabla f(\mathbf{x}^{i-1})) = 0, \quad (8.21)$$

hence
$$c_i = \frac{(\mathbf{x}^{i-1} - \mathbf{x}^i)^{\mathrm{T}}(\nabla f(\mathbf{x}^{i-1}) - \nabla f(\mathbf{x}^i))}{(\mathbf{x}^{i-1} - \mathbf{x}^i)^{\mathrm{T}}(\mathbf{x}^{i-1} - \mathbf{x}^i)}. \tag{8.22}$$

The approximation (8.18) can be enforced to be strictly convex by enforcing $c_i = \max(\beta, c_i)$, with $\beta > 0$ small and prescribed.

8.7.2 Diagonal Hessian matrix approximations

A more general separable approximation is obtained by allowing the Hessian matrix to develop into a diagonal matrix allowing for n coefficients to be solved for. Hence, the approximate Hessian or curvature is of the form $\mathbf{D}_i$, with $\mathbf{D}$ signifying a diagonal matrix. This gives

$$\nabla \tilde{f}^i(\mathbf{x}) = \nabla f(\mathbf{x}^i) + \mathbf{D}_i(\mathbf{x} - \mathbf{x}^i), \tag{8.23}$$

with $D_{i_{jk}}$, $j = k$ unknown for all $j = 1, \dots, n$ and $k = 1, \dots, n$, while $D_{i_{jk}} = 0$ for $j \neq k$.

At $\mathbf{x} = \mathbf{x}^i$, the gradients of the function f and the gradient of the approximation function $\nabla \tilde{f}$ match exactly. Again, the approximate Hessian $\mathbf{D}_i$ of the approximation $\tilde{f}$ is chosen to match additional information. $\mathbf{D}_i$ is obtained by matching the gradient vector at $\mathbf{x}^{i-1}$. Since $\mathbf{D}_i$ has n unknowns that are separable, the n components of the gradient vector are matched exactly,

$$\nabla f(\mathbf{x}^{i-1}) = \nabla f(\mathbf{x}^i) + \mathbf{D}_i(\mathbf{x}^{i-1} - \mathbf{x}^i), \tag{8.24}$$

with each component solved for independently

$$D_{i_{jj}} = \frac{\nabla f_j(\mathbf{x}^{i-1}) - \nabla f_j(\mathbf{x}^i)}{(x_j^{i-1} - x_j^i)}, \; j = 1, \dots, n. \tag{8.25}$$

8.7.3 Symmetric Hessian matrix approximations

Instead of approximating the entire $\mathbf{H}_i$ at every iteration, every iteration can add information to a previous approximation $\mathbf{H}_{i-1}$. Following conventional Quasi-Newton derivations consider the following Hessian update scheme

$$\mathbf{H}_i = \mathbf{H}_{i-1} + \Delta \mathbf{H}_{i-1}. \tag{8.26}$$

The rank of the incremental update $\Delta\mathbf{H}_{i-1}$ depends on the number of difference gradient vectors enforced per iteration. Assuming, $\Delta\mathbf{H}_{i-1} = \mathbf{a}^{i-1}(\mathbf{b}^{i-1})^{\mathrm{T}}$, which is a rank-1 update, then by substituting $\Delta\mathbf{H}_{i-1} = \mathbf{a}^{i-1}(\mathbf{b}^{i-1})^{\mathrm{T}}$ into (8.14) we obtain

$$\mathbf{H}_i\Delta\mathbf{x}^{i-1} = \left(\mathbf{H}_{i-1} + \mathbf{a}^{i-1}(\mathbf{b}^{i-1})^{\mathrm{T}}\right)\Delta\mathbf{x}^{i-1} = \Delta\nabla f^{i-1}, \qquad (8.27)$$

from which $\mathbf{a}^{i-1}$ can be solved

$$\mathbf{a}^{i-1} = \frac{\Delta\nabla f^{i-1}}{(\mathbf{b}^{i-1})^{\mathrm{T}}\Delta\mathbf{x}^{i-1}} - \frac{\mathbf{H}_{i-1}\Delta\mathbf{x}^{i-1}}{(\mathbf{b}^{i-1})^{\mathrm{T}}\Delta\mathbf{x}^{i-1}}. \qquad (8.28)$$

Reconstructing $\mathbf{H}_i$ from (8.26) and (8.28) we obtain

$$\mathbf{H}_i = \mathbf{H}_{i-1} + \left(\frac{\Delta\nabla f^{i-1}}{(\mathbf{b}^{i-1})^{\mathrm{T}}\Delta\mathbf{x}^{i-1}} - \frac{\mathbf{H}_i\Delta\mathbf{x}^{i-1}}{(\mathbf{b}^{i-1})^{\mathrm{T}}\Delta\mathbf{x}^{i-1}}\right)(\mathbf{b}^{i-1})^{\mathrm{T}}, \qquad (8.29)$$

with $\mathbf{b}^{i-1}$ free to be chosen. By choosing $\mathbf{b}^{i-1} = (\Delta\nabla f^{i-1} - \mathbf{H}_{i-1}\Delta\mathbf{x}^{i-1})$ symmetry is enforced

$$\mathbf{H}_i = \mathbf{H}_{i-1} +$$
$$\left(\frac{(\Delta\nabla f^{i-1} - \mathbf{H}_{i-1}\Delta\mathbf{x}^{i-1})(\Delta\nabla f^{i-1} - \mathbf{H}_{i-1}\Delta\mathbf{x}^{i-1})^{\mathrm{T}}}{(\Delta\nabla f^{i-1} - \mathbf{H}_{i-1}\Delta\mathbf{x}^{i-1})^{\mathrm{T}}\Delta x^{i-1}}\right), \qquad (8.30)$$

which yields the symmetric rank-1 update investigated by Conn et al. (1991), subject to starting with an initial symmetric matrix $\mathbf{H}_0$.

8.7.4 Symmetric inverse Hessian matrix approximations

Approximating the inverse Hessian $\mathbf{G}_i$ allows for the search direction to be computed using a matrix-vector multiplication as opposed to solving a linear system of equations. $\mathbf{G}_i$ can be approximated incrementally by adding information per iteration to a previous approximation $\mathbf{G}_{i-1}$. Consider the following inverse Hessian update scheme

$$\mathbf{G}_i = \mathbf{G}_{i-1} + \Delta\mathbf{G}_{i-1}, \qquad (8.31)$$

substituted into (8.14) and restructured to reflect the inverse Hessian, we obtain

$$\Delta\mathbf{x}^{i-1} = (\mathbf{G}_{i-1} + \Delta\mathbf{G}_{i-1})\Delta\nabla f^{i-1}. \qquad (8.32)$$

Assume $\Delta \mathbf{G}_{i-1} = \mathbf{a}^{i-1}(\mathbf{b}^{i-1})^{\mathrm{T}}$, which when substituted into (8.32) gives

$$\Delta \mathbf{x}^{i-1} = \mathbf{G}_{i-1}\Delta \nabla f^{i-1} + \mathbf{a}^{i-1}\left((\mathbf{b}^{i-1})^{\mathrm{T}}\Delta \nabla f^{i-1}\right), \qquad (8.33)$$

from which $\mathbf{a}^{i-1}$ is isolated to obtain

$$\mathbf{a}^{i-1} = \frac{\Delta \mathbf{x}^{i-1}}{(\mathbf{b}^{i-1})^{\mathrm{T}}\Delta \nabla f^{i-1}} - \frac{\mathbf{G}_{i-1}\Delta \nabla f^{i-1}}{(\mathbf{b}^{i-1})^{\mathrm{T}}\Delta \nabla f^{i-1}}. \qquad (8.34)$$

Rewriting (8.31) in terms of only $\mathbf{b}^{i-1}$ gives

$$\mathbf{G}_i = \mathbf{G}_{i-1} + \left(\frac{\Delta x^{i-1}}{(\mathbf{b}^{i-1})^{\mathrm{T}}\Delta \nabla f^{i-1}} - \frac{\mathbf{G}_{i-1}\Delta \nabla f^{i-1}}{(\mathbf{b}^{i-1})^{\mathrm{T}}\Delta \nabla f^{i-1}}\right)(\mathbf{b}^{i-1})^{\mathrm{T}}. \quad (8.35)$$

By choosing $(\mathbf{b}^{i-1})^{\mathrm{T}} = (\Delta \mathbf{x}^{i-1} - \mathbf{G}_{i-1}\Delta \nabla f^{i-1})$ and substituting the result into (8.35) we obtain

$$\mathbf{G}_{i+1} = \mathbf{G}_i + \left(\frac{(\Delta \mathbf{x}^i - \mathbf{G}_i\Delta \nabla f^i)(\Delta \mathbf{x}^i - \mathbf{G}_i\Delta \nabla f^i)^{\mathrm{T}}}{(\Delta x^i - \mathbf{G}_i\Delta \nabla f^i)^{\mathrm{T}}\Delta \nabla f^i}\right), \qquad (8.36)$$

yielding a symmetric update. This specific update was developed by Fletcher and Powell (1963) and is an adaptation of an original procedure first proposed by Davidon (1959).

8.7.5 Non-symmetric inverse Hessian matrix approximations

More generally the Hessian matrix can be approximated as a non-symmetric Hessian matrix approximation. In this section we demonstrate that conventional conjugate gradient directions imply a non-symmetric Hessian matrix. Using the conventional starting point for conjugate gradient directions as outlined in Section 2.3.2, we express a new search direction $\mathbf{u}^i$ as a linear combination of the gradient descent vector computed at the current minimum point $\mathbf{x}^i$, and the previous search direction $\mathbf{u}^{i-1}$. This then gives

$$\mathbf{u}^i = -\nabla f(\mathbf{x}^i) + \beta_i \mathbf{u}^{i-1}, \qquad (8.37)$$

for which we now only have to solve for the scalar β_i such that $\mathbf{u}^i$ is indeed mutually conjugate to the other search directions w.r.t. an assumed matrix $\mathbf{H}_i$. The proposed update formula by Fletcher and Reeves (1964) is given by

$$\beta_i = \frac{\nabla^{\mathrm{T}} f(\mathbf{x}^i) \nabla f(\mathbf{x}^i)}{\nabla^{\mathrm{T}} f(\mathbf{x}^{i-1}) \nabla f(\mathbf{x}^{i-1})}, \tag{8.38}$$

with Polak and Ribiere (1969) proposing an alternative conjugate gradient update

$$\beta_i = \frac{\nabla^{\mathrm{T}} f(\mathbf{x}^i) \left(\nabla f(\mathbf{x}^i) - \nabla f(\mathbf{x}^{i-1}) \right)}{\nabla^{\mathrm{T}} f(\mathbf{x}^{i-1}) \nabla f(\mathbf{x}^{i-1})}. \tag{8.39}$$

Substituting (8.38) into (8.37) we obtain

$$\mathbf{x}^{i+1} = \mathbf{x}^i + \lambda_i \left(-\nabla f(\mathbf{x}^i) + \frac{\nabla^{\mathrm{T}} f(\mathbf{x}^i) \nabla f(\mathbf{x}^i)}{\nabla^{\mathrm{T}} f(\mathbf{x}^{i-1}) \nabla f(\mathbf{x}^{i-1})} \mathbf{u}^{i-1} \right). \tag{8.40}$$

By factoring $\nabla f(\mathbf{x}^i)$ out of the (8.40) we obtain

$$\mathbf{x}^{i+1} = \mathbf{x}^i + \lambda_i \left(-\mathbf{I} + \frac{\nabla \mathbf{u}^{i-1} \nabla^{\mathrm{T}} f(\mathbf{x}^i)}{\nabla^{\mathrm{T}} f(\mathbf{x}^{i-1}) \nabla f(\mathbf{x}^{i-1})} \right) \nabla f(\mathbf{x}^i). \tag{8.41}$$

By comparing (8.16) with (8.41), we see that (8.41) approximates the inverse of the Hessian matrix

$$(\mathbf{H}_i)^{-1} = \lambda_i \left(\mathbf{I} - \frac{\nabla \mathbf{u}^{i-1} \nabla^{\mathrm{T}} f(\mathbf{x}^i)}{\nabla^{\mathrm{T}} f(\mathbf{x}^{i-1}) \nabla f(\mathbf{x}^{i-1})} \right), \tag{8.42}$$

which by inspection reveals that $(\mathbf{H}_i)^{-1}$ is not symmetric, since $\nabla \mathbf{u}^{i-1} \neq \nabla f(\mathbf{x}^i)$, which implies that $\mathbf{H}_i$ is also not symmetric.

8.7.6 Trust Region Methods and Conservatism

Strategies are required to ensure that sequential approximate optimization methods will terminate and converge. A priority therefore is to ensure that the constructed approximation yields a sufficiently accurate solution. Towards this aim Goldfeld et al. (1966) proposed restricting the step size based on the validity of the approximation over a domain, which was later coined by Sorensen (1982) as the well-known trust region methods. As a modern alternative to trust region methods, conservatism was proposed by Svanberg (2002). Conservatism requires each proposed update to be feasible as well as an improvement to the previous iterate. The benefit of both approaches is that strong convergence characteristics of the sequential approximation approaches can be proved albeit for often highly restricted classes of functions.

In an effort to enforce conservatism within the context of gradient-only approaches, Wilke et al. (2010) suggested that the directional derivative of the actual problem at the proposed approximate solution along the update step direction should be negative. At iterate i, the proposed solution $\mathbf{x}^{i*}$ is obtained by taking the update step $\mathbf{x}^{i*} - \mathbf{x}^i$ from the previous solution $\mathbf{x}^i$. This update represents descent of $f(\mathbf{x})$ along the direction $\mathbf{x}^{i*} - \mathbf{x}^i$ if

$$\nabla^{\mathrm{T}} f(\mathbf{x}^{i*})(\mathbf{x}^{i*} - \mathbf{x}^i) \leq \nabla^{\mathrm{T}} \tilde{f}(\mathbf{x}^{i*})(\mathbf{x}^{i*} - \mathbf{x}^i) = 0. \qquad (8.43)$$

Accordingly, any gradient-only approximation may be defined as conservative if (8.43) holds.

This gradient-only definition of conservatism is similar in intent to that of Svanberg's function value based definition that requires that the function value $f(\mathbf{x}^{i*})$ improve on that of the previous iterate. In the gradient-only approach only updates $\mathbf{x}^{i*}$ for which the genuine quality measure,

$$\nabla^{\mathrm{T}} f(\mathbf{x}^{i*})(\mathbf{x}^{i*} - \mathbf{x}^i),$$

is less than or equal to the approximated quality measure

$$\nabla^{\mathrm{T}} \tilde{f}(\mathbf{x}^{i*})(\mathbf{x}^{i*} - \mathbf{x}^i),$$

are accepted. Although no formal proofs are presented here, Wilke et al. (2013b) showed that this definition of conservatism guarantees convergence for certain classes of functions, e.g. smooth convex functions. It is also important to note that for non-smooth and discontinuous functions in general this definition falls short, and is not sufficient to guarantee convergence. It is important to note that, although strong theoretical evidence is lacking, this gradient-only definition of conservatism suffices in general to achieve convergence for practical engineering problems. In sequential approximate optimization, termination and convergence may be affected through this notion of conservatism. Therefore the minimizer

---

Algorithm 8.3 Affecting conservatism in gradient-only sequential approximate optimization using constant diagonal (8.18) Hessian matrix approximations.

Initialization: Given $\mathbf{x}^0$, select the real constant $\epsilon > 0$, initial curvature $c_0 > 0$ and conservatism parameter $\gamma > 1$. Select the maximum number of iterations i_{max}. Set $i := 0$, $l := 0$ and perform the following steps:

1. **Gradient evaluation:** Compute $\nabla f(\mathbf{x}^i)$.

2. **Approximate optimization:** Construct local approximate sub-problem (8.18) at $\mathbf{x}^i$. Solve this subproblem analytically, to arrive at $\mathbf{x}^{i*}$.

3. **Evaluation:** Compute $\nabla f(\mathbf{x}^{i*})$.

4. **Test if $\mathbf{x}^{i*}$ is acceptable:** if (8.43) is satisfied, go to Step 6.

5. **Initiate an inner loop to effect conservatism:**

   (a) Set $l := l + 1$.

   (b) Set $c_i := \gamma c_i$.

   (c) Goto Step 2.

6. **Move to the new iterate:** Set $\mathbf{x}^{i+1} := \mathbf{x}^{i*}$.

7. **Convergence test:** if $\|\mathbf{x}^{i+1} - \mathbf{x}^i\| \leq \epsilon$, OR $i = i_{max}$, stop.

8. **Initiate an additional outer loop:** Set $i := i + 1$ and go to Step 1.

---

of the subproblem $\mathbf{x}^{i*}$ is accepted i.e. $\mathbf{x}^{i+1} := \mathbf{x}^{i*}$ only if $\mathbf{x}^{i*}$ is found to be a gradient-only conservative point. This modification to the gradient-only sequential approximate optimization is listed Algorithm 8.3.

8.8   Gradient-only optimization problem

An important consideration for a holistic understanding of gradient-only approaches, is to understand the characteristics of the designs to which gradient-only strategies converge. This would allow us to better differentiate gradient-only strategies from conventional minimization strategies when step discontinuous functions are considered. We therefore formally define the underlying optimization problem that is consistent with solution strategies that only consider gradient-only information.

Reconsider the derivatives presented in Figures 8.5 (b) and (d), that depict the smooth and piece-wise smooth step discontinuous derivative responses when the same problem is numerically integrated using different numerical strategies. Although conventional interpretations of the smooth derivative function highlights the design with zero slope and the lack thereof for the step discontinuous derivative function, there is a consistent interpretation between the smooth and step discontinuous derivative functions depicted in Figures 8.5 (b) and (d). This interpretation acknowledges that the smooth and step discontinuous derivative functions both change sign from negative to positive only once as λ increases. Therefore, if we define this point as the solution to the optimization problem then we have (i) a unique solution that is defined for both derivative functions based solely on first order information and (ii) the solution defines a minimum when estimated from only first order information as second order (curvature) information is implied by requiring the sign to change from negative to positive with increasing λ.

As illustration of (ii), consider Figure 8.5 (d). The sign change from negative to positive as λ increases is at $\lambda_g^* \approx 0.57$. Consider any point, λ_v to the left of λ_g^* i.e. $\lambda_v < \lambda_g^*$, then the direction is given by $d_v = \lambda_v - \lambda_g^* < 0$. The directional derivative is given by the projection of the derivative $\frac{dE(\lambda_v)}{d\lambda}$ computed at λ_v onto d_v, i.e. by $d_v \frac{dE(\lambda_v)}{d\lambda}$. Since $\frac{dE(\lambda_v)}{d\lambda} < 0$ for $\lambda_v < \lambda_g^*$, the directional derivative is positive. Similarly, the directional derivative for λ_v to the right of λ_g^* i.e. $\lambda_v > \lambda_g^*$ is also only positive. This implies that first order information estimates the function value to only increase irrespective of the direction of departure from λ_g^*. In contrast, when considering Figures 8.5 (a) and (c) it is evident that the function decreases but only as a result of step discontinuities and not because of the trends of the piece-wise smooth sections indicating descent. We define λ_g^* as a strict *non-negative associated gradient (or derivative) projection point* (Wilke et al. (2013b)). This requires the directional derivative at any point λ_v to be positive, where λ_v is in the vicinity of λ_g^*, and the direction defined by $\lambda_v - \lambda_g^*$.

In general, given a real-valued function $f : X \subset \mathbb{R}^n \rightarrow \mathbb{R}$, the general unconstrained gradient-only optimization problem is to find a *non-negative associated gradient projection point* $\mathbf{x}_g^* \in X$ such that for every $\mathbf{u} \in \{\mathbf{y} \in \mathbb{R}^n \mid \|\mathbf{y}\| = 1\}$ there exists a real number $r_u > 0$ for which the following holds:

$$\nabla_A^{\mathrm{T}} f(\mathbf{x}_g^* + \lambda \mathbf{u})\mathbf{u} \geq 0 \ \forall \ \lambda \in (0, r_u].$$

This allows us to determine distinct candidate solutions to an unconstrained gradient-only optimization problem. It is important to note that when multiple candidate solutions exist additional information may be required, after obtaining these solutions using only first order information, in order to uniquely obtain the best solution.

8.9 Exercises

The reader is encouraged to employ a convenient computing environment to complete the exercises. With Python being freely available it is recommended to be used as outlined in Chapter 9.

8.9.1 Consider the Lotka-Volterra system with unknown parameter λ:

$$\frac{dz(z, y, t)}{dt} = (1 - \lambda)z(t) - 0.3z(t)y(t)$$
$$\frac{dy(z, y, t)}{dt} = z(t)y(t) - y(t),$$

with the two initial conditions $z(0) = 0.9$ and $y(0) = 0.9$ integrated using the forward Euler scheme, over an eight second interval, using 50 000 equally spaced time steps. Given $z(8) = 0.722962$ and $y(8) = 1.110567$ plot the sum of the errors squared objective function given in (8.3) for λ between 0 and 1 using 101 equally spaced points.

8.9.2 Consider the Lotka-Volterra system with unknown parameter λ:

$$\frac{dz(z, y, t)}{dt} = (1 - \lambda)z(t) - 0.3z(t)y(t)$$

$$\frac{dy(z, y, t)}{dt} = z(t)y(t) - y(t),$$

with the two initial conditions $z(0) = 0.9$ and $y(0) = 0.9$ integrated using the forward Euler scheme, over an eight second interval, that starts with 50 equally spaced time steps and adding 10 time steps every time the computation is done for a new λ. Given $z(8) = 0.722962$ and $y(8) = 1.110567$ plot the sum of the errors squared objective function given in (8.3) for λ between 0 and 1 using 101 equally spaced points.

8.9.3 Optimize the problem outlined in Exercise 8.9.1 using the Golden section strategy and the gradient-only bisection approach using 100 random starts each. Use the same random starting points for the two line search strategies. Compare the obtained results in terms of the apparent optimal λ, required number of function and derivative evaluations to solve the problem 100 times.

8.9.4 Optimize the problem outlined in Exercise 8.9.2 using the Golden section strategy and the gradient-only bisection approach using 100 random starts each. Use the same random starting points for the two line search strategies. Compare the obtained results in terms of the apparent optimal λ, required number of function and derivative evaluations to solve the problem 100 times.

8.9.5 Critically compare the results obtained in Exercises 8.9.3 and 8.9.4.

8.9.6 Consider the Lotka-Volterra system with unknown parameters λ and δ:

$$\frac{dz(z, y, t)}{dt} = (1 - \lambda)z(t) - 0.3z(t)y(t)$$

$$\frac{dy(z, y, t)}{dt} = \delta z(t)y(t) - y(t),$$

with the two initial conditions $z(0) = 0.9$ and $y(0) = 0.9$ integrated using the forward Euler scheme, over an eight second interval, using 50 000 equally spaced time steps. Given $z(8) = 0.722962$ and $y(8) = 1.110567$ plot the sum of the errors squared objective function given in (8.3) for λ between 0 and 1 and δ between 0.5 and 1.5 using 101 equally spaced points.

8.9.7 Consider the Lotka-Volterra system with unknown parameters λ and δ:

$$\frac{dz(z, y, t)}{dt} = (1 - \lambda)z(t) - 0.3z(t)y(t)$$

$$\frac{dy(z, y, t)}{dt} = \delta z(t)y(t) - y(t),$$

with the two initial conditions $z(0) = 0.9$ and $y(0) = 0.9$ integrated using the forward Euler scheme, over an eight second interval, that starts with 50 equally spaced time steps and adding 10 time steps every time the computation is done for a new (λ, δ) pair. Given $z(8) = 0.722962$ and $y(8) = 1.110567$ plot the sum of the errors squared objective function given in (8.3) for λ between 0 and 1 and δ between 0.5 and 1.5 using 101 equally spaced points.

8.9.8 Optimize the problem outlined in Exercise 8.9.6 using the Golden section strategy and the gradient-only bisection approach using 100 random starts each. Use the same random starting points for the two line search strategies. Compare the obtained results in terms of the apparent optimal λ, required number of function and derivative evaluations to solve the problem 100 times.

8.9.9 Optimize the problem outlined in Exercise 8.9.7 using the Golden section strategy and the gradient-only bisection approach using 100 random starts each. Use the same random starting points for the two line search strategies. Compare the obtained results in terms of the apparent optimal λ, required number of function and derivative evaluations to solve the problem 100 times.

8.9.10 Consider the Lotka-Volterra system with four unknown parameters λ, δ, β and γ :

$$\frac{dz(z, y, t)}{dt} = (1 - \lambda)z(t) - \beta z(t)y(t)$$

$$\frac{dy(z, y, t)}{dt} = \delta z(t)y(t) - \gamma y(t),$$

with the two initial conditions $z(0) = 0.9$ and $y(0) = 0.9$ integrated using the forward Euler scheme, over an eight second interval, that starts with 50 equally spaced time steps. For every 10 evaluations of a $(\lambda, \delta, \beta, \gamma)$ increase the number of time steps by 10. Given $z(8) = 0.722962$ and $y(8) = 1.110567$ solve the problem to find the optimal $(\lambda, \beta, \delta$ and $\gamma)$ using random starting points between 0 and 1 for, λ and β, and between 0.5 and 1.5 for δ and γ.

8.9.11 Construct two third order polynomial approximations of the objective in Exercise 8.9.1. For the first approximation use only zero order information, while for the second approximation use only first order information. Compare the two approximations with each other.

8.9.12 Construct two third order polynomial approximations of the objective in Exercise 8.9.2. For the first approximation use only zero order information, while for the second approximation use only first order information. Compare the two approximations with each other.

8.9.13 Optimize the piece-wise smooth step discontinuous quadratic function given by (7.14) for $n = 2$ using a gradient-only symmetric Hessian matrix approximation. Compare the obtained optimum against the graphical solution of the problem.

8.9.14 Optimize the piece-wise smooth step discontinuous quadratic function given by (7.14) for $n = 4$ using a gradient-only symmetric Hessian matrix approximation.

8.9.15 Compare the symmetric Hessian matrix approximation in Exercises 8.9.13 and 8.9.14 against the actual Hessian of the piece-wise smooth step discontinuous quadratic function.

Chapter 9

PRACTICAL COMPUTATIONAL OPTIMIZATION USING PYTHON

9.1 Introduction to Python

Python is a general purpose computer programming language. An experienced programmer in any procedural computer language can learn Python very quickly. Python is remarkable in that it is designed to allow new programmers to efficiently master programming. The choice of including *Anaconda Python* for application of our mathematical programming concepts is motivated by the fact that *Anaconda Python* supports both symbolic and numerical mathematical operations as part of the installation. Python allows for an intuitive engagement with numerical computations. It is freely available and allows for additional functionality to be developed and extended. All algorithms in this text are made available in Python so as to allow the reader the use of the developed algorithms from the onset. This chapter is not an exhaustive

© Springer International Publishing AG, part of Springer Nature 2018
J.A. Snyman and D.N. Wilke, *Practical Mathematical Optimization,*
Springer Optimization and Its Applications 133,
https://doi.org/10.1007/978-3-319-77586-9_9

treatise on Python and programming in general, but rather the minimum subset of Python required to implement formulated optimization problems and to solve them.

Only selected output of listed codes is presented in this chapter. The premise of withholding the output of some of the Python programs is to encourage an active participation by the reader when reading this chapter.

9.1.1 Installing *Anaconda Python*

Anaconda Python is freely available by following the link [https://www.continuum.io/downloads].

Simply follow the installation instructions and use either Spyder or Jupyter-Notebooks to write and execute your Python programs. All developed programming material is made available as Python files with file extensions `.py` as well as Jupyter-Notebooks with file extensions `.ipynb`.

9.2 Overview

Python is known as an object-orientated programming language which for our purposes implies that every symbol, number, vector, matrix or function we create is an object, where objects have functions that perform specific operations on that object. For example the Python code for computing the following numerical calculation with complex numbers:

```
complex_number = 4 + 9j + 12j - 3
```

returns `1+21j` as an answer and assigns it the name `complex_number`. By typing `complex_number` followed by a dot (.) and pressing the `TAB` key presents the following list of operations:

```
complex_number.conjugate
complex_number.imag
complex_number.real
```

Typing `complex_number.conjugate()` returns 1-21j, whereas the imaginary component is accessible by typing `complex_number.imag` to return the imaginary part of the complex number and `complex_number.real` to return the real part. Hence, an object is both the data as well as relevant functions that operate on it combined into a single package.

Python as a stand-alone language has only limited functionality which *Anaconda Python* extends with additional capability through *modules*. Modules are objects that store functions without any data. Modules need to be imported into memory, whereafter the functionality can be accessed using the . followed by the **TAB** strategy. Consider the following two approaches to import *cos* and *sin* from the module `math`:

```
import math as m
print(m.cos(3.14))
```

or alternatively:

```
from math import cos, sin
print(sin(3.14))
```

Modules may also include submodules that in turn store functions. Consider the function **rand** that is available in the submodule **random** stored inside the module `numpy`:

```
from numpy.random import rand
print(rand())
```

Once a module, submodule, function or object has been loaded into memory, additional information is available via the `help(.)` function. Examples include `help(numpy)` and `help(rand)`. After reading the available help for the **rand** function it is clear that an array of five random numbers is generated by:

```
print(rand(5))
```

A vector of numbers can be stored as a *list* of numbers using square brackets and separating numbers by commas:

```
list_numbers = [1, 4, 9, 25, 36]
```

or as a *numpy* array of numbers by supplying a list of numbers as input to the `numpy.array` function:

```
from numpy import array
array_numbers = array([1, 4, 9, 25, 36])
```

The difference between the list and numpy array is that the list of numbers has limited support for mathematical operations. For example adding two lists results in the two lists being concatenated, while multiplication between two lists is not defined. Multiplying a list by an integer n concatenates the list with it self n times. In turn, a numpy array supports element-wise operations, i.e. multiplying two arrays of equal length results in an array in which corresponding elements are multiplied with each other. Adding two arrays results in corresponding elements being added together. Individual entries in a list or numpy array can be accessed using square brackets that encapsulates the index number that follows the name of the list or array, e.g. `array_numbers[0]` returns the first value stored in the array. Thus the index number corresponding to the first entry in the array is 0, while for the second entry in the array the index number is 1, and for the third entry the index is 2, etc.

The `plot` function in the submodule `matplotlib.pyplot` allows for visualization of both lists and arrays as illustrated by the following two plots:

```
1  import matplotlib.pyplot as plt
2  plt.figure(1)
3  plt.plot(list_numbers)
4  plt.show()
5
6  plt.figure(2)
7  plt.plot(array_numbers)
8  plt.show()
```

where the function `plt.figure(.)` specifies a figure on which a plot can be drawn using `plt.plot(.)`, with the figure then being displayed by `plt.show()`.

The standard Python modules we will be considering in this chapter are

1. `math` - numeric Python using scalars,

2. `sympy` - *sym*bolic *Py*thon,

3. `scipy` - *sci*entific *Py*thon,

4. `numpy` - *num*eric *Py*thon using arrays,

5. `matplotlib` - *Mat*lab *plot*ting *lib*rary for numerical plotting with arrays.

In addition, especially to accompany the new algorithms developed in this book, we make available the Python module `pmo` that can be electronically downloaded from Springer Extra Materials (extras.springer.com/2018).

9.2.1 Basic Arithmetic

1. Multiplication

```
print (3*5)
```

2. Division

```
print (3/5)
```

3. Power

```
print (3**5)
```

9.3 Symbolic Mathematics

In addition to numeric computations, Python allows for symbolic computations to be conducted. This allows for analytical operations such as differentiation and integration to be performed using Python. Consider the one-dimensional quadratic function

$$F(\lambda) = \lambda^2, \tag{9.1}$$

which can easily be visualised using symbolic mathematics in the `sympy` package. We first need to define a variable using the function `symbols`. We can then define a symbolic function that we can then plot over a specified domain using the following code:

```
from sympy import symbols
from sympy.plotting import plot
```

```
4  L = symbols('L')
5  F = L**2
6
7  plot(F,(L,1,6))
```

The result of the two-dimensional visualisation function `plot` is depicted in Figure 9.1.

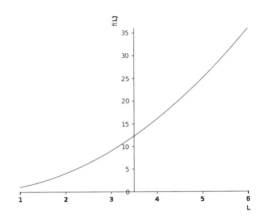

Figure 9.1: Symbolic Python plot of univariate quadratic function

In addition to visualizing univariate functions `sympy` supports the visualization of two-dimensional functions. Consider the two-dimensional Rosenbrock function

$$f(\mathbf{x}) = 100(x_0^2 - x_1)^2 + (x_0 - 1)^2. \tag{9.2}$$

We first need to define variables using the function `symbols` followed by a symbolic representation of the Rosenbrock function that we can then plot over a specified domain using the following code:

```
1  from sympy import symbols
2  from sympy.plotting import plot3d
3
4  x0,x1 = symbols('x0,x1')
5  f = 100*(x0**2 - x1)**2 + (x0-1)**2
6
7  plot3d(f,(x0,-1.5,2),(x1,-0.5,3))
```

The result of the three-dimensional visualisation function `plot3d` is depicted in Figure 9.2.

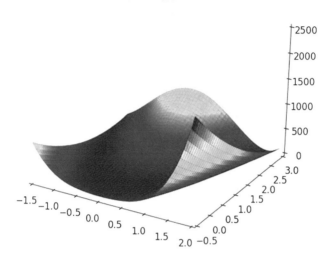

Figure 9.2: Symbolic Python plotted Rosenbrock function

9.3.1 Differentiation

The gradient vector and Hessian matrix of the defined Rosenbrock function can subsequently be computed using the symbolic differentiation function `diff` as follows:

```
from sympy import symbols, diff

x0,x1 = symbols('x0,x1')
f = 100*(x0**2 - x1)**2 + (x0-1)**2

print('First partial derivative',diff(f,x0))
print('Second partial derivative',diff(f,x1))

print('Hessian matrix d/dx0(df/dx0)',diff(f,x0,x0))
```

```
10 print('Hessian matrix d/dx1(df/dx1)',diff(f,x1,x1))
11 print('Hessian matrix d/dx1(df/dx0)',diff(f,x0,x1))
12 print('Hessian matrix d/dx0(df/dx1)',diff(f,x1,x0))
```

Since the Hessian is symmetric for twice continuously differentiable functions, the last two computed elements of the Hessian matrix are equal.

The function `derive_by_array` in `sympy` allows us to compute the gradient vector and Hessian matrix in more compact form as follows:

```
1 from sympy import symbols, diff, derive_by_array,
    pretty_print
2
3 x0,x1 = symbols('x0,x1')
4 f = 100*(x0**2 - x1)**2 + (x0-1)**2
5 x = [x0, x1]
6
7 gradf = derive_by_array(f,x)
8 hessian = derive_by_array(gradf,x)
9
10 print('The gradient vector is:')
11 pretty_print(gradf)
12
13 print('The Hessian matrix is:')
14 pretty_print(hessian)
```

The usage of `pretty_print` allows for the gradient vector and Hessian matrix to be displayed in the more readable format:

```
The gradient vector is:
⎡        ⎛ 2      ⎞                    2              ⎤
⎣400·x₀·⎝x₀  - x₁⎠ + 2·x₀ - 2  - 200·x₀  + 200·x₁⎦
The Hessian matrix is:
⎡        2                           ⎤
⎢1200·x₀  - 400·x₁ + 2   -400·x₀ ⎥
⎢                                    ⎥
⎣     -400·x₀              200    ⎦
```

9.3.2   Numerical Evaluation

The symbolic expression for the Rosenbrock function can numerically be evaluated using the substitution function `subs` as follows:

```
1 print('Rosenbrock function evaluated at x0=5,x1=1 gives',f.
    subs({x0:5,x1:1}))
```

```
2 print('Rosenbrock function evaluated at x0=1,x1=1 gives',f.
      subs({x0:0,x1:0}))
```

The statement `name1:value1,name2:value2` is referred to as a dictionary in Python. The `name` is used to conveniently identify the associated `value`. For example:

```
1 mydict = {'value1':2.73,'value2':3.14}
2 print('The value associated with value2 is',mydict['value2'])
```

prints the value `3.14` to the screen.

Alternatively, the function `lambdify` in `sympy` can be used to construct a numerical function from a symbolic expression, by specifying the symbols in the expression that need to be supplied as input to the function, as follows:

```
1 from sympy import symbols, diff, derive_by_array,
        pretty_print, lambdify

2
3 x0,x1 = symbols('x0,x1')
4 f = 100*(x0**2 - x1)**2 + (x0-1)**2
5 x = [x0, x1]

6
7 gradf = derive_by_array(f,x)
8 hessian = derive_by_array(gradf,x)

9
10 f_numeric = lambdify((x0,x1),f)
11 gradf_numeric = lambdify((x0,x1),gradf)
12 hessian_numeric = lambdify((x0,x1),hessian)

13
14 print('Rosenbrock function evaluated at x0=5,x1=1 gives',
        f_numeric(5,1))

15
16 print('Gradient of the Rosenbrock function evaluated at x0=5,
        x1=1 gives',gradf_numeric(5,1))

17
18 print('Hessian of the Rosenbrock function evaluated at x0=5,
        x1=1 gives',hessian_numeric(5,1))
```

9.3.3 Optimality Criteria

Conveniently the module `sympy` allows us to solve symbolic systems of equations. This capability allows us to find the roots of the gradient vector of a function, i.e. solve optimization problems using an optimality criteria approach.

For example, the necessary condition for a minimum of the Rosenbrock function is computed from the following code:

```
from sympy import symbols, diff, solve

x0,x1 = symbols('x0,x1')

f = 100*(x0**2 - x1)**2 + (x0-1)**2

dfdx0 = diff(f,x0)
dfdx1 = diff(f,x1)

solution = solve([dfdx0,dfdx1],[x0,x1])

print('The gradient vector is zero at the point: ',solution)
```

which gives the solution:

```
('The gradient vector is zero at the point', [(1, 1)])
```

9.4   Numerical Linear Algebra

In addition to symbolic computations in Python, the numerical Python module **numpy** and the scientific Python module **scipy** make a powerful collection of numerical strategies available.

The support for vectors and matrices are limited to **arrays** . Linear algebra operations are accessible via functions within the **numpy** and **scipy** modules. For example, **dot** allows you to compute the inner product between two arrays, subject to the correspondence of the two dimensions of the two arrays.   Also, **solve** under **numpy.linalg** allows for the solution of a linear system of equations. The application of these modules is illustrated by the following numerical examples.

Consider the vector

$$\mathbf{b} = [12.5, 37.5, 120]^{\mathrm{T}},$$

and matrix

$$\mathbf{A} = \begin{bmatrix} 2 & 1 & 0 \\ 1 & 2 & 1 \\ 0 & 1 & 2 \end{bmatrix}.$$

We can compute the solution to $\mathbf{Ax} = \mathbf{b}$ as follows:

```
from numpy import array
from numpy.linalg import solve

b = array([12.5,37.5,120])
A = array([[2,1,0],[1,2,1],[0,1,2]])

x = solve(A,b)

print('The solution is',x)
```

In turn, by multiplying the computed solution $\mathbf{x}$ with the corresponding $\mathbf{A}$ matrix, the solution can be confirmed:

```
from numpy import dot
print('A*x',dot(A,x),'is indeed equal to b')
```

The solution is unique as the number of equations equals the number of unknowns, in addition to $\mathbf{A}$ being full rank.

We can confirm that $\mathbf{A}$ is full rank by computing the rank, determinant or eigenvalues of the matrix:

```
from numpy.linalg import matrix_rank, det, eig

print('Rank of A is',matrix_rank(A))

print('Determinant of A is',det(A))

EigenValues, EigenVectors = eig(A)
print('Eigenvalues of A are',EigenValues)
print('Product of EigenValues are',prod(EigenValues))
```

9.4.1   Overdetermined system of equations

Consider the following overdetermined system of equations $\mathbf{Cx} = \mathbf{d}$ where

$$\mathbf{d} = [12.5, 37.5, 120, 55]^{\mathrm{T}},$$

and the matrix $\mathbf{C}$ given by

$$\mathbf{C} = \begin{bmatrix} 2 & 1 & 0 \\ 1 & 2 & 1 \\ 0 & 1 & 2 \\ 0 & 1 & 1 \end{bmatrix}.$$

We may accept a solution that minimizes some error of the system and can thus be considered superior to any other arbitrary solution. The difference

$$\mathbf{e} = \mathbf{Cx} - \mathbf{d}, \tag{9.3}$$

can be used to compute an appropriate error function by taking the difference squared:

$$\begin{aligned} \mathbf{e}^{\mathrm{T}}\mathbf{e} &= (\mathbf{x}^{\mathrm{T}}\mathbf{C}^{\mathrm{T}} - \mathbf{d}^{\mathrm{T}})(\mathbf{Cx} - \mathbf{d}) \tag{9.4} \\ &= \mathbf{x}^{\mathrm{T}}\mathbf{C}^{\mathrm{T}}\mathbf{Cx} - 2\mathbf{x}^{\mathrm{T}}\mathbf{C}^{\mathrm{T}}\mathbf{d} + \mathbf{d}^{\mathrm{T}}\mathbf{d}. \tag{9.5} \end{aligned}$$

The first order necessary condition for a minimum is obtained by differentiating (9.5) w.r.t. $\mathbf{x}$, which gives the minimum error solution by solving the following linear system:

$$\mathbf{C}^{\mathrm{T}}\mathbf{Cx} = \mathbf{C}^{\mathrm{T}}\mathbf{d}. \tag{9.6}$$

The Python code for solving this linear system is given by

```
1 from numpy import array, dot
2 from numpy.linalg import solve
3
4 d = array([12.5,37.5,120,55])
5 C = array([[2,1,0],[1,2,1],[0,1,2],[0,1,1]])
6
7 CTC = dot(C.transpose(),C)
8 CTd = dot(C.transpose(),d)
9
10 x = solve(CTC,CTd)
11
12 print('The solution that minimizes the error is',x)
```

The quality of the solution can be assessed by

```
1 e = dot(C,x)-d
2 eTe = dot(e,e)
3 print('Cx is',dot(C,x),'which should be',d,'the error squared
      is',eTe)
```

The solution is indeed confirmed to be a minimum since $\mathbf{C}^T\mathbf{C}$ is positive-definite:

```
1 from numpy.linalg import eig
2 EigenValues, EigenVectors = eig(CTC)
3 print('Second order sufficiency: Positive-definite with
       eigenvalues', EigenValues)
```

which confirms the second order necessary condition.

9.4.2 Underdetermined system of equations

Consider the following underdetermined system of equations $\mathbf{Zx} = \mathbf{y}$ that has an infinite number of solutions, and where

$$\mathbf{y} = [12.5, 37.5]^T$$

and the matrix $\mathbf{Z}$ given by

$$\mathbf{Z} = \begin{bmatrix} 2 & 1 & 0 \\ 1 & 2 & 1 \end{bmatrix}.$$

Since we have three unknowns to satisfy only two equations we have an infinite number of solutions. We can formulate an optimization problem, the solution of which gives preference of one solution over others. Enforcing preference for a particular solution is referred to as homogenization. Here we choose to prefer solution vectors with shorter length over those with longer lengths, i.e. that given by the solution to the following optimization problem:

$$\underset{\text{w.r.t. } \mathbf{x}}{\text{minimize}} \frac{1}{2}\mathbf{x}^T\mathbf{x},$$

subject to the constraints:

$$\mathbf{Zx} = \mathbf{y}. \tag{9.7}$$

The Lagrangian for the equality constrained problem is given by

$$\frac{1}{2}\mathbf{x}^T\mathbf{x} + \lambda^T(\mathbf{Zx} - \mathbf{y}), \tag{9.8}$$

from which the necessary KKT conditions follow:

$$\frac{dL}{d\mathbf{x}} = \mathbf{x} + \mathbf{Z}^{\mathrm{T}}\lambda = \mathbf{0}, \tag{9.9}$$

$$\frac{dL}{d\lambda} = \mathbf{Z}\mathbf{x} - \mathbf{y} = \mathbf{0}. \tag{9.10}$$

Consider (9.9) pre-multiplied by $\mathbf{Z}$ to obtain

$$\mathbf{Z}\mathbf{x} + \mathbf{Z}\mathbf{Z}^{\mathrm{T}}\lambda = \mathbf{0}. \tag{9.11}$$

Since $\mathbf{Z}\mathbf{x} = \mathbf{y}$, it follows that we can solve for λ independent of $\mathbf{x}$:

$$\lambda = -(\mathbf{Z}\mathbf{Z}^{\mathrm{T}})^{-1}\mathbf{y}. \tag{9.12}$$

From (9.9) and (9.12) we finally obtain the minimum norm solution

$$\mathbf{x} = \mathbf{Z}^{\mathrm{T}}(\mathbf{Z}\mathbf{Z}^{\mathrm{T}})^{-1}\mathbf{y}. \tag{9.13}$$

Thus the minimum norm solution in Python for the problem is given by

```
from numpy import array, dot
from numpy.linalg import solve

y = array([12.5,37])
Z = array([[2,1,0],[1,2,1]])

ZZT = dot(Z,Z.transpose())
RHS = solve(ZZT,y)

x = dot(Z.transpose(),RHS)

print('The solution with the minimum length is ',x)

print('Zx is ',dot(Z,x),'which corresponds to',y)
```

9.5   Numerical Functions

Python offers two ways to define numerical functions that can represent multidimensional scalar objective and constraint functions. This is in addition to converting symbolic expressions into numerical functions using `lambdify` as discussed in Section 9.3.2.

Algorithm	Identifier	Section	Order of Info.			Constraints		
			0th	1st	2nd	Bnd.	L.	N.L.
Nelder-Mead	'nelder-mead'	2.5	✓					
Powell	'powell'	2.5	✓					
Conj. Grad.	'cg'	2.3.2	✓	✓				
BFGS	'bfgs'	2.4.2.3	✓	✓				
Dogleg Trust	'dogleg'		✓	✓	✓			
Newton CG Trust	'ncg-trust'		✓	✓	✓			
Truncated Newton	'tnc'		✓	✓		✓		
Lim. Mem. BFGS	'l-bfgs-b'		✓	✓		✓		
COBYLA	'cobyla'	App.A	✓	✓				✓
SLSQP	'slsqp'		✓	✓		✓	✓	✓

Table 9.1: Available minimization algorithms for the function `minimize` under `scipy.optimize`

Firstly, using the **def** keyword followed by the name of the function and :. The computation inside the function is defined by the **TAB** indented Python code. The function returns whatever is specified after the **return** keyword. It is important to note, that the input is an array, where the first entry in the array is defined by index 0. For example, the function that defines the two-dimensional Rosenbrock function is given by

```
def rosenbrock(x):
    functionvalue = 100*(x[0]**2 - x[1])**2 + (x[0]-1)**2
    return functionvalue
```

The defined **Rosenbrock** function can then be evaluated at any point as follows:

```
from numpy import array
rosen11 = rosenbrock(array[1,1])
print(rosen11)
```

Secondly, Python offers **lambda** functions to quickly define explicit functions of limited complexity (single line functions) as follows:

```
rosenbrock = lambda x: 100*(x[0]**2 - x[1])**2 + (x[0]-1)**2
```

The lambda constructed **rosenbrock** function can then be evaluated:

```
from numpy import array
rosen00 = rosenbrock(array[0,0])
print(rosen00)
```

9.6 Optimization Algorithms

The scientific Python module `scipy` offers a number of optimization algorithms under `scipy.optimize` using the `minimize` function. The algorithms available through the `minimize` function are presented in Table 9.1, which list the algorithm name, the identifier, the section where it is presented, the order of information that is used and whether it can handle bound (Bnd.), linear (L.) and non-linear (N.L.) constraints.

The `minimize` function requires the following inputs:

```
minimize(fun, x0, args=(), method=None, jac=None, hess=None,
         hessp=None, bounds=None, constraints=(), tol=None,
         callback=None, options=None)
```

The full documentation for `minimize` can be obtained via:

```
1 from scipy.optimize import minimize
2 help(minimize)
```

9.6.1 Unconstrained minimization

Consider the minimization of the already defined two-dimensional Rosenbrock function from the starting point $\mathbf{x}^0 = [5, 3]$ using the BFGS algorithm without supplying the analytical gradient vector:

```
1 x0 = array([5,3])
2 result = minimize(rosenbrock,x0,method='bfgs')
3 print(result)
```

that gives the following result:

```
fun: 6.9100079296177365e-12
hess_inv: array([[ 0.49979588,  0.99962071],
[ 0.99962071,  2.00429513]])
jac: array([ -6.92288171e-05,  3.71612185e-05])
message: 'Desired error not necessarily achieved due to
    precision loss.'
nfev: 632
```

```
nit: 58
njev: 155
status: 2
success: False
x: array([ 0.99999807,  0.99999632])
```

At convergence, the function value is given by the **fun** keyword, the estimated inverse of the Hessian matrix given by **hess_inv** and the gradient
vector by **jac**, with **jac** in reference to the *Jacobian* used to designate
all first order partial derivatives of the function. Additionally, a message
(**message**) is supplied stating that the desired error may not necessarily have been achieved due to precision loss resulting in unsuccessful
convergence as indicated by the **success:false** and also the **status:2**
feedback. Note that the integer related to the **status** feedback is solver
specific, while the accompanied **message** describes the meaning of the
feedback. On closer inspection it is evident that the culprit is the gradient vector at the solution not satisfying the first order optimality criterion to within the default tolerances. Since no analytical expression
for the gradient vector was supplied it had to be computed using finite
differences resulting in the precision loss stated in the accompanied message. The required number of iterations (**nit**) were 58, but the required
number of function evaluations were 632 (**nfev**), which averages to 11
function evaluations per iteration. A total number of 155 gradient esti-
mations (**njev**) were computed using finite differences which largely con-
tributed to the high number of required function evaluations. When the
gradient is not supplied it is estimated by conventional finite differences,
which increasingly affects the associated computational cost adversely
as the dimension of the problem increases.

The number of required function evaluations and numerical accuracy of
the results can be drastically improved on by supplying the Jacobian
function with analytical gradients as follows:

```
def gradient_rosenbrock(x):
    dfdx0 = 400*x[0]*(x[0]**2 - x[1]) + 2*x[0] - 2
    dfdx1 = -200*x[0]**2 + 200*x[1]
    return array([dfdx0,dfdx1])

x0 = array([5,3])
result = minimize(rosenbrock,x0,method='bfgs',jac=
    gradient_rosenbrock)
print(result)
```

which gives the following output:

```
fun: 9.94483975531609e-19
hess_inv: array([[ 0.4998457 ,   0.99969658],
[ 0.99969658,   2.00440351]])
jac: array([  1.09322507e-08,  -4.49452386e-09])
message: 'Optimization terminated successfully.'
nfev: 76
nit: 56
njev: 76
status: 0
success: True
x: array([ 1.,   1.])
```

The required number of function values decreased to 76 over 56 iterations by supplying the analytical gradient for the Rosenbrock function. In addition, the first order optimality criterion is satisfied within the default tolerance at the solution as the gradient vector is much more accurately resolved as indicated by jac.

The function check_grad under scipy.optimize allows for easy verification of analytical gradients by automatically verifying it against built-in finite difference schemes at a specified point:

```
1 point = array([3,2])
2 print('Gradient check: Difference is ',check_grad(rosenbrock,
      gradient_rosenbrock,point))
```

which confirms the user supplied gradient function with the following output:

```
Gradient check: Difference is 6.103515625e-05
```

Instead of analytically computing the gradient vector or relying on Python's built-in finite difference schemes, an accurate numerical finite difference scheme, such as the *complex-step method* see Section 2.3.1.6, can be used to explicitly compute the gradient vector:

```
1 import numpy as np
2 rosenbrock = lambda x: 100*(x[0]**2 - x[1])**2 + (x[0]-1)**2
```

```
3
4  def finitedifference_rosenbrock(x):
5     dfdx = []
6     delta = 1E-20
7
8     for i in range(len(x)):
9        step = zeros(len(x),dtype=np.complex)
10       step[i] = complex(0,delta)
11       dfdx.append(np.imag(rosenbrock(x+step))/delta)
12
13    return array(dfdx)
14
15 x0 = array([5,3])
16 result = minimize(rosenbrock,x0,method='bfgs',jac=
        finitedifference_rosenbrock)
17 print(result)
```

which gives the following output:

```
fun: 1.4496363359720043e-18
hess_inv: array([[ 0.4998082 ,   0.99962102],
[ 0.99962102,   2.00425127]])
jac: array([  9.94697170e-09,  -3.78443943e-09])
message: 'Optimization terminated successfully.'
nfev: 76
nit: 56
njev: 76
status: 0
success: True
x: array([ 1.,   1.])
```

Clearly this only required $76 + 2 \times 76 = 228$ function evaluations instead of the 632 when utilizing the built-in conventional finite difference schemes. As pointed out in Section 2.3.1.6, this illustrates the benefit of not having to resolve the step length in the complex-step method, while still computing numerically accurate sensitivities.

9.6.2 Constrained minimization

Consider the following constrained optimization problem:

$$\text{minimize } f(\mathbf{x}) = 2x_2 - x_1$$
$$\text{subject to}$$
$$g_1(\mathbf{x}) = x_1^2 + 4x_2^2 - 16 \leq 0,$$
$$g_2(\mathbf{x}) = (x_1 - 3)^2 + (x_2 - 3)^2 - 9 \leq 0$$
$$\text{and } x_1 \geq 0 \text{ and } x_2 \geq 0.$$

We aim to solve this problem using sequential least squares quadratic programming SLSQP, that is available in the function minimize.

The objective function is defined keeping in mind that the Python array index numbers start at 0. Hence vector entry x_1 maps to x[0] and x_2 maps to x[1] that defines the objective as follows:

```
1  f = lambda x: 2*x[1] - x[0]
```

Next, the constraint functions need to be constructed inside a dictionary. Before we proceed it is important to note that minimize requires the inequality constraints to be defined in the form $g_i(\mathbf{x}) \geq 0, \ i = 1, \ldots, m$. This differs from the usual convention, used in this book, that states inequality constraints in the form $g_i(\mathbf{x}) \leq 0, \ i = 1, \ldots, m$. Fortunately constraints cast in this form can quickly be rewritten in the required minimize form by multiplying each constraint by -1 to obtain $-g_i(\mathbf{x}) \geq 0, \ i = 1, \ldots, m$.

Each constraint needs to be defined as a dictionary with the following keywords:

1. 'type' that defines the constraint type as either an equality constraint ('eq') or inequality constraint ('ineq'),

2. 'fun' defines the constraint function $h_i(\mathbf{x})$ or $g_i(\mathbf{x})$ depending on the 'type', and

3. optional 'jac' defines the Jacobian of the constraint function.

Multiple constraints are then assembled into a list of dictionaries as follows:

```
1 cons = [{'type': 'ineq',
2      'fun': lambda x:  -x[0]**2 - 4*x[1]**2 + 16},
3    {'type': 'ineq',
4      'fun': lambda x: -(x[0] - 3)**2 - (x[1] - 3)**2 + 9}]
```

In addition the bound constraints are defined per dimension given the upper and lower bounds per list. The lower bound is 0 and the upper bound unspecified. Hence we define per dimension the bound constraints as [0,None], that are assembled into a list of lists for both dimensions as follows:

```
1 bounds = [[0, None], [0, None]]
```

We can now choose an initial starting point and solve the defined problem without supplying analytical gradients:

```
1 x0 = array([1,1])
2 result = minimize(f, x0, constraints=cons, bounds=bounds, method=
       'SLSQP')
3 print(result)
```

to obtain the following output:

```
fun: -3.6527011374767544
jac: array([-1.,  2.])
message: 'Optimization terminated successfully.'
nfev: 24
nit: 6
njev: 6
status: 0
success: True
x: array([ 3.98608288,  0.16669087])
```

By evaluating the constraints the feasibility of the solution can be verified.

Alternatively, analytical gradients can be made available to reduce the computational requirements:

```
def gradf(x):
    return array([-1,2])
```

```
4 def gradg0(x):
5   return array([-2*x[0],-8*x[1]])
6
7 def gradg1(x):
8   return array([-2*(x[0]-3),-2*(x[1]-3)])
9
10 cons = ({'type': 'ineq', 'fun': lambda x:  -x[0]**2 - 4 * x
      [1]**2 + 16, 'jac': gradg0},
11 {'type': 'ineq', 'fun': lambda x: -(x[0] - 3)**2 -(x[1] - 3)
      **2 +9, 'jac': gradg1})
12
13 x0 = array([1,1])
14 result = minimize(f,x0,jac=gradf,constraints=cons,bounds=
      bounds,method='SLSQP')
15 print(result)
```

that gives the following output:

```
fun: -3.6527011374831222
jac: array([-1.,   2.])
message: 'Optimization terminated successfully.'
nfev: 6
nit: 6
njev: 6
status: 0
success: True
x: array([ 3.98608288,   0.16669087])
```

Additional information and example problems are available via:

```
1 from scipy.optimize import minimize
2 help(minimize)
```

9.7   Practical Mathematical Optimization (PMO) Algorithms

The optimization *algorithms presented in this book* are freely available via Springer and distributed under the module name pmo. These novel algorithms are well suited for noisy and discontinuous optimization problems and are implemented to be compatible and consistent with the `scipy.optimize.minimize` framework. The benefit being that once a problem has been constructed it can easily be solved using the already supported algorithms in `minimize` or those supplied with this book, i.e. via the module `pmo`. This supplements the methods available for solving challenging optimization problems.

A summary of the `pmo` algorithms, including their associated function names and reference to relevant sections in this book, are listed in Table 9.2. For each algorithm inside `pmo` a description of the settings to be supplied are available via the `help` function. Previously algorithms were identified within `minimize` by supplying a relevant string identifier for the `method` keyword. The algorithms supplied in `pmo` can be used by specifying the function object for the `method` keyword instead of a string identifier as demonstrated in the following Python code to solve the constrained optimization problem presented in Section 9.6.2:

```
from numpy import array
from scipy.optimize import minimize
import pmo

f = lambda x: 2*x[1] - x[0]

def gradf(x):
    return array([-1,2])

def gradg0(x):
    return array([-2*x[0],-8*x[1]])

def gradg1(x):
    return array([-2*(x[0]-3),-2*(x[1]-3)])

bounds = [[0,None],[0,None]]

cons = ({'type': 'ineq', 'fun': lambda x:  -x[0]**2 - 4 * x
    [1]**2 + 16, 'jac': gradg0},
{'type': 'ineq', 'fun': lambda x: -(x[0] - 3)**2 -(x[1] - 3)
    **2 +9, 'jac': gradg1})

x0 = array([1,1])

result = minimize(f,x0,jac=gradf,constraints=cons,bounds=
    bounds,method=pmo.dynq)
print(result)
```

to obtain the following output:

```
fun: -3.6527010947260279
jac: array([-1,  2])
nfev: 8
nit: 6
njev: 7
success: True
x: array([ 3.98608287,  0.16669089])
```

Additional information on the argument settings is available for each algorithm using the `help` function, e.g. `help(pmo.dynq)`, `help(pmo.etopc)`.

Note that `method=pmo.dynq` sets the keyword `method` equal to the function object for Dynamic-Q, `pmo.dynq` in the `pmo` module. This is in contrast to the string identifier for the built-in algorithms as for example `'SLSQP'`:

```
result = minimize(f,x0,jac=gradf,constraints=cons,bounds=
    bounds,method='SLSQP')
```

which gives the following result:

```
fun: -3.6527011374831222
jac: array([-1.,  2.])
message: 'Optimization terminated successfully.'
nfev: 6
nit: 6
njev: 6
status: 0
success: True
x: array([ 3.98608288,  0.16669087])
```

9.7.1 User defined Algorithms

Included in the `pmo` module is a template code that allows users to implement their own algorithms. The example algorithm supplied within the template merely takes a constant step length along the normalized steepest descent direction until the maximum number of iterations has

Algorithmic Information			Order of Information			Constraints		
Name	Identifier	Section	Zero	First	Second	Bound	Lin.	Nonlin.
LFOPC	'lfopc'	6.2.3	✓	✓		✓	✓	✓
ETOPC	'etopc'	6.5	✓	✓		✓	✓	✓
SQSD	'sqsd'	6.3	✓	✓				
Dynamic-Q	'dynq'	6.4	✓	✓		✓	✓	✓
Snyman-Fatti	'sfglob'	6.6.2	✓	✓				
GO-SSA	'gossa'	8.7.1	✓					
GO-SDA	'gosda'	8.7.2	✓					
GO-BFGS	'gobfgs'	8.7.4	✓					

Table 9.2: Available `pmo` algorithms for the function `minimize` under `scipy.optimize`

been reached, i.e. given $\mathbf{x}^0$,

$$\mathbf{x}^{k+1} = \mathbf{x}^k - \text{constant} \frac{\nabla f(\mathbf{x}^k)}{\|\nabla f(\mathbf{x}^k)\|}, \quad k = 0, 1, 2, \ldots, k_{max}. \qquad (9.14)$$

The function `template` follows the same structure as the standard supported optimizers in `scipy.optimize.minimize`, which allows the algorithms implemented via the template to be used with minimal modifications using `minimize`.

However, before we proceed to the details of the `template` function, some additional understanding of functions in Python is required. Information is supplied to the specified optimizer in `minimize` as keyword inputs with the exception of the objective function and initial guess. For example, in the previous sections the function object that computes the gradient vector was specified as input to `minimize` using the `jac` keyword with appropriate algorithm name as a string. This object is then supplied to the optimizer specified using the `method` keyword using the `jac` keyword. Some algorithms may require specific information to be specified relevant only to that optimizer. Hence the number of required keyword arguments that have to be supplied to `minimize` and that are then passed down to the optimizer may differ from method to method. Python allows us to handle these instances by using a special input to the function namely `**kwargs`, which stands for variable **keyword arguments**. All keyword arguments supplied to the function are stored as a dictionary with the keyword followed by the assigned object during the function call. Here `**kwargs` comes in handy allowing for any number of keyword arguments to be specified, while the algorithm will extract only the required keywords from the dictionary or assign default values when not specified. Consequently, all required information first needs to be extracted from the `kwargs` dictionary or default values assigned when they are not listed. The use of the template is illustrated by the following code:

```
 1 from numpy import ones
 2 from numpy.linalg import norm
 3 from scipy.optimize import OptimizeResult
 4
 5 def template(function,x,**kwargs):
 6 # Begin extraction of keyword information or assign defaults
 7   if 'jac' in kwargs:
 8     gradf = kwargs['jac']
 9   else:
10     print('Jacobian required')
11     return -1
12
13   if 'kmax' in kwargs:
14     kmax = kwargs['kmax']
15   else:
16     kmax = 1000
17
18   if 'xtol' in kwargs:
19     xtol = kwargs['xtol']
20   else:
21     xtol = 1E-8
22
23   if 'steplength' in kwargs:
24     steplength = kwargs['steplength']
25   else:
26     steplength = 0.01
27 # End extraction of keyword information
28
29 # Initialize the iteration counter
30   k = 0
31 # Initialize a fake update in x
32   deltax = x + 2*xtol
33
34 # Conditional loop until maximum number of iterations
35 # have been reached
36   while k <= kmax and np.linalg.norm(deltax) > xtol:
37     gradient  = gradf(x) #Evaluate gradient vector
38 # Update the design vector
39     deltax = -steplength*gradient
40     x = x + deltax
```

```
1 # Increment the iteration counter
2     k = k + 1
3
4     F = function(x)
5 # Return OptimizeResult with selected fields specified
6     return OptimizeResult(fun=F, x=x, nit=k, success=(k > 0))
```

The output is defined by the `scipy.optimize.OptimizeResult` object. The default output has the following thirteen attributes with object types indicated after the colon:

```
Attributes
----------
x : ndarray
The solution of the optimization.
success : bool
Whether or not the optimizer exited successfully.
status : int
Termination status of the optimizer. Its value depends on the
underlying solver. Refer to 'message' for details.
message : str
Description of the cause of the termination.
fun, jac, hess: ndarray
Values of objective function, its Jacobian and its Hessian (if
available). The Hessians may be approximations, see the documentation
of the function in question.
hess_inv : object
Inverse of the objective function's Hessian; may be an approximation.
Not available for all solvers. The type of this attribute may be
either np.ndarray or scipy.sparse.linalg.LinearOperator.
nfev, njev, nhev : int
Number of evaluations of the objective functions and of its
Jacobian and Hessian.
nit : int
Number of iterations performed by the optimizer.
maxcv : float
The maximum constraint violation.
```

By using the **template** code with appropriate new functions users can efficiently expand their collection of optimizers.

9.8   Exercises

9.8.1 Compute the analytical gradient for all the test functions listed in Section 1.6.1 using the *sympy* module.

9.8.2 Confirm that the necessary *first order* condition is satisfied for each test function in Section 1.6.1 at the given solution $\mathbf{x}^*$.

9.8.3 Write a *Python* function that computes the analytical Hessian matrix given the analytical expression for an objective function as a sympy object and list of variables as inputs to the *Python* function.

9.8.4 Compute the analytical Hessian matrices for all the test functions listed in Section 1.6.1.

9.8.5 Confirm that the necessary *second order* condition is satisfied for each test function in Section 1.6.1 at the given solution $\mathbf{x}^*$.

9.8.6 Approximate Powell's badly scaled function in Section 1.6.1 using a first order Taylor series expansion about $\mathbf{x}^0$. Estimate the average accuracy of the approximation for the designs on the unit circle centered around $\mathbf{x}^0$.

9.8.7 Approximate Powell's badly scaled function in Section 1.6.1 using a second order Taylor series expansion about $\mathbf{x}^0$. Estimate the average accuracy of the approximation for the designs on the unit circle centered around $\mathbf{x}^0$.

9.8.8 Solve for all the eigenvalues and vectors for each of the computed $\mathbf{A}$ matrices in Exercise 1.7.6, using the *eig* function in the *numpy.linalg* module.

9.8.9 Setup and solve the constrained problem

$$\max_{\mathbf{x}} f(\mathbf{x}) = \mathbf{x}^T \mathbf{A}\mathbf{x}, \text{ such that } \mathbf{x}^T\mathbf{x} = 1,$$

for each of the computed $\mathbf{A}$ matrices in Exercise 1.7.6.

9.8.10 Critically compare the maximum eigenvalues and associated eigenvectors obtained in Exercise 9.8.8 against the optimal design vectors and function values obtained when maximizing $f(\mathbf{x}) = \mathbf{x}^T \mathbf{A}\mathbf{x}$ in Exercises 9.8.9.

9.8.11 Compute the numerical gradient at $\mathbf{x}^0$ for all the test functions given in Section 1.6.1 using the complex-step method.

9.8.12 Compute the gradient vector at $\mathbf{x}^0$ using the forward, backward, central and complex-step finite difference schemes using appropriate step sizes for each test function in Section 1.6.1. Discuss the accuracy of the various finite difference schemes.

9.8.13 Determine the optimal step size for computing the gradient vector using the forward, backward, central and complex-step finite difference schemes for each test function in Section 1.6. Note that the same step size should be used for all the components of the gradient vector. Compare the optimal step sizes of the various finite difference schemes on the various problems with each other as well as their respective accuracies.

9.8.14 Determine the optimal step size for computing each component of the gradient vector using the forward, backward, central and complex-step finite difference schemes for Wood's function in Section 1.6. Compare the optimal step sizes of the various finite difference schemes on the various components of the gradient vector with each other as well as their accuracies. Determine the additional accuracy of each finite difference scheme when allowing each component to have a different optimal step size.

9.8.15 Plot the univariate function $f(x) = 5x^2 \sin(x)$ between -2 and 2.

9.8.16 Consider the initial value problem presented by Burden et al. (2015):

$$\frac{dz(z,t)}{dt} = z - t^2 + 1,$$

with initial condition $z(0) = 0.5$ integrated using the forward Euler integration scheme with 1 000 equally spaced time steps between 0 and 2 seconds. The forward Euler integration scheme, as outlined by Burden et al. (2015), is given by

$$\tilde{z}^0 = z(0),$$
$$\tilde{z}^{i+1} = \tilde{z}^i + \Delta t \frac{dz(\tilde{z}^i, t^i)}{dt}, \quad i = 0, 1, 2, \ldots$$

where the $\tilde{\ }$ signifies an approximation and superscript i the time step number.

9.8.17 Consider the Lotka-Volterra system (see Section 8.2) with unknown parameter λ:

$$\frac{dz(z, y, t)}{dt} = (1 - \lambda)z(t) - 0.3z(t)y(t)$$

$$\frac{dy(z, y, t)}{dt} = z(t)y(t) - y(t),$$

with the two initial conditions $z(0) = 0.9$ and $y(0) = 0.9$. Integrate the system using the forward Euler integration scheme with 50 000 time steps between 0 and 8 seconds. Given that for the optimal choice of $\lambda = \lambda^*$ that $z(8, \lambda^*) = 0.722962$ and $y(8, \lambda^*) = 1.110567$, construct and plot a scalar error that indicates the difference to the optimal response for values of λ between 0 and 1 using 101 equally spaced points.

9.8.18 How many iterations do you expect the BFGS algorithm to take to minimize the following quadratic function, $f(\mathbf{x}) = \sum_{i=1}^{10} i(x_i - i)^2$? Solve this problem using the *scipy.optimize* implemented BFGS algorithm using 100 random initial designs over the domain $-15 \le x_i \le 15$, $i = 1, \ldots, 10$ and compare the expected against actual number of iterations.

9.8.19 Modify the `template` function in the module `pmo` to implement your own BFGS algorithm. Minimize the quadratic problem in Exercise 9.9.19 and compare the performance of your algorithm with the results in Exercise 9.8.19 for the same initial starting points.

9.8.20 For each computed eigenvector $\mathbf{u}^i$ in Exercise 9.8.8 plot the univariate function $f(\lambda) = (\mathbf{x^0} + \lambda \mathbf{u^i})^T \mathbf{A} (\mathbf{x^0} + \lambda \mathbf{u^i})$ for $0 \le \lambda \le 2$ and $\mathbf{x}^0 = [0, 0, \ldots, 0]$.

Appendix A

THE SIMPLEX METHOD FOR LINEAR PROGRAMMING PROBLEMS

A.1 Introduction

This introduction to the simplex method is along the lines given by Chvatel (1983).

Here consider the *maximization* problem:

$$\text{maximize } Z = \mathbf{c}^T \mathbf{x}$$
$$\text{such that } \mathbf{A}\mathbf{x} \leq \mathbf{b}, \ \mathbf{A} \text{ an } m \times n \text{ matrix} \qquad \text{(A.1)}$$
$$x_i \geq 0, \ i = 1, 2, ..., n.$$

Note that $\mathbf{A}\mathbf{x} \leq \mathbf{b}$ is equivalent to $\sum_{i=1}^{n} a_{ji} x_i \leq b_j, \ j = 1, 2, ..., m.$

Introduce *slack variables* $x_{n+1}, x_{n+2}, ..., x_{n+m} \geq 0$ to transform the

© Springer International Publishing AG, part of Springer Nature 2018

J.A. Snyman and D.N. Wilke, *Practical Mathematical Optimization*,
Springer Optimization and Its Applications 133,
https://doi.org/10.1007/978-3-319-77586-9

inequality constraints to equality constraints:

$$
\begin{aligned}
a_{11}x_1 + \ldots \quad + \quad a_{1n}x_n + x_{n+1} &= b_1 \\
a_{21}x_1 + \ldots \quad + \quad a_{2n}x_n + x_{n+2} &= b_2 \\
\vdots \quad\quad\quad\quad & \\
a_{m1}x_1 + \ldots \quad + \quad a_{mn}x_n + x_{n+m} &= b_m
\end{aligned}
\tag{A.2}
$$

or

$$[\mathbf{A};\mathbf{I}]\mathbf{x} = \mathbf{b}$$

where $\mathbf{x} = [x_1, x_2, ..., x_{n+m}]^T$, $\mathbf{b} = [b_1, b_2, ..., b_m]^T$, and $x_1, x_2, ..., x_n \geq 0$ are the *original decision variables* and $x_{n+1}, x_{n+2}, ..., x_{n+m} \geq 0$ the *slack variables*.

Now assume that $b_i \geq 0$ for all i, To start the process an initial *feasible* solution is then given by:

$$
\begin{aligned}
x_{n+1} &= b_1 \\
x_{n+2} &= b_2 \\
&\vdots \\
x_{n+m} &= b_m
\end{aligned}
$$

with $x_1 = x_2 = \cdots = x_n = 0$.

In this case we have a *feasible origin*.

We now write system (A.2) in the so called standard *tableau* format:

$$
\begin{aligned}
x_{n+1} &= b_1 - a_{11}x_1 - \ldots - a_{1n}x_n \geq 0 \\
x_{n+2} &= b_2 - a_{21}x_1 - \ldots - a_{2n}x_n \geq 0 \\
&\vdots \\
x_{n+m} &= b_m - a_{m1}x_1 - \ldots - a_{mn}x_n \geq 0 \\
Z &= c_1x_1 + c_2x_2 + \ldots + c_nx_n
\end{aligned}
\tag{A.3}
$$

The left side contains the *basic variables*, in general $\neq 0$, and the right side the *nonbasic variables*, all $= 0$. The last line Z denotes the objective function (in terms of *nonbasic* variables).

In a more general form the *tableau* can be written as

$$
\begin{aligned}
x_{B1} &= b_1 - a_{11}x_{N1} - \ldots - a_{1n}x_{Nn} \geq 0 \\
x_{B2} &= b_2 - a_{21}x_{N1} - \ldots - a_{2n}x_{Nn} \geq 0 \\
&\vdots \\
x_{Bm} &= b_m - a_{m1}x_{N1} - \ldots - a_{mn}x_{Nn} \geq 0 \\
Z &= c_{N1}x_{N1} + c_{N2}x_{N2} + \ldots + c_{Nn}x_{Nn}
\end{aligned}
\tag{A.4}
$$

The $\geq$ at the right serves to remind us that $x_{Bj} \geq 0$ is a necessary condition, even when the values of x_{Ni} change from their zero values.

$\mathbf{x}_B =$ vector of basic variables and $\mathbf{x}_N =$ vector of nonbasic variables represent a *basic feasible solution.*

A.2 Pivoting to increase the objective function

Clearly if any $c_{Np} > 0$, then Z increases if we increase x_{Np}, with the other $x_{Ni} = 0, i \neq p$. Assume further that $c_{Np} > 0$ and $c_{Np} \geq c_{Ni}, i = 1, 2, ..., n$, then we decide to increase x_{Np}. But x_{Np} can not be increased indefinitely because of the constraint $\mathbf{x}_B \geq \mathbf{0}$ in *tableau* (A.4). Every entry i in the tableau, with $a_{ip} > 0$, yields a constraint on x_{Np} of the form:

$$
0 \leq x_{Np} \leq \frac{b_i}{a_{ip}} = d_i, \quad i = 1, 2, ..., m.
\tag{A.5}
$$

Assume now that $i = k$ yields the strictest constraint, then let $x_{Bk} = 0$ and $x_{Np} = d_k$. Now $x_{Bk}(= 0)$ is the *outgoing* variable (out of the base), and $x_{Np} = d_k(\neq 0)$ the *incoming* variable. The k-th entry in tableau (A.4) changes to

$$
x_{Np} = d_k - \sum_{i \neq k} \bar{a}_{ki}x_{Ni} - \bar{a}_{kk}x_{Bk}
\tag{A.6}
$$

with $\bar{a}_{ki} = a_{ki}/a_{kp}, \quad \bar{a}_{kk} = 1/a_{kp}.$

Replace x_{Np} by (A.6) in each of the remaining $m - 1$ entries in tableau (A.4) as well as in the objective function Z. With (A.6) as the first entry this gives the *new tableau* in terms of the new basic variables:

$$
x_{B1}, x_{B2}, \ldots, x_{Np}, ..., x_{Bm} \text{ (left side) } \neq 0
$$

and nonbasic variables:

$$x_{N1}, x_{N2}, \ldots, x_{Bk}, \ldots, x_{Nn} \text{ (right side)} = 0.$$

As x_{Np} has increased with d_k, the objective function has also increased by $c_{Np}d_k$. The objective function (last) entry is thus of the form

$$Z = c_{Np}d_k + c_{N1}x_{N1} + c_{N2}x_{N2} + \cdots + c_{Nm}x_{Nm}$$

where the x_{Ni} now denotes the new nonbasic variables and c_{Ni} the new associated coefficients. This completes the first *pivoting iteration*.

Repeat the procedure above until a tableau is obtain such that

$$Z = Z^* + c_{N1}x_{N1} + \cdots + c_{Nm}x_{Nm} \text{ with } c_{Ni} \leq 0, \quad i = 1, 2, \ldots, m.$$

The optimal value of the objective function is then $Z = Z^*$ (no further increase is possible).

A.3 Example

maximize $Z = 5x_1 + 4x_2 + 3x_3$ such that

$2x_1 + 3x_2 + x_3 \leq 5$
$4x_1 + x_2 + 2x_3 \leq 11$
$3x_1 + 4x_2 + 2x_3 \leq 8$
$x_1, x_2, x_3 \geq 0.$

This problem has a *feasible origin*. Introduce slack variables x_4, x_5 and x_6 and then the *first tableau* is given by:

I:

x_4	$=$	5	$-$	$2x_1$	$-$	$3x_2$	$-$	x_3	≥ 0	$x_1 \leq 5/2(s)$
x_5	$=$	11	$-$	$4x_1$	$-$	x_2	$-$	$2x_3$	≥ 0	$x_1 \leq 11/4$
x_6	$=$	8	$-$	$3x_1$	$-$	$4x_2$	$-$	$2x_3$	≥ 0	$x_1 \leq 8/3$
Z	$=$			$5x_1$	$+$	$4x_2$	$+$	$3x_3$	

Here $5 > 0$ and $5 > 4 > 3$. Choose thus x_1 as *incoming variable*. To find the outgoing variable, calculate the constraints on x_1 for all the entries (see right side). The strictest (s) constraint is given by the first entry, and thus the *outgoing variable* is x_4. The first entry in the next tableau is

$$x_1 = \tfrac{5}{2} - \tfrac{3}{2}x_2 - \tfrac{1}{2}x_3 - \tfrac{1}{2}x_4.$$

Replace this expression for x_1 in all other entries to find the next tableau:

$$
\begin{aligned}
x_1 &= \tfrac{5}{2} - \tfrac{3}{2}x_2 - \tfrac{1}{2}x_3 - \tfrac{1}{2}x_4 & &\geq 0 \\
x_5 &= 11 - 4(\tfrac{5}{2} - \tfrac{3}{2}x_2 - \tfrac{1}{2}x_3 - \tfrac{1}{2}x_4) - x_2 - 2x_3 & &\geq 0 \\
x_6 &= 8 - 3(\tfrac{5}{2} - \tfrac{3}{2}x_2 - \tfrac{1}{2}x_3 - \tfrac{1}{2}x_4) - x4x_2 - 2x_3 & &\geq 0 \\
Z &= 5(\tfrac{5}{2} - \tfrac{3}{2}x_2 - \tfrac{1}{2}x_3 - \tfrac{1}{2}x_4) + 4x_2 + 3x_3 &
\end{aligned}
$$

After simplification we obtain the *second tableau* in standard format:

II:

$$
\begin{array}{rclcrcrcrcll}
x_1 &=& \tfrac{5}{2} &-& \tfrac{3}{2}x_2 &-& \tfrac{1}{2}x_3 &-& \tfrac{1}{2}x_4 & \geq 0 & x_3 \leq 5 \\
x_5 &=& 1 &+& 5x_2 & & &+& 2x_4 & \geq 0 & \text{no bound} \\
x_6 &=& \tfrac{1}{2} &+& \tfrac{1}{2}x_2 &-& \tfrac{1}{2}x_3 &+& \tfrac{3}{2}x_4 & \geq 0 & x_3 \leq 1(s) \\
\hline
Z &=& \tfrac{25}{5} &-& \tfrac{7}{2}x_2 &+& \tfrac{1}{2}x_3 &-& \tfrac{5}{2}x_4 & &
\end{array}
$$

This completes the first iteration. For the next step it is clear that x_3 is the incoming variable and consequently the outgoing variable is x_6. The first entry for the next tableau is thus $x_3 = 1 + x_2 + 3x_4 - 2x_6$ (3-rd entry in previous tableau).

Replace this expression for x_3 in all the remaining entries of tableau II. After simplification we obtain the *third tableau*:

III:

$$
\begin{array}{rclcrcrcrcl}
x_3 &=& 1 &+& x_2 &+& 3x_4 &-& 2x_6 & \geq 0 \\
x_1 &=& 2 &-& 2x_2 &-& 2x_4 &+& x_6 & \geq 0 \\
x_5 &=& 1 &+& 5x_2 &+& 2x_4 & & & \geq 0 \\
\hline
Z &=& 13 &-& 3x_2 &-& x_4 &-& x_6 &
\end{array}
$$

In the last entry all the coefficients of the nonbasic variables are negative. Consequently it is not possible to obtain a further increase in Z by increasing one of the nonbasic variables. The optimal value of Z is thus $Z^* = 13$ with

$$x_1^* = 2 \; ; \; x_2^* = 0 \; ; \; x_3^* = 1.$$

Assignment A.1

Solve by using the simplex method:

maximize $z = 3x_1 + 2x_2 + 4x_3$ such that

$x_1 + x_2 + 2x_3 \leq 4$
$2x_1 + 3x_3 \leq 5$
$2x_1 + x_2 + 3x_3 \leq 7$
$x_1, x_2, x_3 \geq 0.$

A.4 The auxiliary problem for problem with infeasible origin

In the previous example it is possible to find the solution using the simplex method only because $b_i > 0$ for all i and an initial solution $x_i = 0$, $i = 1, 2, ..., n$ with $x_{n+j} = b_j$, $j = 1, 2, ..., m$ was thus feasible, that is, the origin is a feasible initial solution.

If the LP problem does not have a feasible origin we first solve the so called *auxiliary problem*:

Phase 1:

$$\text{maximize } W = -x_0$$
$$\text{such that } \sum_{i=1}^{n} a_{ji}x_i - x_0 \leq b_j, \quad j = 1, 2, ..., m \qquad (A.7)$$
$$x_i \geq 0, \quad i = 0, 1, 2, ..., n$$

where x_0 is called the new artificial variable. By setting $x_i = 0$ for $i = 1, 2, ..., n$ and choosing x_0 large enough, we can always find a feasible solution.

The original problem clearly has a feasible solution if and only if the auxiliary problem has a feasible solution with $x_0 = 0$ or, in other words, the original problem has a feasible solution if and only if the optimal value of the auxiliary problem is zero. The original problem is now solved using the simplex method, as described in the previous sections. This solution is called *Phase* 2.

A.5 Example of auxiliary problem solution

Consider the LP:

maximize $Z = x_1 - x_2 + x_3$ such that

$2x_1 - x_2 + x_3 \le 4$
$2x_1 - 3x_2 + x_3 \le -5$
$-x_1 + x_2 - 2x_3 \le -1$
$x_1, x_2, x_3 \ge 0.$

Clearly this problem does not have a feasible origin.

We first perform *Phase* 1:

Consider the *auxiliary problem*:

maximize $W = -x_0$ such that

$2x_1 - x_2 + 2x_3 - x_0 \le 4$
$2x_1 - 3x_2 + x_3 - x_0 \le -5$
$-x_1 + x_2 - 2x_3 - x_0 \le -1$
$x_0, x_1, x_2, x_3 \ge 0.$

Introduce the slack variables x_4, x_5 and x_6, which gives the tableau (not yet in standard form):

x_4	$=$	4	$-$	$2x_1$	$+$	x_2	$-$	$2x_3$	$+$	x_0	$\ge$	0	$x_0 \ge -4$

$$
\begin{array}{rclcrcrcrcrcccl}
x_4 & = & 4 & - & 2x_1 & + & x_2 & - & 2x_3 & + & x_0 & \ge & 0 & x_0 \ge -4 \\
x_5 & = & -5 & - & 2x_1 & + & 3x_2 & - & x_3 & + & x_0 & \ge & 0 & x_0 \ge 5(s) \\
x_6 & = & -1 & + & x_1 & - & x_2 & + & 2x_3 & + & x_0 & \ge & 0 & x_0 \ge 1 \\
W & = & & & & & & & & - & x_0 & & &
\end{array}
$$

This is not in standard form as x_0 on the right is not zero. With $x_1 = x_2 = x_3 = 0$ then $x_4, x_5, x_6 \ge 0$ if $x_0 \ge \max\{-4; 5; 1\}$.

Choose $x_0 = 5$, as prescribed by the second (strictest) entry. This gives $x_5 = 0, x_4 = 9$ and $x_6 = 4$. Thus x_0 is a basic variable ($\ne 0$) and x_5 a nonbasic variable. The *first standard tableau* can now be write as

I:

x_0	$=$	5	$+$	$2x_1$	$-$	$3x_2$	$+$	x_3	$+$	x_5	$\geq$	0	$x_2 \leq \frac{5}{3}$

$$x_0 = 5 + 2x_1 - 3x_2 + x_3 + x_5 \geq 0 \quad \Big| \quad x_2 \leq \tfrac{5}{3}$$
$$x_4 = 9 \qquad\quad - 2x_2 - x_3 + x_5 \geq 0 \quad \Big| \quad x_2 \leq \tfrac{9}{2}$$
$$x_6 = 4 + 3x_1 - 4x_2 + 3x_3 + x_5 \geq 0 \quad \Big| \quad x_2 \leq 1(s)$$
$$\overline{W = -5 - 2x_1 + 3x_2 - x_3 - x_5}$$

Now apply the simplex method. From the last entry it is clear that W increases as x_2 increases. Thus x_2 is the *incoming variable*. With the strictest bound $x_2 \leq 1$ as prescribed by the third entry the *outgoing variable* is x_6. The second tableau is given by:

II:

$$x_2 = 1+ \;\; 0.75x_1+ \;\; 0.75x_3+ \;\; 0.25x_5- \;\; 0.25x_6\geq \;\; 0 \quad \text{no bound}$$
$$x_0 = 2- \;\; 0.25x_1- \;\; 1.25x_3+ \;\; 0.25x_5- \;\; 0.75x_6\geq \;\; 0 \quad x_3 \leq \tfrac{8}{5}(s)$$
$$x_4 = 7- \;\; 1.5x_1- \;\; 2.5x_3+ \;\; 0.5x_5+ \;\; 0.5x_6\geq \;\; 0 \quad x_3 \leq \tfrac{14}{5}$$
$$\overline{W = -2+ \;\; 0.25x_1+ \;\; 1.25x_3- \;\; 0.25x_5- \;\; 0.75x_6}$$

The new incoming variable is x_3 and the outgoing variable x_0. Perform the necessary pivoting and simplify. The next tableau is then given by:

III:

$$x_3 = 1.6 - 0.2x_1 + 0.2x_5 + 0.6x_6 - 0.8x_0 \geq 0$$
$$x_2 = 2.2 + 0.6x_1 + 0.4x_5 + 0.2x_6 - 0.6x_0 \geq 0$$
$$x_4 = 3 - x_1 \qquad\qquad - x_6 + 2x_0 \geq 0$$
$$\overline{W = \qquad\qquad\qquad\qquad - x_0}$$

As the coefficients of x_0 in the last entry is negative, no further increase in W is possible. Also, as $x_0 = 0$, the solution

$$x_1 = 0; \;\; x_2 = 2.2; \;\; x_3 = 1.6; \;\; x_4 = 3; \;\; x_5 = 0; \;\; x_6 = 0$$

corresponds to a feasible solution of the original problem. This means that the first phase has been completed.

The initial tableau for *Phase 2* is simply the above tableau III without the x_0 terms and with the objective function given by:

$$
\begin{aligned}
Z &= x_1 - x_2 + x_3 \\
&= x_1 - (2.2 + 0.6x_1 + 0.4x_5 + 0.2x_6) + (1.6 - 0.2x_1 + 0.2x_5 + 0.6x_6) \\
&= -0.6 + 0.2x_1 - 0.2x_5 + 0.4x_6
\end{aligned}
$$

in terms of nonbasic variables.

Thus the initial tableau for the original problem is:

x_3	$=$	1.6	$-$	$0.2x_1$	$+$	$0.2x_5$	$+$	$0.6x_6$	≥ 0	no bound
x_2	$=$	2.2	$+$	$0.6x_1$	$+$	$0.4x_5$	$+$	$0.2x_6$	≥ 0	no bound
x_4	$=$	3	$-$	x_1			$-$	x_6	≥ 0	$x_6 \leq 3(s)$
Z	$=$	-0.6	$+$	$0.2x_1$	$-$	$0.2x_5$	$+$	$0.4x_6$		

Perform the remaining iterations to find the final solution (the next incoming variable is x_6 with outgoing variable x_4).

Assignment A.2

Solve the following problem using the *two phase* simplex method:

maximize $Z = 3x_1 + x_2$ such that

$$
\begin{aligned}
x_1 - x_2 &\leq -1 \\
-x_1 - x_2 &\leq -3 \\
2x_1 + x_2 &\leq 4 \\
x_1, x_2 &\geq 0.
\end{aligned}
$$

A.6   Degeneracy

A further complication that may occur is degeneracy. It is possible that there is more than one candidate outgoing variable. Consider, for example, the following tableau:

$$
\begin{array}{rcrcrcrcrcr|l}
x_4 & = & 1 & & & & & - & 2x_3 & \geq & 0 & x_3 \leq \frac{1}{2}(s) \\
x_5 & = & 3 & - & 2x_1 & + & 4x_2 & - & 6x_3 & \geq & 0 & x_3 \leq \frac{1}{2}(s) \\
x_6 & = & 2 & + & x_1 & - & 3x_2 & - & 4x_3 & \geq & 0 & x_3 \leq \frac{1}{2}(s) \\
\hline
Z & = & & & 2x_1 & - & x_2 & + & 8x_3 & & &
\end{array}
$$

With x_3 the incoming variable there are three candidates, x_4, x_5 and x_6, for outgoing variable. Choose arbitrarily x_4 as the outgoing variable. Then the tableau is:

$$
\begin{array}{rcrcrcrcrcr|l}
x_3 & = & 0.5 & & & & & - & 0.5x_4 & \geq & 0 & \text{no bound on } x_1 \\
x_5 & = & & - & 2x_1 & + & 4x_2 & + & 3x_4 & \geq & 0 & x_1 \leq 0(s) \\
x_6 & = & & & x_1 & - & 3x_2 & + & 2x_4 & \geq & 0 & x_1 \leq 0(s) \\
\hline
Z & = & 4 & + & 2x_1 & - & x_2 & - & 4x_4 & & &
\end{array}
$$

This tableau differs from the previous tableaus in one important way: two *basic* variables have the value *zero*. A basic feasible solution for which one or more of the basic variables are zero, is called a *degenerate* solution. This may have bothersome consequences. For example, for the next iteration in our example, with x_1 as the incoming variable and x_5 the outgoing variable there is no increase in the objective function. Such an iteration is called a degenerate iteration. Test the further application of the simplex method to the example for yourself. Usually the stalemate is resolved after a few degenerate iterations and the method proceeds to the optimal solution.

In some, very exotic, cases it may happen that the stalemate is not resolved and the method gets stuck in an infinite loop without any progress towards a solution. So called cycling then occurs. More information on this phenomenon can be obtained in the book by Chvatel (1983).

A.7 The revised simplex method

The revised simplex method (RSM) is equivalent to the ordinary simplex method in terms of tableaus except that matrix algebra is used for the calculations and that the method is, in general, faster for large and sparse systems. For these reasons modern computer programs for LP problems always use the RSM.

We introduce the necessary terminology and then give the algorithm for an iteration of the RSM. From an analysis of the RSM it is clear that the algorithm corresponds in essence with the tableau simplex method. The only differences occur in the way in which the calculations are performed to obtain the incoming and outgoing variables, and the new basic feasible solution. In the RSM two linear systems are solved in each iteration. In practice special factorizations are applied to find these solutions in an economic way. Again see Chvatel (1983).

Consider the LP problem:

$$\text{maximize } Z = \mathbf{c}^T \mathbf{x} \tag{A.8}$$

such that $\mathbf{Ax} \leq \mathbf{b}$, $\quad \mathbf{A}$ $m \times n$ and $\mathbf{x} \geq \mathbf{0}$.

After introducing the slack variables the constraints can be written as:

$$\tilde{\mathbf{A}}\mathbf{x} = \mathbf{b}, \quad \mathbf{x} \geq 0 \tag{A.9}$$

where $\mathbf{x}$ includes the slack variables.

Assume that a basic feasible solution is available. Then, if $\mathbf{x}_B$ denotes the m basic variables, and $\mathbf{x}_N$ the n nonbasic variables, (A.9) can be written as:

$$\tilde{\mathbf{A}}\mathbf{x} = [\mathbf{A}_B \mathbf{A}_N] \begin{bmatrix} \mathbf{x}_B \\ \mathbf{x}_N \end{bmatrix} = \mathbf{b}$$

or

$$\mathbf{A}_B \mathbf{x}_B + \mathbf{A}_N \mathbf{x}_N = \mathbf{b} \tag{A.10}$$

where $\mathbf{A}_B$ is an $m \times m$ and $\mathbf{A}_N$ an $m \times n$ matrix.

The objective function Z can be written as

$$Z = \mathbf{c}_B^T \mathbf{x}_B + \mathbf{c}_N^T \mathbf{x}_N \tag{A.11}$$

where $\mathbf{c}_B$ and $\mathbf{c}_N$ are respectively the basic and nonbasic coefficients.

It can be shown that A_B is always non-singular. It thus follows that

$$\mathbf{x}_B = \mathbf{A}_B^{-1}\mathbf{b} - \mathbf{A}_B^{-1}\mathbf{A}_N \mathbf{x}_N. \tag{A.12}$$

Expression (A.12) clearly corresponds, in matrix form, to the first m entries of the ordinary simplex tableau, while the objective function entry is given by

$$
\begin{aligned}
Z &= \mathbf{c}_B^T \mathbf{x}_B + \mathbf{c}_N^T \mathbf{x}_N \\
&= \mathbf{c}_B^T \left(\mathbf{A}_B^{-1} \mathbf{b} - \mathbf{A}_B^{-1} \mathbf{A}_N \mathbf{x}_N \right) + \mathbf{c}_N^T \mathbf{x}_N \\
&= \mathbf{c}_B^T \mathbf{A}_B^{-1} \mathbf{b} + \left(\mathbf{c}_N^T - \mathbf{c}_B^T \mathbf{A}_B^{-1} \mathbf{A}_N \right) \mathbf{x}_N.
\end{aligned}
$$

Denote the basis matrix $\mathbf{A}_B$ by $\mathbf{B}$. The complete tableau is then given by

$$
\frac{\mathbf{x}_B = \overbrace{\mathbf{B}^{-1}\mathbf{b}}^{\mathbf{x}_B^*} - \mathbf{B}^{-1}\mathbf{A}_N\mathbf{x}_N \geq \mathbf{0}}{Z = \underbrace{\mathbf{c}_B^T\mathbf{B}^{-1}\mathbf{b}}_{Z^*} + \left(\mathbf{c}_N^T - \mathbf{c}_B^T\mathbf{B}^{-1}\mathbf{A}_N \right)\mathbf{x}_N} \tag{A.13}
$$

We now give the RSM in terms of the matrix notation introduced above. A careful study of the algorithm will show that this corresponds exactly to the tableau method which we developed by way of introduction.

A.8   An iteration of the RSM

(Chvatal, 1983)

Step 1: Solve the following system:

$\mathbf{y}^T\mathbf{B} = \mathbf{c}_B^T$. This gives $\mathbf{y}^T = \mathbf{c}_B^T\mathbf{B}^{-1}$.

Step 2: Choose an incoming column. This is any column $\mathbf{a}$ of $\mathbf{A}_N$ such that $\mathbf{y}^T\mathbf{a}$ is less than the corresponding component of $\mathbf{c}_N$.

$\left(\text{See (A.13)}: Z = Z^* + (\mathbf{c}_N^T - \mathbf{y}^T\mathbf{A}_N)\mathbf{x}_N\right)$

If no such column exists, the current solution is optimal.

Step 3: Solve the following system:

$\mathbf{B}\mathbf{d} = \mathbf{a}$. This gives $\mathbf{d} = \mathbf{B}^{-1}\mathbf{a}$.

(From (A.13) it follows that $\mathbf{x}_B = \mathbf{x}_B^* - \mathbf{d}t \geq \mathbf{0}$, where t is the value of the incoming variable).

Step 4: Find the largest value of t such that

$$
\mathbf{x}_B^* - t\mathbf{d} \geq \mathbf{0}.
$$

If no such t exists, then the problem is unbounded; otherwise at least one component of $\mathbf{x}_B^* - t\mathbf{d}$ will equal zero and the corresponding variable is the outgoing variable.

Step 5: Set the incoming variable, (the new basic variable), equal to t and the other remaining basis variables

$$\mathbf{x}_B^* := \mathbf{x}_B^* - t\mathbf{d}$$

and exchange the outgoing column in $\mathbf{B}$ with the incoming column $\mathbf{a}$ in $\mathbf{A}_N$.

Bibliography

D.M. Allen. The relationship between variable selection and data augmentation and a method for prediction. *Technometrics*, 16:125–127, 1974.

L. Armijo. Minimization of functions having Lipschitz continuous first partial derivatives. *Pacific Journal of Mathematics*, 16:1–3, 1966.

D.V. Arnold. *Noisy Optimization With Evolution Strategies*. Springer US, 2002.

J.S. Arora. *Introduction to Optimal Design*. McGraw-Hill, New York, 1989.

J.S. Arora. *Introduction to Optimum Design*. Elsevier Academic Press, San Diego, second edition, 2004.

J.S. Arora, O.A. El-wakeil, A.I. Chahande, and C.C. Hsieh. Global optimization methods for engineering applications: A review. *Structural and Multidisciplinary Optimization*, 9:137–159, 1995.

J. Barhen, V. Protopopescu, and D. Reister. TRUST: A deterministic algorithm for global optimization. *Science*, 276:1094–1097, 1997.

M.S. Bazaraa, H.D. Sherali, and C.M. Shetty. *Nonlinear Programming, Theory and Algorithms*. John Wiley, New York, 1993.

R.E. Bellman. *Dynamic programming*. Princeton University Press, 1957.

D.P. Bertsekas. Multiplier methods: A survey. *Automatica*, 12:133–145, 1976.

© Springer International Publishing AG, part of Springer Nature 2018
J.A. Snyman and D.N. Wilke, *Practical Mathematical Optimization,*
Springer Optimization and Its Applications 133,
https://doi.org/10.1007/978-3-319-77586-9

D.P. Bertsekas. *Convex Optimization Algorithms*. Athena Scientific, first edition, 2015.

F.H. Branin. Widely used convergent methods for finding multiple solutions of simultaneous equations. *IBM Journal of Research and Development*, 16:504–522, 1972.

F.H. Branin and S.K. Hoo. A method for finding multiple extrema of a function of n variables. In F.A. Lootsma, editor, *Numerical Methods of Nonlinear Optimization*, pages 231–237, London, 1972. Academic Press.

R.L. Burden, J.D. Faires, and A.M. Burden. *Numerical Analysis*. Cengage Learning, 2015.

H.-S. Chung and J.J. Alonso. Using gradients to construct response surface models for high-dimensional design optimization problems. In *39th AIAA Aerospace Sciences Meeting and Exhibit*, 2001.

V. Chvatel. *Linear Programming*. W.H. Freeman & Co Ltd, New York, 1983.

A.R. Conn, N.I.M. Gould, and Ph.L. Toint. Convergence of quasi-Newton matrices generated by the symmetric rank one update. *Mathematical Programming–Series B*, 50:177–195, 3 1991.

G. Dantzig. *Linear Programming and Extensions*. Princeton University Press, Princeton, 1963.

W.C. Davidon. Variable metric method for minimization. R&D Report ANL-5990 (Rev.), US Atomic Energy Commission, Argonne National Laboratories, 1959.

E. De Klerk and J.A. Snyman. A feasible descent cone method for linearly constrained minimization problems. *Computers and Mathematics with Applications*, 28:33–44, 1994.

L.C.W. Dixon and G.P. Szegö. The global optimization problem: An introduction. In L.C.W. Dixon and G.P. Szegö, editors, *Towards Global Optimization 2*, pages 1–15, Amsterdam, 1978. North-Holland.

L.C.W. Dixon, J. Gomulka, and G.P. Szegö. Towards a global optimization technique. In L.C.W. Dixon and G.P. Szegö, editors, *Towards Global Optimization*, pages 29–54, Amsterdam, 1975. North-Holland.

L.C.W. Dixon, J. Gomulka, and S.E. Hersom. Reflections on the global optimization problem. In L.C.W. Dixon, editor, *Optimization in Action*, pages 398–435, London, 1976. Academic Press.

N. Dyn, D. Levin, and S. Rippa. Numerical procedures for surface fitting of scattered data by radial functions. *SIAM Journal on Scientific and Statistical Computing*, 1986:639–659, 1986.

J. Farkas and K. Jarmai. *Analysis and Optimum Design of Metal Structures*. A.A. Balkema Publishers, Rotterdam, 1997.

A.V. Fiacco and G.P. McCormick. *Nonlinear Programming*. John Wiley, New York, 1968.

R. Fletcher. An ideal penalty function for constrained optimization. *IMA Journal of Applied Mathematics (Institute of Mathematics and Its Applications)*, 15:319–342, 1975.

R. Fletcher. *Practical Methods of Optimization*. John Wiley, Chichester, 1987.

R. Fletcher and M.J.D. Powell. A rapidly convergent descent method for minimization. *The Computer Journal*, 6:163–168, 1963.

R. Fletcher and C.M. Reeves. Function minimization by conjugate gradients. *The Computer Journal*, 7:149–154, 1964.

C. Fleury. Structural weight optimization by dual methods of convex programming. *International Journal for Numerical Methods in Engineering*, 14:1761–1783, 1979.

A. Forrester, A. Sóbester, and A. Keane. *Engineering Design via Surrogate Modelling: A Practical Guide*. Wiley, New York, NY, USA, 2008.

R. Franke. Scattered data interpolation: Tests of some method. *Mathematics of Computation*, 38(157):181–200, 1982.

J.C. Gilbert and J. Nocedal. Global convergence properties of conjugate gradient methods. *SIAM Journal on Optimization*, 2:21–42, 1992.

S.M. Goldfeld, R.E. Quandt, and H.F. Trotter. Maximization by quadratic hill-climbing. *Econometrica*, 34:541–551, 1966.

N. Gould, D. Orban, and P. Toint. Numerical methods for large-scale nonlinear optimization. *Acta Numerica*, 14:299–361, 2005.

A.O. Griewank. Generalized descent for global optimization. *Journal of Optimization Theory and Applications*, 34:11–39, 1981.

A.A. Groenwold and J.A. Snyman. Global optimization using dynamic search trajectories. *Journal of Global Optimization*, 24:51–60, 2002.

A.A. Groenwold, L.F.P. Etman, J.A. Snyman, and J.E. Rooda. Incomplete series expansion for function approximation. *Structural and Multidisciplinary Optimization*, 34:21–40, 2007.

R.T. Haftka and Z. Gürdal. *Elements of Structural Optimization*. Kluwer, Dortrecht, 1992.

R.L. Hardy. Multiquadric equations of topography and other irregular surfaces. *Journal of Geophysical Research*, 76:1905–1915, 1971.

R.L. Hardy. Research results in the application of multiquadric equations to surveying and mapping problems. *Surveying and Mapping*, 35:321–332, 1975.

R.L. Hardy. Theory and applications of the multiquadric-biharmonic method 20 years of discovery 1968–1988. *Computers & Mathematics with Applications*, 19:163–208, 1990.

M.R. Hestenes. Multiplier and gradient methods. *Journal of Optimization Theory and Applications*, 4:303–320, 1969.

D.M. Himmelblau. *Applied Nonlinear Programming*. McGraw-Hill, New York, 1972.

W. Hock and K. Schittkowski. *Lecture Notes in Economics and Mathematical Systems. No 187: Test examples for nonlinear programming codes*. Springer-Verlag, Berlin, Heidelberg, New York, 1981.

C. Homescu and I.M. Navon. Optimal control of flow with discontinuities. *Journal of Computational Physics*, 187:660–682, 2003.

N. Karmarkar. A new polynomial time algorithm for linear programming. *Combinatorica*, 4:373–395, 1984.

W. Karush. Minima of functions of several variables with inequalities as side conditions. Master's thesis, Department of Mathematics, University of Chicago, 1939.

J. Kennedy and R.C. Eberhart. Particle swarm optimization. In *Proceedings of the IEEE International Conference on Neural Networks*, pages 1942–1948, 1995.

A.I. Khuri and S. Mukhopadhyay. Response surface methodology. *Wiley Interdisciplinary Reviews: Computational Statistics*, 2:128–149, 2010.

J. Kiefer. Optimal sequential search and approximation methods under minimum regularity conditions. *SIAM Journal of Applied Mathematics*, 5:105–136, 1957.

N. Klier, S. Shallcross, and O. Pankratov. Asymptotic discontinuities in the RKKY interaction in the graphene Bernal bilayer. *Physical Review B*, 90:245118, 2014.

R.J. Kohavi. A study of cross-validation and bootstrap for accuracy estimation and model selection. In *Proceedings of the 14th International Joint Conference on Artificial Intelligence–Volume 2*, IJCAI'95, pages 1137–1143, San Francisco, CA, USA, 1995. Morgan Kaufmann Publishers Inc.

S. Kok and J.A. Snyman. A strongly interacting dynamic particle swarm optimization method. *Journal of Artificial Evolution and Applications*, 2008:1–9, 2008.

H.W. Kuhn and A.W. Tucker. Nonlinear programming. In J. Neyman, editor, *Proceedings of the Second Berkeley Simposium on Mathematical Statistics and Probability*. University of California Press, 1951.

S. Lauridsen, R. Vitali, F. van Keulen, R.T. Haftka, and J.I. Madsen. Response surface approximation using gradient information. In *Fourth World Congress on Structural and Multidisciplinary Optimization*, Dalian, China, 2002.

S. Lucidl and M. Piccioni. Random tunneling by means of acceptance-rejection sampling for global optimization. *Journal of Optimization Theory and Applications*, 62:255–277, 1989.

J. Lyness and C. Moler. Numerical Differentiation of Analytic Functions. *SIAM Journal of Numerical Analysis*, 4:202–210, 1967.

A.I. Manevich. Perfection of the conjugate directions method for unconstrained minimization problems. In G.I.N. Rozvany and N. Olhoff, editors, *Proceedings of the Third World Congress of Structural and Multidisciplinary Optimization*, Buffalo, New York, 1999. WCSMO.

J.R.R.A. Martins, P. Sturdza, and J.J. Alonso. The connection between the complex-step derivative approximation and algorithmic differentiation. *39th AIAA Aerospace Sciences Meeting and Exhibit*, 921:1–11, 2001.

J.R.R.A. Martins, P. Sturdza, and J.J. Alonso. The complex-step derivative approximation. *ACM Transactions on Mathematical Software*, 29:245–262, 2003.

R.E. Meyer. A simple explanation of the Stokes phenomenon. *SIAM Review*, 31:435–445, 1989.

J.J. Moré and D.J. Thuente. Line search algorithms with guaranteed sufficient decrease. *ACM Transactions on Mathematical Software*, 20:286–307, 1994.

M.D. Morris, T.J. Mitchell, and D. Ylvisaker. Bayesian design and analysis of computer experiments: Use of derivatives in surface prediction. *Technometrics*, 35:243–255, 1993.

J.A. Nelder and R. Mead. A simplex method for function minimization. *The Computer Journal*, 7:308–313, 1965.

J. Nocedal and S.J. Wright. *Numerical Optimization*. Springer-Verlag, New York, 1999.

P.Y. Papalambros and D.J. Wilde. *Principles of Optimal Design, Modeling and Computation*. Cambridge University Press, Cambridge, 2000.

E. Polak and G. Ribiere. Note on the convergence of methods of conjugate directions. *Revue Francaise d'Informatique et de Recherche Operationnelle*, 3:35–43, 1969.

M.J.D. Powell. An efficient method for finding the minimum of a function of several variables without calculating derivatives. *The Computer Journal*, 7:155–162, 1964.

S.S. Rao. *Engineering Optimization, Theory and Practice*. John Wiley, New York, 1996.

J.B. Rosen. The gradient projection method for nonlinear programming, Part I, Linear constraints. *SIAM Journal of Applied Mathematics*, 8:181–217, 1960.

J.B. Rosen. The gradient projection method for nonlinear programming, Part II, Nonlinear constraints. *SIAM Journal of Applied Mathematics*, 9:514–532, 1961.

A. Sarkar, A. Chatterjee, P. Barat, and P. Mukherjee. Comparative study of the Portevin-Le Chatelier effect in interstitial and substitutional alloy. *Materials Science and Engineering: A*, 459:361–365, 2007.

F. Schoen. Stochastic techniques for global optimization: A survey of recent advances. *Journal of Global Optimization*, 1:207–228, 1991.

N.Z. Shor, K.C. Kiwiel, and A. Ruszcaynski. *Minimization methods for non-differentiable functions*. Springer-Verlag New York, Inc., New York, NY, USA, 1985.

C. Smith and M. Gilbert. Application of discontinuity layout optimization to plane plasticity problems. *Proceedings of the Royal Society A*, 463:2461–2484, 2007.

J.A. Snyman. A new and dynamic method for unconstrained minimization. *Applied Mathematical Modelling*, 6:449–462, 1982.

J.A. Snyman. An improved version of the original leap-frog method for unconstrained minimization. *Applied Mathematical Modelling*, 7:216–218, 1983.

J.A. Snyman. Unconstrained minimization by combining the dynamic and conjugate gradient methods. *Quaestiones Mathematicae*, 8:33–42, 1985.

J.A. Snyman. The LFOPC leap-frog algorithm for constrained optimization. *Computers and Mathematics with Applications*, 40:1085–1096, 2000.

J.A. Snyman. A gradient-only line search method for the conjugate gradient method applied to constrained optimization problems with severe noise in the objective function. *International Journal for Numerical Methods in Engineering*, 62:72–82, 2005.

J.A. Snyman and L.P. Fatti. A multi-start global minimization algorithm with dynamic search trajectories. *Journal of Optimization Theory and Applications*, 54:121–141, 1987.

J.A. Snyman and A.M. Hay. The spherical quadratic steepest descent (SQSD) method for unconstrained minimization with no explicit line searches. *Computers and Mathematics with Applications*, 42:169–178, 2001.

J.A. Snyman and A.M. Hay. The Dynamic-Q optimisation method: An alternative to SQP? *Computers and Mathematics with Applications*, 44:1589–1598, 2002.

J.A. Snyman and S. Kok. A reassessment of the snyman-fatti dynamic search trajectory method for unconstrained global optimization. *Journal of Global Optimization*, 43:67–82, 2009.

J.A. Snyman and N. Stander. A new successive approximation method for optimal structural design. *AIAA Journal*, 32:1310–1315, 1994.

J.A. Snyman and N. Stander. Feasible descent cone methods for inequality constrained optimization problems. *International Journal for Numerical Methods in Engineering*, 39:1341–1356, 1996.

J.A. Snyman, N. Stander, and W.J. Roux. A dynamic penalty function method for the solution of structural optimization problems. *Applied Mathematical Modelling*, 18:453–460, 1994.

D.C. Sorensen. Newton's Method with a Model Trust Region Modification. *SIAM Journal on Numerical Analysis*, 19:409–426, 1982.

M. Squire and G. Trapp. Using Complex Variables to estimate derivatives of real functions. *SIAM Review*, 40:110–112, 1998.

N. Stander and J.A. Snyman. A new first order interior feasible direction method for structural optimization. *International Journal for Numerical Methods in Engineering*, 36:4009–4025, 1993.

R. Storn and K. Price. Differential evolution – a simple and efficient heuristic for global optimization over continuous spaces. *Journal of Global Optimization*, 11:341–359, 1997.

G. Strang. *Computational Science and Engineering.* Wellesley-Cambridge Press, 2007.

K. Svanberg. A globally convergent version of MMA without line search. In G.I.N. Rozvany and N. Olhoff, editors, *Proceedings of the First World Conference on Structural and Multidisciplinary Optimization*, Goslar, Germany, 1995. Pergamon Press.

K. Svanberg. The MMA for modeling and solving optimization problems. In K. English, K. Hulme, and E. Winer, editors, *Proceedings of the Third World Conference on Structural and Multidisciplinary Optimization*, Buffalo, New York, 1999. WCSMO.

K. Svanberg. A class of globally convergent optimization methods based on conservative convex separable approximations. *SIAM Journal on Optimization*, 12:555–573, 2002.

H. Theil and C. van de Panne. Quadratic programming as an extension of conventional quadratic maximization. *Management Science*, 7:1–20, 1961.

A. Törn and A. Zilinskas. *Global Optimization: Lecture Notes in Computer Science*, volume 350. Springer-Verlag, Berlin, 1989.

G.N. Vanderplaats. *Numerical Optimization Techniques for Engineering Design.* Vanderplaats R & D, Colorado Springs, 1998.

D.J. Wales and J.P.K. Doye. Global optimization by basin-hopping and the lowest energy structures of lennard-jones clusters containing up to 110 atoms. *Journal of Physical Chemistry A*, 101:5111–5116, 1997.

G.R. Walsh. *Methods of Optimization.* John Wiley, London, 1975.

D. N. Wilke and S. Kok. Numerical sensitivity computation for discontinuous gradient-only optimization problems using the complex-step method. *Blucher Mechanical Engineering Proceedings*, 1:3665–3676, 2014.

D.N. Wilke. How to get rid of discontinuities when constructing surrogates from piece-wise discontinuous functions. In J. Herskovits, editor, *5th International Conference on Engineering Optimization*, Iguassu Falls, Brazil, 2016.

D.N. Wilke, S. Kok, and A.A. Groenwold. The application of gradient-only optimization methods for problems discretized using non-constant methods. *Structural and Multidisciplinary Optimization*, 40:433–451, 2010.

D.N. Wilke, S. Kok, and A.A. Groenwold. Relaxed error control in shape optimization that utilizes remeshing. *International Journal for Numerical Methods in Engineering*, 94:273–289, 2013a.

D.N. Wilke, S. Kok, J.A. Snyman, and A.A. Groenwold. Gradient-only approaches to avoid spurious local minima in unconstrained optimization. *Optimization and Engineering*, 14:275–304, 2013b.

D.A. Wismer and R. Chattergy. *Introduction to Nonlinear Optimization. A Problem Solving Approach*. North-Holland, New York, 1978.

P. Wolfe. Convergence conditions for ascent methods. *SIAM Review*, 11:226–235, 1969.

P. Wolfe. Convergence conditions for ascent methods. II: Some corrections. *SIAM Review*, 13:185–188, 1971.

J.A. Yorke and Jr. W.N. Anderson. Predator-Prey patterns (Volterra-Lotka equations). *Proceedings of the National Academy of Sciences*, 70, 1973.

R. Zieliński. A statistical estimate of the structure of multiextremal problems. *Mathematical Programming*, 21:348–356, 1981.

Index

© Springer International Publishing AG, part of Springer Nature 2018
J.A. Snyman and D.N. Wilke, *Practical Mathematical Optimization*,
Springer Optimization and Its Applications 133,
https://doi.org/10.1007/978-3-319-77586-9

Printed in the United States
By Bookmasters